Automotive Innovation

Automotive Innovation
The Science and Engineering behind
Cutting-Edge Automotive Technology

Patrick Hossay

CRC Press
Taylor & Francis Group
Boca Raton London New York

CRC Press is an imprint of the
Taylor & Francis Group, an **Informa** business

CRC Press
Taylor & Francis Group
6000 Broken Sound Parkway NW, Suite 300
Boca Raton, FL 33487-2742

International Standard Book Number-13: 978-1-138-61176-4 (Hardback)

Library of Congress Cataloging-in-Publication Data

Names: Hossay, Patrick, 1964- author.
Title: Automotive innovation : the science and engineering behind cutting-edge automotive technology / Patrick Hossay.
Description: First edition. | Boca Raton, FL : CRC Press/Taylor & Francis Group, [2020]
Identifiers: LCCN 2019009155 | ISBN 9781138611764 (hardback)
Subjects: LCSH: Automobiles—Technological innovations—Popular works. | Automobiles—Design and construction—Popular works.
Classification: LCC TL240 .H655 2020 | DDC 629.2—dc23
LC record available at https://lccn.loc.gov/2019009155

Visit the Taylor & Francis Web site at
http://www.taylorandfrancis.com

and the CRC Press Web site at
http://www.crcpress.com

Contents

Preface

Cars have changed radically over the past few decades, and the pace of change is only accelerating. Innovations in engine design, fuel systems, digital control, advanced transmissions, and a range of other technologies have fundamentally redefined the powertrain. And advanced electronic control systems, active chassis control, driver assistance systems, not to mention electrified drivetrains, advanced batteries, and new lightweight materials are allowing us to profoundly reimagine what is possible. It can be tough to keep up. And that's the point of this book.

These exciting changes and innovations in automotive technology are complex, but they don't need to be intimidating. Fundamentally, the same basic principles are at work, whether you're looking at a Model T or a Tesla. The laws of science and mechanical principles haven't changed. Admittedly, the engineering particulars have become more involved, and there are a lot more of them. But, in the end, all of the technology in the most advanced vehicles can be understood in principle by anyone with a basic grasp of science and mechanics.

So, think of this book as a primer, a basic survey of the new automotive landscape with an eye toward a timely orientation to the most interesting innovations and the most promising technological advances out there. One aim of this work is to provide a solid introductory text for an undergraduate course. In particular, the idea is to fill the gap between a vocational-based automotive repair text and an advanced engineering text. In fact, this book grew out of an undergraduate survey course on automotive technology and design. Finding a useful text for this sort of course has always been difficult, as nearly all introductory texts in automotive technology focus on vocational training for mechanics; and the only other alternative is often a technically dense engineering text, a rather intimidating introduction to the field. This text is aimed at the midpoint: a true introductory survey of the science and engineering in automotive technology that allows a generally informed reader to develop an understanding of the principles, trends, and challenges in automotive technology and the possible directions of future developments.

With this in mind, the aim is to keep this work accessible and engaging. A useful orientation to the field should be readable and stimulating for students, mechanics, automotive enthusiasts, and anyone else who may have an interest in cars, technology, innovation, or engineering. This stuff is really amazing, exciting, and frequently ingenious; but all-too-often, the amazing stuff is buried under layers of engineering terminology and daunting computations that can thoroughly snuff out the flame of enthusiasm in the uninitiated. Cars can be really exiting; a book on them should be too.

So, this book will be useful for students of automotive engineering and technology that need an orientation to the field. This book should also be useful to a seasoned automotive technician trying to stay on top of a rapidly changing field, or a newly minted mechanic who needs a general orientation to the near future of the automobile, and even an automotive enthusiast who just wants to better understand how recent technological changes come together.

With luck this text will inspire budding engineers and maybe even motivate a few mechanics and gear heads to dig deeper, continue to explore the field, and perhaps even choose to take the next step and select a career that allows them to contribute to redefining the future of the automobile. This is truly a golden age in automotive design, a time when the future seems up for grabs, and a new possibilities have become not just feasible but likely.

Remember, this book is intended as a primer. You don't need a deep background in automotive technology to keep up. But a basic understanding of science and the fundamentals of mechanics won't hurt. Each topic and each chapter begin at the beginning, the basic principles that underpin the technology. Subsequently, the chapters move on to the ideas and engineering that define some of the most exciting innovations in the field; and in the end, each chapter addresses some of the most promising advances for the near future. In sum, the chapters offer a basic lay of the land, an orientation to the technology that is reshaping that field and plenty of real-world examples of remarkable automotive innovations.

Since the idea is to keep this book accessible, approachable, and short, this text has defined limits. It is fundamentally about cars, the current cars on the road now, and the likely changes that will define the cars on the road tomorrow. It's not about the automotive industry more generally, or the future of transportation infrastructure, manufacturing, or policy. Nor is it a detailed examination of research in science or engineering. So, automotive-related innovations that could one day reshape vehicles by remaking transportation infrastructure, such as alternative fuels, fuel cells, or intelligent transportation systems (ITS), are interesting, but that's really not what this book is about. Likewise, this is a primer; so, for a complete presentation of the advanced engineering techniques and computations needed to design these systems, you'll need to look elsewhere. In short, both the scope and depth of this text are intentionally focused. This is an introductory survey of contemporary automotive innovations for readers with a basic mechanical and science background.

With all of this in mind, this book addresses four principle areas: first, the technology of the combustion-based automobile on the road now and in the near future, addressed in the first three chapters; second, the technology of the electrified drivetrain that's increasingly present now and very likely to become dominant in the near future, addressed in the subsequent three chapters; third, innovations in chassis and body design, which are covered in Chapters 7 and 8; and lastly, a basic introduction to the sensor and navigation technology that enables advanced driver assistance systems and the possibilities for self-driving cars, addressed in the final chapter.

So, within this broad framework, Chapter 1 begins with the basics of the internal combustion engine and quickly moves on to review recent innovations in ignition management, advanced fuel delivery, combustion chamber design, and moves through to the basic principles of advanced low-temperature combustion possibilities and ingenious new engine designs. Chapter 2 then builds on this foundation with an examination of the digital control technology that has redefined the internal combustion engine, from variable valve timing and lift to variable intakes, as well as promising developments in more advanced active control mechanisms that enable precise on the fly changes in just about every aspect of the engine, defying the tradeoffs and limits engineers once faced when designing automotive engines. Chapter 3 then connects these technologies to the road by examining the rest of the powertrain, beginning with the basic principles of gearing and moving through to advanced innovations in transmission design including continuously variable transmissions, automatic manual transmission, dual-clutch systems, torque vectoring, and even advances in future tire design, where the rubber literally meets the road.

This then paves the way for an exploration of electrification of the drivetrain. Chapter 4 begins with a general introduction to electric motors and their performance advantages and challenges, with a particular focus on brushless DC and induction AC motors and control technology. It ends with an introduction to some of the more promising advances in motor design that may represent the electric machines of future automobiles. Chapter 5 examines the electrified powertrain, beginning with hybrid vehicle engineering, discussing varying hybrid drive architectures as well as the nature of regenerative braking and recent energy recovery innovations. It also explores electric vehicle technology, and the challenges and possibilities for future EVs. Energy storage technology is examined in Chapter 6, from basic battery science to promising developments in advanced battery chemistry and design.

The subsequent two chapters explore advanced vehicle design beyond the powertrain, beginning with a basic discussion of vehicle structure and handling and moving onto advanced suspension, active chassis control, new materials, and crashworthiness. Vehicle aerodynamics is examined in Chapter 8, again beginning with basic concepts of airflow and bluff bodies and moving onto an examination of recent applications such as air curtains, active shutters, ground effect management, and other advanced aerodynamic innovations.

The last chapter examines advanced driver assistance systems. This includes a discussion of sensing technology now in use, such as LIDAR, SONAR, and RADAR, as well as the benefits and challenges of applying these technologies to advanced vehicle control and driver assistance features, such as lane keeping, active cruise control, and crash avoidance. The chapter moves on to discuss the potential for a more extensive incorporation of advanced vehicles into roadway control technology, exploring V2X possibilities, the challenges of advanced driver assistance and autonomous vehicles, as well a basic introduction to the artificial intelligence needed for such systems.

In the end, the hope is that this book will help get the reader up to speed, oriented to the basic science and technology that defines the modern automobile, and its likely future.

Like any book that dares to offer a sweeping survey of a field, it's likely that a few points have been missed. Reader's comments are very welcome and can help improve possible future versions of this text. It is also certain that this book benefited greatly from the advice of my colleagues. In particular, I'd like to thank Justine Ciraolo, Jason Shulman, Marc Richard, and most especially Kristina Lawyer for their very helpful and thoughtful suggestions. Of course, any errors are entirely my own. I would also like to thank the array of component suppliers and carmakers that agreed to provide images and insights for this text and supporting materials. They are identified thorough the text, and I am most grateful.

Author

Professor Patrick Hossay heads the Energy Studies and Sustainability programs at Stockton University where he teaches courses in automotive technology, green vehicle innovations, and energy science. He is also an experienced aircraft and automotive mechanic, and enjoys restoring classic cars and motorcycles.

1

Bringing the Fire

It makes sense to start our exploration of automotive innovation with what has been the heart of the automobile for more than a century: the internal combustion engine. And it makes sense to start a look at the internal combustion engine with the heart of the process: combustion. The very idea of combustion is usually taken for granted, as a notion that is self-explanatory. We all know what combustion is, it's when something explodes or burns, right? But, thankfully, that's not exactly what's taking place in your engine. We're going to need a more precise understanding of exactly what and how something is burned in an engine before we can understand the complete workings of internal combustion.

So, what exactly is combustion? Put in general terms, it's a chemical reaction that converts organic material to carbon dioxide while releasing heat energy. The process is called rapid oxidation because it's defined by a fast reaction with oxygen. So, at the most basic level, a chemist would write down a combustion reaction like this:

$$C + O_2 + heat \rightarrow CO_2 + heat$$

Very simply, this means organic materials made of carbon (C), like wood or paper or in our case gasoline, react with oxygen (O_2) with added heat to create carbon dioxide gas (CO_2) and more heat.

This already allows for some very useful observation: first, combustion needs oxygen in a most fundamental way. You can think of a car's engine as a large air pump, drawing in massive amounts of air, delivering it through a combustion process that consumes the oxygen, and pushing the deplete product out the other end. In fact, to burn a single gallon of gasoline completely, an engine needs to draw in nearly 9,000 gallons of air. This need to provide a ready supply of oxygen for combustion is a critical criterion and definitive challenge of any engine's performance.

The second observation is that combustion is a *process*, not an instantaneous event. The heat resulting from the reaction, on the right side of the equation above, is the same heat that feeds the continuation of the reaction on the left. So, what we get in the engine is absolutely not an explosion, but a rapid combustive expansion that defines a wave of pressure, more like a push than a bang. This wave is often called a **flame front**, fed by the combustion of gasoline, and propagating a swell of pressure that can be converted by the engine into rotational energy. If we can accelerate that flame front, while keeping it controlled and even, we can achieve greater power from combustion. But, if we lose control of the combustion process, and perhaps even get more than one flame front from multiple points of ignition, the result is closer to an explosion, reducing performance while potentially harming the engine. Before we see how this all plays out in a typical engine, we'll need to understand a bit more about the chemical that fuels this process—gasoline.

What Is Gasoline?

The equation above is a bit too simple. After all, we're not burning single-carbon molecules. We're burning gasoline. Which raises the question, what exactly is gasoline? Gasoline is a 'hydrocarbon', which means it is a long chain of carbon connected to hydrogen. Essentially, the longer the chain, the more potential energy it contains. Crude oil comes in chains that could be dozens of carbon atoms long. But long molecules like this, while they have plenty of potential energy in them, are thick, hard to pump, and hard to burn. Think of tar. So, we need to refine these long, heavy hydrocarbons into shorter chains of lighter hydrocarbons that are easier to burn—that's gasoline (Image 1.1).

To make long carbon chains of crude oil into smaller chains that can be burned by our engine, we need to break them apart into smaller chains, a process called **cracking**. Cracking entails putting these long crude oil molecules in a tall tank, applying heat, pressure, and a chemical catalyst to encourage a reaction, and breaking them up into smaller chains of carbon. The sweet spot for our purposes is 4–12 atoms long for gasoline or 13–20 atoms for diesel. So, gasoline is not really a single chemical compound, it is an irregular mixture of different lighter and heavier compounds, with chemical characteristics that vary.

As we start thinking about putting that gasoline in our engine, it's useful to have a sense of how the fuel performs, or more precisely how easily it ignites. Not all fuels are the same. Some might not ignite very easily at normal pressure and temperature. This is the case with diesel, for example. Others might be more volatile and ignite easily, like butane lighter fluid. If a fuel is heated or put under pressure, it might even ignite on its own, called **autoignition**. If this happens before we want it to in an engine, the result can be very bad.

To help us manage this potential problem, we've developed a way to compare the tendency of a fuel to ignite on its own when put under pressure. To do this, we compare it to a set standard, a defined hydrocarbon chain length that we can use to measure and compare other hydrocarbons. Since the molecules in gasoline are typically 4–12 carbon atoms long, we can use a strand that's eight atoms long as a good comparison. Because a hydrocarbon with eight carbon atoms is called octane, we call this the **octane rating**, and use octane's ability to withstand compression before spontaneously igniting as a guide. If a given fuel can be squeezed to 85% of the value of octane before autoignition, we give it an octane rating of 85. If we can squeeze it more, to say 105% of octane, we give it a rating of 105. Pretty simple.

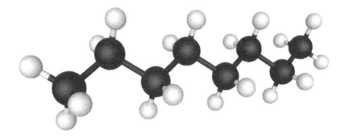

IMAGE 1.1
Hydrocarbon.

A chain of carbon atoms (black) and hydrogen atoms (white) make up a molecule of gasoline. This particular molecule has eight carbon atoms, so it's a molecule of 'octane'. Crude oil is comprised of much longer molecules that must be broken, or 'cracked', into smaller components to make gasoline.

So, the commonly held misbelief that higher octane fuel contains more energy or power is simply not correct.

Moving beyond the simple notion of an idealized perfect combustion, what does the combustion of gasoline really look like? Since gasoline is actually a mixture of many different hydrocarbons, the answer's tricky. But for the sake of simplicity, let's pretend it's all pure octane. A chemist would write the resulting equation like this:

$$\underset{\text{25 units of air (oxygen and nitrogen)}}{25\left[O_2 + 3.76N_2\right]} + \underset{\text{gasoline}}{2C_8H_{18}} \rightarrow \underset{\text{carbon dioxide}}{16CO_2} + \underset{\text{water}}{18H_2O} + \underset{\text{nitrogen from the air}}{94N_2} + \text{heat}$$

This looks complicated, but it's really not. Notice the eight under the 'C' on the left; that tells us it's octane. Simple chemistry tells us that a complete reaction, in which every molecule of gasoline is combined with the right amount of oxygen, would take place with a mixture ratio of 14.7 to 1. This means that under normal conditions 1 g of gasoline can combine with 14.7 g of air to produce a perfect oxidation reaction. This ratio is called the **stoichiometric ratio**. It's often represented by the Greek letter lambda—λ. When the mixture has more fuel than the stoichiometric mix, λ is less than 1, and when there's more air than the stoichiometric ratio, λ is greater than 1.

But this ideal ratio isn't always ideal. For example, providing a **rich mixture**, with more fuel than the 14.7:1 ratio, can offer more power or easier ignition, since there is more gasoline available. At cold start-up, for example, an engine might run more smoothly with a ratio of 12:1 (or $\lambda = 0.8$) until it warms up. Or under high loads or high acceleration, we may also want a richer mixture to provide more power. Alternatively, when driving on a highway without much need for power, we would be using much more gas than we need with such a rich mixture, so we could use a **lean mixture** with less gasoline to improve fuel economy, say 24:1 (or $\lambda = 1.55$).

An additional complication is that we're not providing pure oxygen to this reaction, we're proving air. While air is about 21% O_2, it's 78% nitrogen, or N_2. So far, we've ignored the nitrogen because it's not part of the oxidizing reaction. But, if the heat and pressure of combustion rise too high, as a result, for instance, of heavy torque demand, nitrogen molecules break apart and combine with oxygen, usually producing nitrogen oxide (NO), but also a bit of NO_2, collectively called **nitrogen oxides** or NO_x. Nitrogen oxides are a primary cause of smog and when combined with water in the atmosphere, can form nitric acid, a cause of acid rain. Similarly, if we're providing more fuel than can be effectively burned, some gas will slip through without burning; we call these unburned hydrocarbons (UHCs). An ongoing challenge then is to keep the heat of combustion controlled and the mixture correct to minimize these undesirable pollutants.

The Engine

The basic components of an internal combustion engine are simple: a cylinder that's just a long tube closed on one side and open on the other, and a piston that can slide up and down in that cylinder. These two components define the combustion chamber. We add a couple of valves that open and close to let us deliver air and fuel into the cylinder and let exhaust out. And we can add a means of mechanical connection to the bottom of the piston, so when it moves up and down, it causes a shaft to rotate (after all, what we're after is rotation) (Image 1.2).

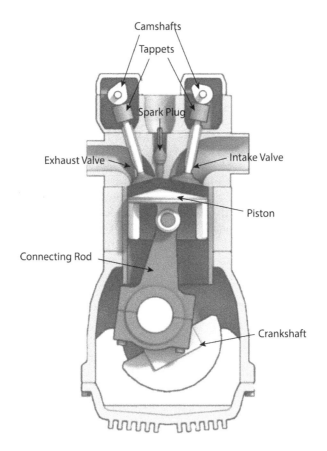

IMAGE 1.2
Basic engine components.

A basic internal combustion engine has some key primary components: a piston, a cylinder, a cylinder head, valves in the head, a connecting rod, and a crankshaft. In this case, our simple engine has dual overhead cams that operate the valves.

Image: Richard Wheeler

The piston is connected to a connecting rod by a pin (a piston pin), and the connecting rod is connected to an off-center, or eccentric, lobe on a shaft. When the piston goes up and down, the shaft rotates. Rotating that crankshaft is the whole purpose of the engine. So, all we need to do now is make the piston go up and down; that's where combustion comes in. A combustion event in the cylinder when the piston is all the way up will push the piston down and turn the crankshaft. To keep that process going, in an even and regular cycle so that the shaft spins evenly and with power, we've defined a **four-stroke** process.

The Four Strokes

The basic up and down movement of the piston is defined in four strokes; that's why we call it a 'four-stroke' engine. Each stroke is one full movement of the piston up or down. So, in four strokes, the piston has gone up and down twice. Let's look at each of these strokes.

We can begin with the **intake stroke**. During this stroke, the piston moves from the top of the cylinder to the bottom with the intake valve open. This allows an air and fuel mixture to be drawn into the combustion chamber as the crankshaft turns a half rotation. We call that incoming air–fuel mixture a **charge** (Image 1.3).

As the crankshaft continues to turn, the intake valve closes and the piston is pushed up, compressing the charge. This is called the **compression stroke**. This adds pressure to the fuel–air mixture, resulting in a rise in heat, preparing the fuel for combustion.

With the fuel–air mixture compressed, and the piston nearing the top of the compression stroke, a spark plug is used to create a small electric arc that ignites the mixture. With combustion initiated, a pressure wave begins at the spark plug and rapidly travels through the combustion chamber. This defines a **flame front** that pushes strongly down on the piston, adding energy to the rotation of the crankshaft. This is the power stroke. Later, we'll see that we can design engines differently to control and define the propagation of that flame front through the combustion chamber; in fact, this is a critical element of current engine research and innovation (Image 1.4).

At the end of the power stroke, the crankshaft has built rotational momentum that will push the piston back up. The exhaust valve opens, and as the piston moves up, it pushes out all burnt charge from the combustion chamber, defining the **exhaust stroke** and allowing the process to begin again. The high-energy exhaust gas moves quickly, initially passing the exhaust valves at more than 1,500 mph, and marking a significant variation from the relatively slow-moving incoming charge during the intake stroke. This is why intake valves are often larger than exhaust valves, as the slow-moving intake air needs a larger passage to fill the chamber (Image 1.5).

As these four strokes repeat themselves, they define the basic operation of the internal combustion, four-stroke engine. The fundamentals are pretty simple: suck in a charge,

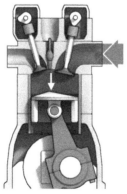
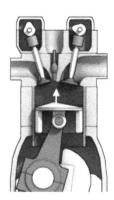

Intake Stroke   Compression Stroke

IMAGE 1.3
Intake and compression stroke.

As the piston moves down the cylinder, the combustion chamber expands, drawing in an air–fuel mixture, or charge, through the open intake valve as the exhaust valve remains closed. Subsequently, as the air–fuel mixture is compressed, both pressure and temperature rise, preparing the fuel–air charge for combustion. This is why the octane rating of our fuel is so important. If the octane rating is too low, the fuel might autoignite before we're done with the compression stroke, we call that preignition. Or if the rating is too high, the fuel might not quite be ready for ignition when we want it to be.

Image: Richard Wheeler

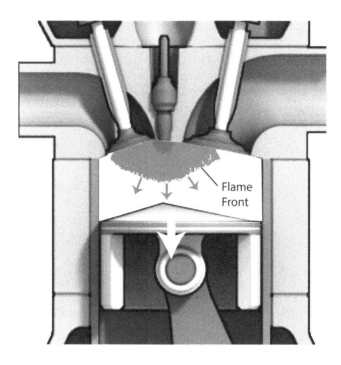

IMAGE 1.4
Flame front.

The flame front can be visualized as an arc of combustion that propagates away from the point of ignition at the spark plug and travels through the charge in the chamber, quickly increasing the heat and pressure of the cylinder and pushing the piston downward. A moving wall of combustion propagated by heat, such as this is, is called deflagration. Defining the character and speed of this deflagration, and the way it travels through the combustion chamber, is a central concern of internal combustion innovation.

Image: Richard Wheeler

squeeze it so it's ready for combustion, ignite it and release energy in the form of heat and pressure, and clear the cylinder to start the process again.

A diesel engine operates in basically the same manner, but with a few important differences that are worth noting. Diesel fuel is less volatile than gasoline, and thus harder to ignite. So, to achieve ignition, we need higher pressure and higher temperature. So, diesel engines achieve a much higher pressure during the compression stroke, which also produces a much higher resulting temperature called the **heat of compression.** While gasoline engines reduce volume by about ten to one on average, diesel engines double that, compressing volume by a ratio of 15–20 to 1. (We'll talk about the important of compression ratios in the next chapter.)

At this high pressure, diesel fuels will spontaneously ignite. So, to get a precise ignition point, we can't have the fuel in the cylinder during the compression stroke or it's likely to ignite unpredictably. Consequently, in the intake stroke, the engine takes in only air, the diesel fuel is then injected into the combustion chamber at the end of the compression stroke, just before the piston reaches the top of its stroke, or **top dead center** (TDC). The tremendous heat causes the injected spray to vaporize and ignite quickly, and a rapid pressure rise occurs by the time the piston reaches just past the top of the stroke.

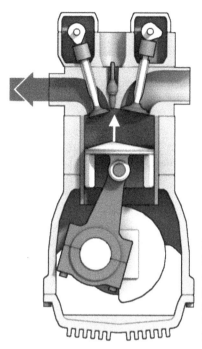

IMAGE 1.5
Exhaust stroke.

The angular momentum of the engine helps push the piston back up the cylinder. With the exhaust valve now open, this expels the burnt air–fuel mixture, allowing the cylinder to begin the process again with another intake stroke.

Image: Richard Wheeler

Unfortunately, spraying diesel fuel late in the compression stage means the fuel mixture will remain uneven in the chamber when ignition begins, with some lean and some rich areas. Those rich areas cause more soot (called **particulate mater** or PM) and the lean regions cause more nitrogen oxides than a gasoline engine. This typically means a clean burning diesel requires more complex (and expensive) emissions treatment than its gasoline counterpart.

Still, a typical diesel engine has an advantage in economy and torque. Relatively slow burning diesel operates at a lower speed range than gasoline, but can compensate with a longer piston travel, or **stroke**. It is able to take advantage of the resulting leverage to produce more rotational force or **torque**. A gasoline engine operates with a shorter stroke, but higher engine speed, or rotations per minute (rpm), capacity due to the faster burn rate of gasoline. It's often assumed that the greater power and efficiency of diesel is fully explained by the higher compression ratio. And certainly higher compression allows for higher thermal efficiency. But, a tremendous advantage in efficiency comes from the fact that varying the injection of the fuel can control engine speed, so diesel engines do not need to throttle the air supply to control engine speed like gasoline engines do. So, unlike conventional gasoline engines, diesels do not expend energy by drawing air through a restricted intake channel at low throttle speeds, a power reduction called **pumping loss**. This and the higher thermal efficiency explain why a diesel engine can typically achieve about 30% greater efficiency than a gasoline counterpart. With a lower rpm range, and a heartier construction to handle the increased pressure, diesel engines also tend to have a longer life than their gasoline counterparts. Of course, as diesel engines are made smaller and faster to suite automobile applications, and as gasoline engines borrow ideas from their diesel brethren (more on this later), these distinctions are less absolute than they once were.

The Engine Comes Together

So, we now have a basic sense of the principal components needed to make all this come together: cylinders, pistons, valves, connecting rods, a crankshaft, ignition, combustion, momentum, and all the rest. Let's look more closely at the components we need to make this real.

The development of angular momentum keeps the piston moving between power strokes and the crankshaft rotating evenly. To help maintain this momentum, we put a large disk on the crankshaft called a flywheel. This disk with a large radius will increase what physicists call the **moment of inertia** of the moving parts. Because we want to do this with the minimal mass necessary (so we don't slow down the engine), flywheels tend to be thin disk with a large radius. The inertia is defined by the mass and the radius squared ($i = mr^2$). So, a small increase in radius can allow a larger decrease in mass. The flywheel will also provide a contact point for the transmission and a convenient engagement gear for the electric starter. While the flywheel allows for smoother operation due to the increased moment of inertia, for the same reason it also takes more energy to get the engine spinning. So, a common performance upgrade is to swap out a heavy flywheel for a lighter one that will allow for faster engine acceleration.

Because the flywheel rotates at a near-constant speed, and pistons apply torque to the rest of the crankshaft with each firing impulse, the shaft can whip slightly due to the torsional force of each crank throw. On longer, six- and eight-cylinder engines, the result can throw off the valve timing mechanism at the front end of the crankshaft where the torsional oscillation is most severe, or even results in mechanical failure due to severe vibration. To prevent this, a smaller flywheel with a rubber hub is attached to the front end of longer crankshafts. While the flywheel favors a constant speed, the rubber serves to absorb irregular rotational energy of the shaft, operating as a torsional oscillation dampener, or **harmonic balancer**, to absorb potentially damaging oscillations (Image 1.6).

IMAGE 1.6
Crankshaft and flywheel.

The large flywheel attached to the right end of the crankshaft increased the assembly's moment of inertia, thus improving it's rotational momentum. Along with the harmonic balancer on the left, this produces more balanced and even operation.

Image: Jaguar MENA

Besides inertia, the other element that keeps our engine functioning smoothly is the fact that most engines have more than one piston. So, when one piston is in the compression stroke, and so not producing any rotational energy, another piston is in the power stroke, pushing the crankshaft around. Engines can have any number of cylinders, from 1 to as many as 12 or more. In fact, Cadillac produced a concept car in 2003 called the Sixteen that had, you guessed it, 16 cylinders. The cylinders are made to begin ignition in succession at even intervals, called the **firing order**. This allows for the cylinders to work together keeping the crankshaft rotating more uniformly, rather than abruptly jerking with each power stroke or slowing between strokes. It can also allow us to ensure that two adjacent cylinders aren't firing at the same time, since they'd both be trying to draw in a charge simultaneously, potentially causing one to draw fuel and air away from the other, a problem called **induction robbery**.

Holding all this together is the main structure of the engine, called the **engine block**. An engine block can be made of cast iron, aluminum, or more advanced materials, but it's basic character is the same: a large solid structure that contains the cylinders and holds these key components in place. The bottom of the block is often called the **crankcase** because it holds the crankshaft. The cylinders are usually located in the upper portion of the block. The valves are contained in a separate component, called the **cylinder head**, which attaches to the top of the block and completes the combustion chamber.

There are, of course, multiple configurations of cylinders and so multiple shapes to engine blocks. The common V8 gets its name from the V-like shape defined by the cylinders. This is a common shape because it allows a larger number of cylinders to fit in a relatively compact package. Alternatively, with fewer cylinders, it is possible to line up the cylinders in an in-line or 'straight' configuration and still fit it under the hood. Of course, you could line up eight cylinders in a 'straight 8' configuration, and enjoy greater power and smooth operation, if you have the space under the hood, but that would require a really long hood. Smaller cars might opt for an in-line four cylinder. If the engine is cooled by air, rather than a fluid cooling system, there's an advantage to having the cylinders on opposing sides of the block to achieve maximum airflow over each cylinder. This 'horizontally opposed' configuration was used for many years by Volkswagen in its air-cooled Beetles and buses. It's the type of engine that powered the ridiculously cool Porsche 911 for more than three decades. And is still the norm in air-cooled small aircraft engines. In short, there are many variations on cylinder configurations, each presenting trade-offs in size, fit, power, and other characteristics that define the appropriate engine for any given application.

Because the engine block contains the combustion chambers, it has to absorb a lot of heat. To help manage the engine's temperature, the block is designed with small channels that allow coolant to flow throughout the block (Image 1.7). The coolant that flows through these **cooling jackets** then flows through the radiator, allowing for the dissipation of excess heat and proper engine temperature regulation. Similarly, in order to both control temperature and reduce friction and wear, oil is used to provide a lubricating film between moving parts. The bottom of the block is fitted with an oil pan, to contain this oil as well as an oil pump to properly circulate the oil under pressure throughout the engine.

To minimize wear and heat buildup in the block, and to keep all the moving parts in their proper location, all the spots where moving parts would rub against each other are designed to minimize friction. The walls of the combustion chamber are a key example. They must not only endure wear and remain dimensionally stable at high temperatures but also provide minimal friction. Precisely defined patterns of small scratches, called **cross-hatch** by mechanics and **micro-asperities** by engineers (who like to sound smart),

IMAGE 1.7
Cooling the engine.

This cut-away clearly shows the cylinders as well as the oil passages and coolant jackets that channel through the block around the cylinders to provide cooling and lubrication.

provide a surface that is better at allowing oil to cling to the combustion walls and reducing friction with the piston. This texturing of the surface allows film lubrication to produce **hydrodynamic lift**. Recent research indicates that dimpling may provide similar advantages, so look for that as a possible future innovation.

Other junctions of moving parts, for example where the block holds the crankshaft or camshaft, use specially designed fittings. Because these fittings 'bear' the force and friction required to hold the component in place, they are called **bearings**. The crankshaft is held in place with **main bearings**, with the specialized bearings meant to hold the crankshaft from sliding forward called **thrust bearings**. Similarly, the connecting rod is linked to the crankshaft with **rod bearings**. Oil is drawn by the oil pump from the pan and pumped through channels in the block and into the crankshaft to all these bearings, so that the moving parts do not actually rub against each other but are suspended by a pressurized layer of lubricating oil, called **hydrodynamic lubrication**. Oil is also delivered to coat the stems of the valves, and other moving parts in the head. It is estimated that on average 10%–30% of engine output is lost to internal friction; so, reduction of friction is a key element of performance and efficiency innovations in modern engines.

As mentioned, the head serves to close off the upper end of the chamber, creating a tight seal with the block, to ensure a sealed cylinder. It must also define a tight seat for the valves, allowing them to open and close freely, while ensuring the valves seal tightly when closed (Image 1.8). It also holds the spark plug in place and defines the basic geometry of the upper portion of the combustion chamber. As we will see later, variations in the shape

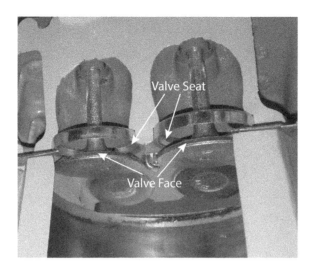

IMAGE 1.8
Valve seats.

The mating surface, or **valve face**, of intake and exhaust valves are precisely machined to provide a tight seal with the **valve seat**. Precisely defined angles can be cut into the face to ensure this.

of the combustion chamber can have a major impact on the performance of the engine. Because the combustion chamber needs to ensure a pressure tight seal, a **head gasket** composed of multiple thin layers of metal is used at the connection of the block and head. If this gasket fails, it is possible coolant from the coolant jackets in the block could get into the cylinders. Since coolant cannot be compressed, this can lead to catastrophe for an engine. When the piston comes up in the compression stroke and is met by an incompressible fluid, something's got to give.

Valve Train

The key to this whole thing working is having the valves open and close at the right time, this is called **valve timing**. It's pretty clear that the system only works if the intake valve is actually open during the intake stroke, or the exhaust valve is open during the exhaust stroke. Just as obvious, if the intake valve were open during the power stroke, the result would not be good. To ensure that everything operates in harmony, we connect the valves to the rotation of the crankshaft through a smaller shaft called a camshaft (Image 1.9).

The camshaft has a series of eccentric lobes, one for each valve. As the shaft rotates, the lobes will raise and lower each valve a defined amount, called lift, causing them to open and close (Image 1.10).

The camshaft is connected to the crankshaft so that it rotates in harmony with the crankshaft and overall engine operation. Since each valve needs to open and close once with each four-stroke cycle, the camshaft will need to turn once each time the crankshaft turns twice to complete a four-stroke cycle. The connection can be made with simple gears, chain, or a belt. The gears and chain are durable, but they can make an engine nosier (Image 1.11). The belt is quieter, but it's more susceptible to wear and needs to be replaced more frequently.

IMAGE 1.9
Camshafts.

The eccentric lobes on the cams of the camshaft define the movement, or 'lift', and timing of the valves.

Source: ThyssenKrupp Presta Chemnitz GmbH

IMAGE 1.10
The valve train.

When the camshaft is located above the valves, it's called an overhead cam. As the camshafts rotate, they push on the valve stems, causing the valves to open and close in harmony with the cylinder's strokes.

Image: Volvo

The amount each valve rises is tied to the lift defined by the camshaft and is critical to the operation of the engine. A small lift will create a small opening, and so allow less charge to enter the chamber or less exhaust to exit the chamber. A larger opening will increase the ability of the engine to 'breath' by allowing more charge in and more exhaust out. But this could potentially outstrip the capacity of the engine, or adversely affect fuel economy or emission by allowing more fuel into the combustion chamber than can be

IMAGE 1.11
Advanced chain drive.

Timing belts typically offer reduced noise when compared to timing chains, but must be replaced before failure, as the failure of a timing belt can mean a valve is extended into the combustion chamber when the piston moves up. If the piston hits the open valve in its upward stroke, the results won't be good for either one. This advanced chain drive by BorgWarner attempts to let you have your cake and eat it too offering the reliability of a chain and the performance of a belt. The system promises improved fuel economy with an inverted 'silent tooth' chain.

Image: BorgWarner

effectively burned. It's not uncommon for gearheads to swap stock 'cams' for performance cams that have a more aggressive profile (more lift); and this can allow an engine to produce more power. Sometimes this makes sense, sometimes not so much.

Defining the Combustion Chamber

As previously noted, the heart of the internal combustion engine is combustion; and the shapes and materials that define the combustion chamber in turn define the combustion itself and fundamentally define the performance of any engine. Like just about any other component of an automobile, designing the elements of a combustion chamber has been a game of trade-offs. You want components that are light but also strong, and those two things don't always go together. You want parts that are free moving when cool but also when very hot, again at times a tough requirement. Nevertheless, new technologies are

IMAGE 1.12
Bringing it all together.

This 1.5-liter, three-cylinder, direct-injection engine by Volvo demonstrates modular design, aimed at enabling greater powertrain options while still benefiting from economies of scale in production. As advanced as it is, the basic components remain the same. Like any engine, it has a block, head, valves, valve train, and crankcase. You can't see the pistons, but they're in there.

Image: Volvo

making these compromises less compromising, offering new possibilities that sometimes allow us to have our cake and eat it too.

Innovations in design and material have even changed something as fundamental as the block itself. The block may seem like a fairly simple component at first glance, but look again. It needs to be strong enough to define the combustion chambers and hold the engine componentry together under tremendous loads. It must be capable of being formed and manufactured precisely, to allow for connecting points, coolant and lubricating channels that permeate the block and allow for proper lubrication and cooling of the engine. It needs to be able to withstand and transfer intense heat, and endure high loads. And, it needs to be as light as possible. Because of the relative mass of the block, even a small percentage change in weight would mean a significant change in the overall engine weight.

Happily, advances in materials and manufacturing are redefining this once-simple component. A generation or two ago, blocks were nearly universally made of cast iron, a durable material that had the strength to hold engine parts together, absorbed high temperatures, and offer a workable, machinable material at low cost. But this is changing. Aluminum–silicon alloys, magnesium, and advanced composites are changing the way we think about engine blocks.

In fact, aluminum alloys have been used in engine blocks increasingly since the 1970s, but not without some challenges. The block is more than simply a container for the other components of the engine; it defines the major surface area of the combustion chamber.

So, it must be thermally stable and able to endure significant wear. Until recently, aluminum blocks required the use of a steel sleeve to handle the wear of the piston sliding against the cylinder. Because this steel sleeve adds to the weight and size of the engine, numerous alternatives have been used, with varying success. Most of these approaches try to remove the aluminum near the surface of the cylinder electrochemically; this leaves a hardwearing silicon layer to define a more capable cylinder wall without adding the weight of a steel sleeve. While the relative merits of various techniques that accomplish this are debated, all present some challenges. Another option might be the installation of a removable cylinder liner that is designed to have coolant flow around it, what's called a **wet liner**. While popular with a few French manufacturers, these still present a significant increase in engine weight and size, diminishing the intended aim of using an aluminum block. Recent work with low-friction plasma metal coating may signal the possibility of aluminum or magnesium blocks with no steel liner, making it much lighter and smaller. The use of this new technology on the GT500 Shelby Mustang cut the engine weight and actually used recycled engines to provide the raw material.[1] Called Plasma Transferred Wire Arc (PTWA) technology, the process blows a fine mist of molten steel plasma onto the cylinder walls to create a hard-wearing finish without a cylinder liner.

As strong as aluminum but lighter, magnesium shows real promise as an engine material. This is not a new idea; Volkswagen and Porsche manufactured engine blocks out of magnesium back in the 1960s, but not without challenges. In fact, more recently, when BMW chose to use magnesium in the engine block of its N53 engine, they had to couple it with an aluminum insert forming the cylinders and coolant channels. A newer magnesium alloy, with the catchy name AMC-SC1, is designed specifically for engine blocks and may offer a block that is lighter than aluminum but with greater strength and similar manufacturing requirements and cost.[2] This isn't exactly a fast-track technology; but don't be surprised if you see more magnesium powertrain components in the future.

Perhaps more promising are advanced metal composites or **metal matrix composites** (MMCs). Combining a metal-binding agent, or **matrix**, as the major component with a reinforcing ceramic, organic, or another metal can allow us to precisely engineer specific characteristics into materials to suit certain purposes. **Compacted graphite cast iron** (CGI) is a promising example of this in the search for a superior block material. CGI has been used previously in the manufacturing of brake components that are strong, lightweight, low wear, and able to transfer heat better than previous materials. Since these are all desirable characteristics of an engine block, the fit seems right. Also called vermicular graphite iron, CGI used thicker graphite particles than exist in typical cast iron (all iron contains some graphite). These particles define thick tentacles of graphite that create a tight interwoven bond with the surrounding iron matrix. So, while more dense than aluminum, the resulting superior strength of CGI means it can be made thinner, resulting in a competitive weight with greater thermal conductivity and excellent internal dampening. The result is a block that is more compact, up to 75% stronger than gray iron and five times more fatigue resistant than aluminum.[3] Recent advances in manufacturing may allow for the more widespread use of CGI and similar MMCs, as some have seen limited use due to low machinability. But high cost is the number one barrier.

---

[1] "Ford Developed High-Tech Plasma Process that can Save an Engine from the Scrapyard while Reducing CO_2 Emission." *Ford Media Center* December, 2015.

[2] C.J. Bettles et al., AMC-SC1: A New Magnesium Alloy Suitable for Powertrain Applications. *SAE Technical Papers*, March, 2003.

[3] P.K. Mallick, Advanced materials for automotive applications: An overview. In J. Rowe (ed) *Advanced Materials in Automotive Engineering*. Woodhead Publishing, Cambridge, 2012, 5–27.

Some have explored a far more surprising composite material for an engine block: carbon fiber. Basically a plastic matrix with long strands of engineered carbon added for strength, carbon fiber is usually associated with low heat and modest strength. (The base material is plastic, after all.) However, a carbon fiber block for specialized racing has seen some preliminary testing.[4] This combination of plastic reinforced with carbon strands is many times lighter than iron, and about half the weight of an aluminum block. It would be an exceptional material for a block if it could handle the heat and force. If these issues can be worked out, we may see composite materials used for bocks in production cars sometime in the future. And if not, this still gives us an impressive example of the sorts of innovations being pursued. We'll look more closely at advanced metals and composites in Chapter 8.

Pistons

The piston is clearly a central component in any engine. After all, the piston defines the core function of the engine, converting combustion into rotation. So, even small variations in the shape, weight, and design of this component alter the performance of an engine in fundamental ways. Today's engines are smaller, hotter, and faster than their predecessors. And today's piston needs to endure tremendous pressure—over 6 tons of force every two-tenths of a second at 6,000 rpm—and intense heat—regularly over 500°F—and convert it as efficiently as possible into linear mechanical movement. And it must be engineered precisely enough to make an exact fit with the cylinder wall, but not so tight that it can't slide. In addition it needs to maintain this precision fit under significant heat variations. If it's not obvious, achieving all of this is not easy.

At the dawn of the automobile, pistons were large and made of cast iron. With the metallurgy of the time, only a bulky chunk of iron was thought capable of absorbing the force and wear inside an engine. However, early on it was recognized that iron's poor ability to shed heat led to high engine head temperatures and a resulting problem in the combustion chamber as the air fuel mixture expanded with the heat. This expansion meant a less-dense fuel–air charge and a resulting reduction in the power the engine could produce. So, by the 1920s, aluminum was the common material for pistons since it has more than three times iron's ability to move heat.

To make this a bit tougher, like any oscillating or rotating engine component, piston weight is a critical element of performance. Because changing the momentum of fast-moving parts requires significant energy, any weight saved on moving parts such as the crankshaft, rods, valves and pistons is far more definitive than the same weight reduction on the block or other stationary parts. At high rpm, a typical piston can accelerate to 50 mph, back to zero, and back to 50 mph in the opposite direction in less than a millisecond. Heavy parts take more energy and time to get moving and come to a stop, so putting these parts on a diet is good for acceleration, maximum operating speed, and fuel economy.

As a result, today's pistons are smaller, lighter, and more durable and precisely engineered than ever. More precise manufacturing has allowed for more exact and complex head shapes, stronger asymmetric lower walls to the piston, called skirts, and much

---

[4] D. Sherman, "Is This the Engine of the Future? In-depth with Matti Holtzberg and His Composite Engine Block". *Car and Driver* May 6, 2011.

thinner walls, all while maintaining strength (Image 1.13). Similarly, reducing the piston mass and improved alloys has also addressed the challenge of thermal expansion, allowing much tighter tolerance, or closer fit, within the cylinder.

Managing piston heat is no easy trick. The piston is exposed to the full heat of combustion most directly, and unlike the block, doesn't have a large mass to help absorb and dissipate that heat. What's more, because all the parts of these lighter and thinner pistons don't heat up at the same rate, thermal expansion also requires exact asymmetries and tappers that allow the maintenance of precise clearances even as extreme and uneven heat is applied.

Already thinner and lighter, these new pistons need to be made stronger as well. With rising engine speeds and pressure, ring grooves are hardened with anodizing or laser hardening to improve strength. The fitting of cast iron or steel insert to reinforce the top ring groove, a practice often used in diesel engines, is now being considered for aluminum gasoline pistons. Taking another cue from diesel pistons, high-performance gasoline pistons are now being manufactured as two fused parts to allow for cooling channels in the piston that can help move heat from the head to the ring pack. Sloshing oil in these cooling galleries creates a cooling effect, sometimes called **cocktail shaker cooling**, which can dramatically reduce piston thermal loading.

Designing a high-performance piston requires that we also think about the friction and wear on piston surfaces. The sliding of pistons against the cylinder walls is the largest bearing surface in the engine that is not fed by pressurized oil, and this alone can account for 5% to nearly 8% of your fuel use.[5] This friction loss is as great as the crankshaft and valve train friction loss combined. So, to reduce energy loss from friction, skirts are now

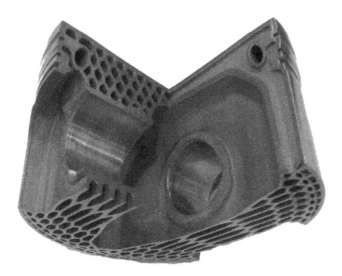

IMAGE 1.13
Advanced piston manufacturing.

Future piston structures may be defined with 3D printing. IAV Automotive Engineering is developing this piston with a honeycomb lattice to define the piston's structure. IAV reports exceptional strength, about 20% less weight than its conventional counterpart, and reduced thermal expansion.

---

[5] C. Kirner, J. Halbhuber, B. Uhlig, A. Oliva, S. Graf and G. Wachtmeister, Experimental and simulative research advances in the piston assembly of an internal combustion engine. *Tribology International* 99, 2016, 159–168.

printed with a thin graphite-infused, low-friction patch. A recently developed piston coating that includes graphite, molybdenum disulfide, and carbon fiber promises a full 10% friction reduction over an uncoated piston (Image 1.14).[6]

Even with this increasing sophistication and precision, the aluminum piston is now facing a challenge from a high-tech version of its old iron rival made of new high-strength steel alloys for diesel passenger car applications. Because of the greater strength of this new steel, the piston can be made smaller and lighter. In particular, the distance from the piston pin to the upper surface of the piston, called the crown, can be shortened. This allows for less mass and potentially a longer connecting rod and thus a larger displacement in the same overall size engine. Or, with the same displacement, a shorter block could be used, allowing a significant reduction in engine size and weight. Because steel offers lower heat conductivity than aluminum, the steel piston must make use of cooling galleries, which significantly complicates the manufacturing process since the piston must be made in two parts. Because steel expands less with heat than aluminum, the challenge of accommodating thermal expansion is eased a bit. Whether gasoline engine applications of an advanced steel piston could be developed is still to be seen. We'll look more closely at advanced steel in Chapter 8.

Another piston option that has been touted by producers of aftermarket performance parts but not by OEM producers is ceramic-coated piston crowns. The idea is to produce a strong, insulating coating on the top of pistons. Zirconium ceramic fits the bill well, as it provides good insulation and thermal expansion behavior that matches the metal and thus allows for an enduring bond.[7] The hope is that this insulation will help maintain the heat energy in the cylinder, increasing thermal efficiency; and evidence indicates this is

IMAGE 1.14
Steel piston.

Aluminum pistons are now common in all production automobiles, with cast aluminum common in fleet vehicles and hypereutectic or stronger forged aluminum used in performance applications, but pistons manufactured from new steel alloys such as this are set to change this for diesel engines. This monotherm steel piston made by Mahle was the first made for light passenger cars.

Source: Mahle

---

[6] M. Ross, "Pumped Up: Piston Evolution." *Engine Technology International.com* March, 2015.
[7] D. Das, K.R. Sharma and G. Majumdar, Review of emission characteristics of low heat rejection internal combustion engines. *International Journal of Environmental Engineering and Management* 4 (4), 2013, 309–314.

the case.[8] However, the process can be expensive, the gains are modest, and the resulting increased combustion temperature also tends to produce higher NO_x emission, making this a problematic adoption for production cars.[9]

As goes the piston, so goes the connecting rod. Like the piston, the rod has been exposed to increasing pressure and heat. Performance rods are often made of forged steel, though aluminum alloy rods are not unusual. Titanium offers extremely light and strong performance rods. And because of a rod's simpler design requirements, it's easier and more cost-effective to make rods out of titanium than pistons. Aluminum MMCs also offer some promise. In fact, because it's not exposed to the same extreme temperatures as the piston, the connecting rod offers many promising avenues for weight reduction. Newer research is examining the use of composite materials for connecting rods, with Lamborghini exploring the possibility of a carbon fiber rod that is half the weight of a conventional steel rod.[10]

Similarly, the design of piston rings has evolved over time. Rings provide the seal against the cylinder wall, but in defining that tight fit, they also account for a majority of the friction in the engine. Classically, an upper ring provided a tight seal, called the **compression ring**, and a lower ring ensured that engine oil from the crankcase didn't make its way in the combustion chamber, called an **oil control ring** or scraper. This lower ring allows excess oil to be cleaned off the cylinder walls and returned to the crankcase through oil drain holes in the ring groove, leaving only a slight lubricating film, enough to allow movement of the piston without excessive fouling of the combustion chamber. There are multiple variations in the number, cross-sectional design, and the placement of the rings, though generally automobile manufacturers have moved to smaller **ring packs** to reduce friction loss, with some recent compression rings less than a millimeter thick, less than half the thickness of typical rings a decade ago.[11] Ceramic coating is also used to reduce ring friction and wear, and new application technologies are significantly reducing the cost of this once-expensive option (Images 1.15 and 1.16).[12]

The Head

The head, and the valves it contains, define the last remaining wall of the combustion chamber. The head must be engineered precisely enough to define the **top end** of the combustion chamber, ensure a tight seal with the block, and provide precision-machined **valve seats** that are able to maintain an exact fit at high temperatures.

---

[8] K.S. Mahajan and S.H. Deshmukh, Structural and thermal analysis of piston. *International Journal of Current Engineering and Technology* 5 (June), 2016, 22–29. Available at http://inpressco.com/category/ijcet; K. Thiruselvam, Thermal barrier coatings in internal combustion engine. *Journal of Chemical and Pharmaceutical Sciences* 7, 2015, 413–18; and A. Sh. Khusainov, A.A. Glushchenko, Theoretical prerequisites for lowering piston temperature in internal combustion engines. *International Conference on Industrial Engineering, ICIE 2016 Procedia Engineering* 150, 2016, 1363–1367.

[9] D. Das, K.R. Sharma and G. Majumdar, Review of emission characteristics of low heat rejection internal combustion engines. *International Journal of Environmental Engineering and Management* 4 (4) 2013, 309–314. Available at http://www.ripublication.com/ ijeem.htm

[10] D. Undercoffler, "Lambo Expands Carbon-fiber Footprint." *Automotive News* July 4, 2016.

[11] V.W. Wong and S.C. Tung, Overview of automotive engine friction and reduction trends—Effects of surface, material, and lubricant-additive technologies. *Friction* 4(1), 2016, 1–28.

[12] L. Kamo, P. Saad, W. Bryzik and M. Mekari, Ceramic coated piston rings for internal combustion engines. *Proceedings of WTC2005 World Tribology Congress III* September 12–16, 2005, Washington, DC, USA.

IMAGE 1.15
Advanced piston.

Pistons have advanced significantly. Ring packs have become much smaller and slipperier than they once were. And cooling channels, such as the one in this Federal-Mogul Elastothermic piston, can reduce piston crown temperatures by about a fifth.

Source: ©2018 Federal-Mogul LLC

IMAGE 1.16
Piston coatings.

Ceramic coatings can reduce surface friction by a third. This advanced coating by FederalMogul includes solid lubricants and carbon fibers, offering reduced friction and significantly improved wear.

Source: ©2018 Federal-Mogul LLC

Perhaps the most noticeable variation in the design of the head is how many valves are integrated into the combustion chamber and in exactly what way. While the classic engine had one intake and one exhaust valve, by the early 1980s this convention was regularly challenged in an effort to allow greater flow in and out of the cylinder. In fact, in the mid-1980s, an experimental Maserati V6 2.0-liter turbo used six valves per cylinder, defining a very tight fit. While six-valve heads are decidedly not common, four valves per cylinder certainly are. Particularly with the greater demands being placed on smaller engines, multivalve heads accommodate higher rpm, allowing the engine to breath at a much faster rate. Because the incoming charge is cooler than the high-heat, high-energy exhaust, a common variation is to increase the size or number of the intake valves specifically. This allows more area for the relatively cool, heavy and slower-moving intake charge to enter the cylinder. There's reason for caution, however. More valves are not always better and can complicate the valve train and constrain head design options. Similarly, the common performance upgrade of installing oversized valves in hopes of producing more horsepower can be a fool's errand, as the valve size is frequently not the performance bottleneck.

The larger question is how do we integrate these valves into the head, and what geometry will we use to shape the upper combustion chamber? There are three primary considerations at play in answering this question: First, maintaining the high energy and full emulsification of the incoming fuel–air mixture, principally by producing an agitated flow into the chamber that keeps the kinetic energy of the incoming charge high. Second, intensifying the energy of the vaporized charge in the last stage of compression to prepare the charge for combustion, we call this **squish**. And, third, removing the exhausted charge in the chamber quickly and completely after combustion to allow it to be replaced with a fresh charge, we call this **scavenging**. Let's look at each of these.

A priority is to ensure a highly energized flow into the cylinder as this helps produce a desirable and fast-moving flame front upon combustion. The placement of the valves and shape of the piston are vital here. Positioning the valves so that the intake occurs obliquely, with both an end-over-end rotation, called **tumble**, and a spiral rotation of the mixture, called **swirl**, is the goal. Without this energy, the charge might lose the integrity of its targeted mixture profile, and the resulting flame front from a low energy charge may move too slowly for a high-speed engine (Image 1.17).

A clean and fast flame front is also encouraged by shaping the piston and head to define a desired squish pattern. **Squish** is the final squeezing of the charge in the last bit of the compression stroke. Ideally this injects a burst of energy into the fuel–air mixture. With contemporary engines designed to make maximum use of this effect, the area between the

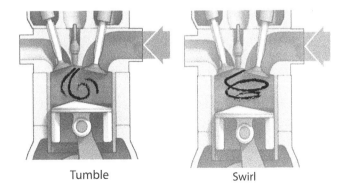

Tumble   Swirl

IMAGE 1.17
Swirl and tumble.

Proper swirl and tumble patterns are key variables in the vaporization and emulsification of charge delivery in a high-performing engine. Producing a high-energy charge promotes a faster flame front that can meet the needs of high-speed performance.

head and piston at TDC, called the **clearance volume**, is tiny by the standards of past generations. Modern engine pistons come within a single millimeter of the head. This brings the charge into an unstable and highly energized state, greatly speeding the rate of combustion and allowing the piston to absorb as much of the combustion energy as possible. The chamber shape and piston head are designed to make maximum use of this energy. But there's a trick to this: you need to avoid preignition and maintain low emission of NO_x which tend to increase with rising ignition temperature. We'll look at just how that's done in the next chapter.

Scavenging is yet another consideration. The shape and dynamics that keep an incoming charge energized and turbulent can also be used the help eject the exhaust out of the cylinder. One method is allowing for some **valve overlap**, a short period when the exhaust valve hasn't yet completely closed and the intake valve begins to open. This can allow for **cross-flow** thru the combustion chamber, with some of the energy of the incoming air being used to push the remaining exhaust out of the chamber. The effect can be facilitated by placing the valves opposed to each other so that flow from one to the other drafts through the cylinder.

So, how do we shape the head to ensure that we achieve all these aims? As you might expect, there's no single answer. Over the years, there have been any number of combustion chamber designs. Early 'flat heads', for example, had valves either side by side (called a T head) or across from each other (L head) that faced upward and were adjacent to the cylinder. Once common, these configurations are now obsolete, as they can't offer the sort of compression and control desired in modern, high-performance automobiles. More recently, engines have tended toward some variation of three head designs. The first two utilize a so-called I-head that mounts valves facing the piston directly and can thus allow the upward facing valve stems to connect with one or two camshafts in the head, thus called an **overhead cam**. These can commonly entail either an inverted cup-shaped chamber that has earned it the name **bathtub** chamber, or an angled design with one side higher than the other, forming a **wedge** chamber.

A **heart-shaped** chamber is a common variation on the bathtub design that is defined by two squish regions, typically a large circle at the spark plug and a smaller one on the opposing end. The result forms a crescent or heart shape. The spark plug in the center of the head favors a desirable flame front. However, having the valves close to each other makes heat transfer a problem, limiting octane tolerance of the engine (Image 1.18).

A third common configuration is the **hemispherical head**. This may be the most well-known head design, largely thanks to Dodge's promotion of its 'hemi' engine. The rounded chamber top offers good geometry for turbulence and facilitates the use of opposing valve placement and so allows excellent cross-flow scavenging. In addition, because the valves are on opposing sides of the head, they are more thermally isolated. This helps keep the valves cooler and so helps avoiding knocking. A variant of this design can accommodate multiple valves and is called a **pentroof**, or penta, design. Because it must provide multiple valve seats, it tends to have flatter sides.

New innovations in the material and manufacturing have also helped improve valve design. Stainless steel valves are now common and available in various alloys that can improve hardness and heat resistance. Recently developed titanium valves can be 40% lighter than stainless steel valves. As previously discussed, the weight of oscillating components can be definitive to performance. Valve weight in particular can be a limiting factor on maximum engine speed. Valves that are even just a slight bit lighter can significantly reduce the energy needed to operate the valves and so allow for more aggressive valve lift. This can mean significant rewards in engine response and speed. Sodium-filled

IMAGE 1.18
Chamber designs.

Just about every manufacturer has used some version of a wedge design (upper left). With the valves aligned side by side on the long wall of an asymmetric wedge, the spark plug is placed on the opposing short side. The walls of this design allow for advantageous tumble. And the intense push of the charge from the narrow to the full side of the wedge at the late point of compression offers desirable squish. The heart-shaped head (upper right) represents a common head design. Note the close proximity of the valves.

The bathtub chamber (lover left) defines a symmetrical head shape that can place valves upright or at slight opposing angles. The hemispherical head (lower right) may be the most well-known head design, largely thanks to Dodge's promotion of its 'hemi' engine with excellent cross-flow scavenging and good thermal isolation for the valves, the design deserves some praise.

valves offer another interesting example. As the sodium liquefies with heat, it shifts back and forth through the stem of the valve, transferring heat away from the head, another example of shaker cooling.

Ignition

With the combustion chamber defined, and the charge primed and ready for ignition, the obvious question arises: when do we 'light 'er up'? The ignition of the charge has to come at the right time, not too soon, since this might cause the cylinder pressure to rise too high before the piston completes the compression stroke; and not too late, since this might mean we'll lose some of our ability to absorb the combustion energy before the end of the power stroke.

Because ignition timing needs to change while the engine operates, it's not quite as simple as the timing of the valve and piston movement. These two parts can move in synch with no trouble; every turn of the crankshaft needs half a turn of the camshaft, no matter what engine speed, no matter what driving conditions. But, for ignition timing, the situation changes. As previously noted, combustion is a process. The flame front initiated by the spark plug is not instantaneous, it takes time to pass through the combustion chamber, and that time varies as a result of the energy in the charge and the richness of the mixture. Leaner air–fuel ratios define a slower-moving flame front. High energy due to intense compression or high heat can speed the flame front. And, of course, the engine moves at dramatically different speeds, sometimes rotating at an idle, say about 800 rpm, and sometimes at as much as 6,000 or more rpm. Because we need the flame front to develop it's push at just the right time, this means we need to adjust the ignition event to account for engine speed and mixture, so that the main push of the flame front occurs when the piston is early in the downward stroke and can absorb the energy effectively.

Consequently, to ensure proper timing of the flame front, we need to initiate ignition before the piston reaches the top of the compression stroke, or TDC. This is called **ignition advance**. If we didn't have an advance, and the spark plug ignited at TDC, the main push of combustion would occur late in the power stroke, and the piston would reach the bottom of the stroke before the combustion force was complete. The faster the engine is rotating, the worse this effect would get, since the flame front moves at a relatively fixed rate and won't speed up to match engine speed. We measure the advance of the ignition in degrees of crankshaft rotation before TDC, or BTDC, and so talk about advance in number of degrees.

This means that at low rpm, we might want the spark plug to fire 15° BTDC. But at high rpm, we might need ignition to be 30° or more BTDC, to allow the expanding combustion gasses to fill the chamber and provide a maximum amount of power to the piston well before the piston reaches the end of the power stroke (Image 1.19). Older cars had a simple mechanism that could allow for two timing settings, one for low rpm and one for high rpm. New cars allow for much better ignition timing because the engine's operations are computer controlled (lots more on this in the next chapter). Ideally, we'd like to advance the ignition timing as much as we can, to allow us to harvest as much of the energy from the power stroke as possible (Image 1.20). The problem is, if we advance the timing too much, we can create or exacerbate engine knocking.

Knocking

In a perfect engine, the arcing of the spark plug will ignite a critical **flame kernel** that will in turn begin the propagation of an even flame front and a resulting fast and powerful but smooth push against the piston head. An abrupt, explosive blow, even though powerful, won't do. Think of this like a playground merry-go-round. To keep it moving, you need even and firm pushes. Imagine trying to get that merry-go-round spinning by slamming it with a sledgehammer. When this sort of uncontrolled or uneven combustion happens in an engine, we call it **knocking**. This is a general term that actually refers to two things that can go wrong: **preignition** and **detonation**.

Preignition, as the name implies, is when the ignition process starts early through autoignition, before the spark plug fires. To fully understand what causes this, we need

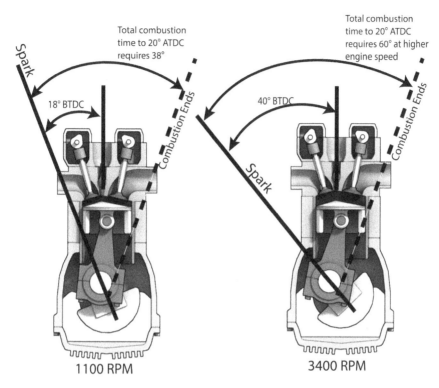

IMAGE 1.19
Ignition advance.

Ensuring that combustion takes place just after the piston reaches top dead center (TDC) requires having initial ignition take place well before TDC, and be adjusted to accommodate engine speed. The faster the engine speed, the greater the ignition advance required.

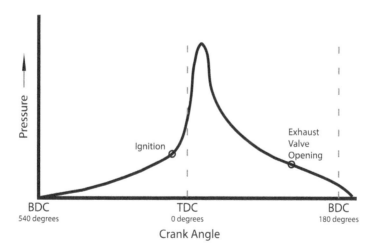

IMAGE 1.20
Combustion chamber pressure.

Ideally, maximum combustion chamber pressure is reached just after TDC, about 15° or so, allowing for maximum recovery of combustion energy by the piston. To ensure proper scavenging, the exhaust valve opens (EVO) before combustion pressure has completely dropped.

to recall a key characteristic of gasoline—octane rating. Remember that the octane rating gives us a sense of the fuel's tolerance to compression. A high octane rating means the fuel can be compressed more without fear of triggering autoignition, and a low octane rating indicates that the fuel is more likely to spontaneously ignite under relatively lower pressure. So, if an engine is run on a fuel with an octane rating that is too low, the heat and pressure caused by the compression stroke could trigger autoignition before the spark plug fires. This might be made more likely if the spark plugs or valves get overheated, adding excess heat to the compression stroke. This preignition is clearly not a good thing. If ignition happens too early, the force of combustion could actually push against the piston while it's still in the compression stroke, effectively trying to turn the engine against its direction of rotation.

A more common form of knocking is detonation. Unlike preignition, detonation does not begin before the spark plug fires. Instead, the spark-initiated ignition increases the heat and pressure in the combustion chamber, causing spontaneous autoignition in at least one other location in the cylinder (Image 1.21). So, combustion in the chamber happens in more than one location at a time, and instead of having a singular kernel that leads to a unified and even push, we have multiple points of ignition that lead to a more disorganized combustion and a much less even push against the piston. The resulting violent combustion is a frequent cause of damage to bearings and pistons.

This threat of detonation is the core challenge of ignition timing. One cause of detonation could be excessive ignition advance. If the spark plug fires too early, the cylinder pressure will increase while the combustion chamber is still relatively large and before the flame front can travel through the chamber, leading to one or more points of unburned

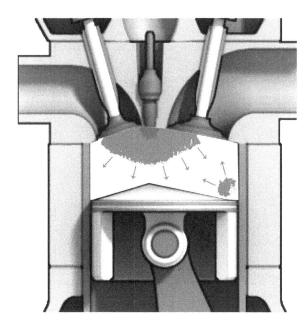

IMAGE 1.21
Knocking.

Desired ignition is defined by a single ignition point and flame front in the combustion chamber. Knocking can be caused by preignition, defined by the autoignition of the fuel–air mixture before the spark plug fires, or detonation, defined by multiple ignition points caused by the increased heat and pressure resulting from spark ignition.

fuel igniting on their own as the chamber's pressure and heat increase beyond the fuel's autoignition point. The solution is to back off on the ignition advance, and allow the piston to travel a bit more before initializing combustion. But, if you back off too much, as discussed, you'll lose the ability to harvest some of the energy from combustion. The balance can be tricky.

A central variable in this effort is how much we compress the charge. This is defined by the engine's **compression ratio**, defining the amount the air is squeezed prior to combustion. This ratio is defined by the proportion of the volume of the combustion chamber when it's at it's largest (bottom dead center) to the volume at its smallest (top dead center). We called the smallest volume **clearance volume** (most often written V_c), the change in volume caused by piston movement is called **displacement volume** (V_d), and so the total volume when the piston is at BDC is the two volumes combined. That means the compression ratio is $(V_c + V_d)/V_c$.

Taking what we've already discussed into account, we can see that defining an engine's compression ratio is a balancing act too. A higher compression ratio can allow us to draw more power from combustion. This could enable us to gain significant improvements in fuel economy or power, as we're able to draw a much greater amount of energy from a given amount of fuel. But, as the compression ratio goes up, so does the need for higher octane fuel that can withstand the higher pressure without autoignition; and, if we compress the charge too much, we invite preignition and detonation. Preignition is more likely when the engine is operating at high load since the engine temperature goes up. So, a high-performance engine might have a compression ratio (CR) 8:1 to avoid knocking at high engine loads, while an engine targeting high efficiency might be closer to 13:1. With advanced control discussed in the next chapter and appropriate fuel, compression ratios as high as 14:1 are possible. By comparison, diesel engines start at 14:1 and go to about 23:1.

This adds complexity to much of what we've discussed so far, since how we manage heat, engine speed, and knocking are all shaped by the compression ratio. For example, a higher compression ratio will increase engine heat, but cooling the piston temperature with galleries could reduce the threat of detonation and thus allow a higher compression ratio. Or a richer mixture can speed the rate of combustion, but it could also cool the combustion chamber and may reduce the possibility of detonation.

Fuel Delivery

Delivering fuel and air to the combustion chamber can be tricky too. To be ready for combustion, the fuel must be atomized (suspended in small bits), emulsified (fully blended with air), and vaporized (changed into a stable and consistent gas). As we discussed previously, without adequate access to oxygen, gasoline will not burn. So, the delivery of fuel and the delivery of air are irrevocably linked. Complicating this somewhat is our desire to avoid the production of nitrogen oxides. NO_x is formed when combustion temperatures are high enough to break apart the nitrogen in the incoming air supply. As a result, **exhaust gas recirculation** (EGR) has been used since the 1970s to return a small portion of the exhaust gases to the combustion chamber in the next intake stroke as a way of cooling the temperature of combustion and reducing the formation of these undesirable pollutants.

Early engines did not offer any real opportunity for precise control of the mixture. The engine that powered the Wright brother's first flight, for example, simply spilled fuel on a manifold and used the engine's heat to help vaporize it. And into the 1980s, automobiles utilized **carburetors** that channeled the engine's incoming air through a narrow passage, a venturi. The resulting vacuum simply sucked up gas from an adjacent small reservoir. The system worked adequately, but the mixture was imperfect, and would not meet the needs of today's engines. Carburetors were replaced by more precisely controlled spray nozzles for gasoline, **fuel injection**, by the start of the 1990s. The early systems simply sprayed fuel through a nozzle into the central throttle that replaced the carburetor, called **throttle body injection** (TBI). In the end, TBI was not too different from the carbureted systems they supplanted.

Both of these systems, the TBI and carburetors, presented an inherent challenge. Because the air and fuel were mixed at the start of the intake manifold, and had to travel through the intake manifold channels to get into the chamber, it was always a challenge to ensure the mixture remained fully emulsified and atomized as it made its way to the cylinder. Suspended fuel particles are heavier than the surrounding air, and thus inclined to settle if the flow calms or slows. So, this was particularly a problem at low engine speeds when induction flow slowed. To address this, the manifold had to be designed to ensure continued high-energy turbulent flow to maintain suspension of the fuel particles. Intuitively, you'd want the path of an intake manifold to be as smooth and large as possible so that plenty of air can reach the cylinders with minimal resistance. But, a large open intake would mean low-energy, slow-moving air, which could easily cause the fuel to drop out of the mixture. Similarly, if you need to ensure the air–fuel mixture remains energized, smooth walls and wide turns are not your friend. You want sharp edges, rough walls and sharp corners to stimulate turbulence. However, such a design also meant high flow resistance, and at higher engine speed this meant a significant power loss. So, the result was a significant compromise in high-end power to ensure continued emulsification of the air–fuel mixture at low engine speeds.

Current systems, however, get around this challenge by adding fuel just before the cylinder, in the case of **port injection**, or in the cylinder itself, with **direct injection**. This eliminates an extended run of fuel–air mixture through a manifold, and thus allows for more precise and independent control of both air and fuel. Manifolds for a port or direct-injected engine carried only air and so could be made smooth and streamlined. This reduced pumping losses and so improved efficiency. An open chamber at the start of the induction system can act as an air reservoir to slow the incoming air and allow a resulting pressure increase. This air reservoir, or **plenum**, can dampen pressure fluctuations and provide higher density air to the inlet channels that feed each cylinder with air, called **intake runners**.

The most common fuel system, port injection, sprays the fuel on the backside of the valve just outside the cylinder. Designing the spray pattern can ensure desirable atomization and a spray and inlet geometry that promotes a beneficial swirl and tumble pattern. The result is a fuel flow that is far more precisely defined than previous carburetor or centralized injection systems. While older designs sprayed fuel continuously, newer sequential port fuel injection provides pulses timed with the intake cycle.

More recently, direct injection offers a notable uptick in precision and control. Borrowing once again from the diesel world, rather than mixing the fuel and air before it enters the cylinder, **gasoline direct injection** (GDI) injects a precise spray of fuel in the cylinder itself. The basic idea has been around for a while, but lack of capable control mechanisms and the challenge of high-pressure precision injection have made this idea problematic and expensive until very recently. The resulting control over the amount and pattern of the fuel delivery offers substantial advantages in fuel economy and performance. Like diesel engines, GDI provides a major decrease in pumping loss at low speed by providing a very lean charge and avoiding the energy loss resulting from drawing intake air through a narrow throttle. Fuel pulses can be altered for a lean burn when acceleration and torque demand is low, a stoichiometric burn for moderate driving, or a rich burn for cold starts or when increased power is demanded.

Direct injection not only allows for precise mixture control, but it can also allow us to more easily vary the distribution of the mixture throughout the charge, defining rich and lean areas of the combustion chamber, though not all GDI systems are capable of this. It is this characteristic that allows for a very lean operation without jeopardizing ignition (Image 1.22). A charge with a consistent air–fuel mixture throughout its volume is called a **homogeneous charge**, and this is the rough characteristic of a conventional engine's intake. The challenge is that a very lean homogeneous charge may not provide enough fuel to initiate combustion at the spark plug. But the richness required to initiate combustion may be excessively rich at low-load conditions and may invite knocking. So, our desire to provide a much more lean charge at low loads can be achieved with a **stratified charge**, defined by a cloud of rich mixture around the sparkplug tapering to a leaner mixture as you move through the chamber. The ability to lean the mixture outside of the initial ignition region significantly reduces the threat of knocking, allows for improved efficiency, and so is a major potential benefit of GDI.

Achieving a precise stratified charge is no easy task. Injection takes place in the later stage of the compression stroke, allowing a very short amount of time to deliver and vaporize the charge. The required injection speed and pressure to make this happen requires a high-pressure fuel delivery that can provide fuel at nearly 1,500 psi, some 30 times the 40–70 psi typically present in a port injection fuel manifold. Shaping the distribution of

IMAGE 1.22
Advanced ignition systems.

Very lean operation or extra EGR can improve efficiency, but it can also make achieving a reliable ignition of the charge challenging. Advanced ignition systems can help address this challenge. For example, BorgWarner's EcoFlash system utilizes high-frequency ignition (HFI) to produce an extended corona discharge. The system can ignite the hydrocarbons throughout the combustion chamber virtually simultaneously, allowing for the precise and reliable ignition of a very lean charge. Comparing the corona image on the left to a conventional sparkplug on the right gives a clear sense of the difference.

Image: BorgWarner

this high-pressure injection is also a challenge, and has generally been accomplished with three approaches: a **wall-guided** process that uses a specifically designed piston head and the chamber wall to shape the injection spray; an **air-guided** process that relies on swirl and tumble pattern; or a **spray-guided** process, using precisely defined mist from an injector adjacent to the plug to place a rich cloud directly at the plug location.[13] While any direct-injection system will have some characteristics of each of these approaches, each system typically adopts one as the primary determinant of chamber dynamics.

A wall-guided, or geometry-guided, approach has the advantage of allowing the greatest amount of time between injection and ignition, since injection begins much earlier in the compression stroke. Typically, the injector is located at the side of the chamber near the intake valves, with the spark plug at the center of the chamber. Proximity to the intake valves allows the fuel spray to mix with the incoming air more easily, and helps keep the injector cooler. A specially designed piston head receives the injection spray and shapes it in the combustion chamber as it moves upward (Image 1.23). Often a deep bowl piston head is used. The challenge is that this piston geometry can make the attainment of a homogeneous charge more difficult. And, the spraying of fuel directly on the piston and cylinder walls can lead to incomplete vaporization, or **wetting**, and resulting higher emissions of CO and unburnt hydrocarbons.[14] As a result, this approach is not typically used alone.

An air-guided system is similar, though the swirl of the intake is more definitive in shaping the charge, so the design of the inlet ducts becomes more critical and the shape of

IMAGE 1.23
Piston crown.

A wall-guided GDI system relies on the interaction of the injector spray pattern and piston head geometry to shape the incoming fuel charge. The shape of the piston crown in this Buick Enclave engine is designed to receive the injector spray.

---

[13] G. Fiengo, A. di Gaeta, A. Palladino, and V. Giglio, *Common Rail System for GDI Engines.* Springer Briefs in Electrical and Computer Engineering. Springer, London, 2013.
[14] G. Fiengo, A. di Gaeta, A. Palladino, and V. Giglio, *Common Rail System for GDI Engines.* Springer Briefs in Electrical and Computer Engineering. Springer, London, 2013.

the piston head less so. The intake ports are specifically designed to develop the targeted swirl and tumble pattern, often oriented more vertically and using baffles or narrowing to define a distinct tumble pattern that will help concentrate the fuel mixture at the spark-plug. In some systems with two intake valves, only one value operates in stratified mode, to allow for a more defined, high-velocity pattern. The challenge is that while the piston head geometry is not as vital to defining the charge geometry and so piston wetting is reduced, a flow-guided system does not fully resolve the performance and efficiency limits at high load previously mentioned.

These challenges can be largely met with more recently developed spray-guided GDI systems. This approach locates the injector close to the spark plug in the center of the chamber and can therefore define more precise spray dynamics (Image 1.24). This allows for a more reliable and symmetrical fuel distribution and more effective utilization of the air charge, and so results in the best efficiency and performance at differing engine speeds. Better injector control can also eliminate wetting and allow for a much higher level of EGR, as the rich cloud around the sparkplug will avoid the misfire otherwise associated with a higher level of exhaust gas in the chamber. However, the injector location next to the spark plug presents some challenges as well. It can cause undesirable heating of the injector and

IMAGE 1.24
Mazda SkyActiv fuel injection.

Mazda uses a notable piston cavity to define a stratified air–fuel mixture around the spark plug. This allows delayed ignition timing during warm-up to speed emissions system warming and helps avoid the flame front contacting the piston head too soon.

Image: Mazda

lead to plug fouling and soot formation. The soot deposit can also lead to disturbed injector orifices and thus compromised spray patterns. So, while a spray-defined system can offer greater efficiency, it also requires not only a robust sparkplug that can accommodate the thermal stress of repeated cooling and heating, but a very precise control of the fuel delivery (Image 1.25).

The needed precise control is provided by what is called a **piezoelectric injector**. This injection technology utilizes a crystalline material that we will see a lot of in advanced automotive components. Within the crystalline structure are opposing positive and negative charges, called dipoles. When the crystal is exposed to an electric charge these dipoles react, and the material physically expands. In this case, this expansion can be used with a levering mechanism that amplifies the movement to open and close an injector orifice. The result is far more precise control of the fuel spray, with pulses as short as a few milliseconds. Classic **solenoid injectors** utilize electromagnets to actuate injection pulses. While cost-effective, this technology has a longer response time and is not capable of graduated control, making it difficult to provide the precision spray needed for GDI systems. By contrast, varying the voltage applied to piezo injector allows for variation in the extent of the reaction and regulated control of the injector opening. The result is a highly responsive very precise delivery of fuel that can accurately control the volume of fuel injection, from 1 to 150 mg/pulse. This also enables the delivery of multiple pulses per cycle, called **split injection**, and can enable very lean operation.[15]

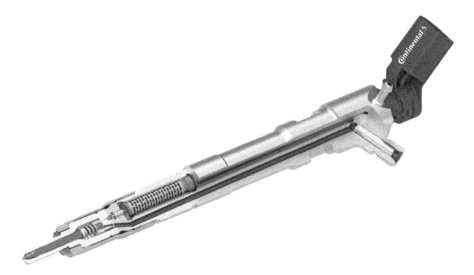

IMAGE 1.25
The piezoelectric injector.

This piezo common rail injector by Continental offers an impressive example of a cutting-edge piezoelectric application. With an ability to handle pressures over 36,000 psi (2,500 bar), a piezo-servo drive, and self-adjusting servo valve and needle control, this unit can provide extremely precise fuel regulation, variable injection patterns and up to ten injections per cycle. All of this means more precise control, which translates to improved fuel economy and reduced emissions.

Image: Continental AG

---

[15] C. Park, S. Kim, H. Kim, S. Lee, C. Kim and Y. Moriyoshi, Effect of a split-injection strategy on the performance of stratified lean combustion for a gasoline direct-injection engine. *Proceedings of the Institute of Mechanical Engineers: Journal of Automotive Engineering* 225 (10), 2011, 1415–1426.

Split injection can enable a small preinjection during the intake stroke, for example, which can help cool the combustion chamber. This allows for the exploitation of what is called the **heat of vaporization**, defined by the heat absorbed by the fuel as it transitions from suspended liquid droplets to a vapor. As the atomized fuel in this preinjection pulse vaporizes it absorbs heat energy, not unlike the heat removed from your body as perspiration evaporates. This cooling increases the density of the air in the cylinder and more importantly helps avoid knock by cooling the combustion chamber.[16] The remaining fuel is delivered later with a longer pulsation. In some diesel engines, piezoelectric injectors can provide up to seven injections per cycle.

The use of high-pressure direct injection does not exclude the use of lower pressure port injection; in fact, the two can be used to compliment each other. So, while direct injection offers a cooler combustion chamber and so helps avoid knocking, a port injection system can cool the incoming air and thus allow for its increased density and a resulting improved air intake efficiency (see the next chapter for volumetric efficiency). This allows the best of both worlds. The cooling effect of GDI on the combustion chamber can reduce knocking and potentially allow for a higher compression ratio for greater efficiency. And by allowing more time for the fuel to vaporize, port injection can reduce some of the challenges of exclusive direct injection. Direct injection can deliver fuel at high engine loads, when chamber cooling is important, and a combination can be used at lower loads to help keep injectors clear and efficiency high.

Combining DI and PI can also avoid some of the challenges of direct injection alone. The need to deliver some 50 times the fuel pressure required by port injection systems can result in increased engine noise. In addition, while not universal, there is potential for more pronounced carbon buildup on some GDI engines. By injecting fuel directly on the backside (or tulip) of the valves, port injection actually helped keep valves clean. This effect is lost when the fuel is injected directly in the cylinder. More critically, during rapid acceleration, a GDI system will be challenged to transition quickly from low-end stratified charge to a rich homogeneous charge as the engine accelerates. Combining GDI with PI can help address each of these issues. The combination can also allow for other innovations. For example, the Toyota's D-4S engine couples this with slits on the side of the injector that are periodically charged to blow the carbon off the outside of the injector as an automated self-clearing mechanism.[17]

Mercedes has been using a cutting-edge form of multiple method fuel delivery, called Turbulent Jet Injection (TJI) in its F1 cars since 2014; Ferrari is looking to follow suit soon. TJI replaces the spark plug with a jet injector inside a small ignition chamber (Image 1.26). While the main combustion chamber is supplied with the majority of the charge, about 3% of the fuel is injected in this small adjacent chamber, providing an excessively rich mixture around the plug. Ignition triggers a rapid flame jet streaming through precisely defined nozzles into the main combustion chamber. The fuel is ignited simultaneously at several points in the top chamber, causing more effective combustion, more power, and lower emission while avoiding knocking. This allows for an ultra-lean burn, with a fuel–air ratio exceeding 29:1 ($\lambda = 2$), and theoretically approaching 60:1. Several manufacturers have worked over the past two decades to define an ultra-lean burning engine that would be highly efficient and clean. The availability of precision fuel injection has now made that possible.

---

[16] S.P. Chincholkara and Dr. J.G. Suryawanshib, Gasoline direct injection: An efficient technology. *5th International Conference on Advances in Energy Research*, ICAER 2015, December 15–17, 2015, Mumbai, India *Energy Procedia* 90 2016, 666–672.

[17] C. Schweinsberg, "Toyota Advances D4S with Self-Cleaning Feature on Tacoma." *Wardsauto.com* August 27, 2015.

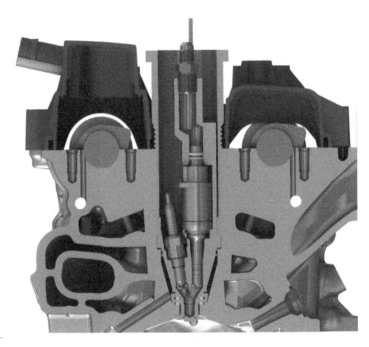

IMAGE 1.26
Turbulent jet injection (TJI).

The TJI system replaces the spark plug with a small precombustion ignition chamber that can trigger a rapid flame jet ignition into the main combustion chamber.

Source: Mahle

Low-Temperature Combustion

Similar technology is being explored to pursue the possibility of what is called **low-temperature combustion** (LTC). The idea is a gasoline-fueled engine that operates with compression ignition like a diesel, and so burns at a much lower temperature. The goal is to reduce the heat of combustion so that we can dramatically reduce the generation of harmful NO_x caused by excessive heat, or PM caused by excessive richness, and reap higher fuel efficiency. Some of these aims are achieved in current production cars with EGR, recirculating about 10% of the exhaust gas back into the combustion chamber to cool the combustion chamber. This is a well-established technology. But a promising route for significant further improvement is to design a gasoline engine that can capture the sort of thermal efficiency that has long existed in diesel engines, while avoiding the emissions challenges of diesel fuel. Because gasoline offers a much lower autoignition temperature than diesel, this sort of design could only be possible with the precise fuel and combustion control offered by modern fuel systems.

Homogeneous charge compression ignition (HCCI) is being examined as a possible path to this sort of ultra-efficient combustion. A lean, well-mixed fuel–air charge in the intake stroke is ignited by the heat of compression rather than a spark plug. Again, the hope is to achieve the efficiency of a diesel engine with the emission of a gasoline engine. With rising pressure and even mixture distributed throughout the combustion chamber, ignition takes place as a spontaneous reaction throughout the cylinder, producing a rapid and distributed pressure rise rather than a flame front. The low temperature and fast reaction means very low production of NO_x.

A key to making this work is in the fuel. In HCCI, all fuel has to fully vaporize prior to the reaction, or remaining liquid droplets will lead to undesirable emissions. Even more importantly, fuel characteristics are critical to ignition timing in HCCI, since the system relies on spontaneous autoignition, without the timing assistance of either a spark plug or a timed injector. So, in HCCI fuel chemistry plays a determinant role. Since the charge is definitively uniform, variations in fuel mixture or vaporization distribution that shaped spark ignition engines do not exist. As you might guess, typical pump gasoline is generally not precise enough for such a system. The imprecise autoignition point of pump gas makes it hard to achieve a predictable ignition, particularly at low load. Instead, a precisely engineered fuel with a well-defined octane rating is needed. All of this means HCCI needs very accurate control. And HCCI experiments in the laboratory, where control is relatively easy, have at times lead to more CO and more unburned fuel in the emissions.[18] A turbocharger, which we will discuss in the next chapter, can help produce more complete combustion by providing more air to the combustion chamber to help ensure that all the fuel is completely burned. But, ensuring a complete and controlled HCCI burn remains a challenge.[19]

The challenge is that we need to provide a mixture that is rich enough to support autoignition, but not so rich that it can't all be efficiently burned once combustion has started. A solution might come from **stratified charge compression ignition** (SCCI). At partial loads, the mixture can be leaned to ensure efficiency. If combined with compression ignition technology, the result could be highly efficient.

The trick is that we need to be able to ensure a desirable burn under varying engine conditions and loads. The perfect compression ignition engine really needs a fuel that is somewhere between gasoline and diesel, and the precise point changes as the engine load changes. A diesel engine defines a precise ignition point by injecting the fuel late in the process, timed to generate the targeted ignition sequence. But, this generates plenty of NO_x and soot. Gasoline typically produces less NO_x and soot, but ignition through compression is hard to control. Current research in what's called **Reactively Controlled Compression Ignition** (RCCI) tries to bridge this by providing a primary charge of gasoline through port injection and a smaller secondary injection of diesel through direct injection. Fuel reactivity could even potentially be precisely controlled with multiple injections of diesel to produce a specific gradient of fuel mixtures in the squish region and outward. Or an alternative fuel with a more defined octane rating might be used to more precisely determine the ignition point.[20] The result is high efficiency and low emissions, possibly the best of both worlds.

---

[18] A.A. Hairuddin, A.P. Wandel and T. Yusaf, An introduction to a homogeneous charge compression ignition engine. *Journal of Mechanical Engineering and Sciences* 7 (December), 2014, 1042–1052; and P. Kumar and A. Rehman, Homogeneous charge compression ignition (HCCI) combustion engine—A review. *IOSR Journal of Mechanical and Civil Engineering* 11 (6) Ver. II, 2014, 47–67.

[19] S. Saxena and I.D. Bedoya, Fundamental phenomena affecting low temperature combustion and HCCI engines, high load limits and strategies for extending these limits. *Progress in Energy and Combustion Science* 39, 2013, 457–488; M.M. Hasan and M.M. Rahman, Homogeneous charge compression ignition combustion: Advantages over compression ignition combustion, challenges and solutions. *Renewable and Sustainable Energy Reviews* 57 (May), 2016, 282–291; and M. Izadi Najafabadi and N.A. Aziz, Homogeneous charge compression ignition combustion: Challenges and proposed solutions. *Journal of Combustion* 57 (May), 2013, 1–14.

[20] S. Saxena and I.D. Bedoya, Fundamental phenomena affecting low temperature combustion and HCCI engines, high load limits and strategies for extending these limits. *Progress in Energy and Combustion Science* 39, 2013, 457–488; and S.L. Kokjohn, R.M. Hanson, D.A. Splitter and R.D. Reitz, Fuel reactivity controlled compression ignition (RCCI): A pathway to controlled high-efficiency clean combustion. *International Journal of Engine Research* 12 Special issue paper 209.

If this sounds like science fiction, it's not. Ford has looked at an ethanol boost 'Bobcat' Engine to possibly replace its 6.7 liter diesel in its super duty truck. The ethanol boosting system (EBS) runs as a typical port injection gasoline engine at modest loads. But to meet high-load demands, the engine injects E85 ethanol from a separate tank. With the high octane and cooling capacity of E85, the engine can accommodate more compression and produce more power on demand. The engine isn't ready for prime time, but Ford and others are looking for ways they can provide all the power and torque of diesel, without all the expensive diesel emissions treatment equipment. Just as impressive, Mazda is placing a variant of an HCCI engine in its next-generation Mazda 3, defining the first production car with a form of gasoline compression ignition. Its supercharged SkyActiv-X engine will offer an ultra-lean burn mode that can decrease emission by nearly a third, the manufacturer claims (Image 1.27).[21]

IMAGE 1.27
Spark-controlled compression ignition.

Mazda's SkyActiv-X engine offers what the carmaker is calling a spark-controlled compression ignition (SPCCI) engine. The aim is to combine the reliability and reduced emissions of spark ignition with the fuel economy of compression ignition. While the idea of spark-controlled compression ignition may seem oxymoronic, the basic principle is a sort of hybrid of spark and compression ignition that can offer improved efficiency by enabling a lean mixture to burn more evenly and reliably. Ignition is initiated by the spark plug, which raises the cylinder pressure so that the remaining lean charge in the cylinder ignites via compression. Under conditions when compression ignition is not ideal, the engine can shift back to traditional spark ignition.

Image: Mazda

---

[21] A. Stoklosa, "Driving Mazda's Next Mazda 3 with Its Skyactiv-X Compression-Ignition Gas Engine." *Car and Driver* September, 2017.

None of these innovations would have been possible a generation ago. All of them require a more sophisticated engine management system that can precisely determine the vehicle's driving conditions and engine parameters, and deliver fuel exactly as needed. Taking into account torque demand, engine speed, and other operating parameters, the engine management system can alter the fuel and type of charge to suit changing conditions. A homogeneous charge could be used for improved emissions throughout the range. At a modest torque range, a simple stratified charge can be implemented, adding EGR at low torque range to improve emissions. When more power is desired, a duel injection could occur, defining a homogeneous charge in most of the cylinder and a localized enriched charge near the plug. This sort of control capacity has fundamentally altered the possibilities for the internal combustion engine, and is the subject of the next chapter.

2

The End of Compromise

For a century, automotive engine design has been a game of compromise. Designing an engine for high engine speed meant it would be rough and inefficient at low speed. An engine designed for low-end torque meant limited high-speed performance. Similarly, an intake designed for efficient cruising often meant compromised acceleration. Or a high compression ratio meant for efficient cruising could invite knocking in high-end operation. However, the enhanced control capacity made possible with more sophisticated micro-processors and advanced materials and design are quickly making such compromises a thing of the past. Increasingly, we are able to change the fundamental characteristics of the engine while it's running, to effectively remake the engine into what we want when we want it.

Advanced Digital Control

A fundamental element that has made this possible is the use of increasingly powerful computer systems that manage and actually modify the basic characteristics of the engine while it operates. This has allowed for an ever-increasing level of control and innovation. These dedicated computer systems, or **embedded systems**, function with an array of actuators, sensors, microprocessors, controllers and instruments to replace and enhance mechanical functions with more precise and smarter electronics. So, they add considerable functionality and capacity to today's automobile, but they also add complexity.

While enhanced digital control has reset the bar for automotive performance and rede-fined what's possible, it's also worth noting a few challenges as we begin. Automotive digital control technology has developed irregularly, and its evolution is far from com-plete. Multiple stand-alone control units marked the early adoption of embedded systems, and soon progressed to an array of microcontrollers, sensors, and associated functions. Like the broader systems overall, the software has developed incrementally, without a clear overarching architecture or plan.[1] The result is somewhat of a technological patch-work quilt in the modern car. Each manufacturer defines a distinct control architecture, with their own priorities, approaches, and functions. A typical automobile manufactured today may have well over 100 **electronic control units** (ECUs), with multiple intercon-necting buses that allow devices to communicate with each other. Because each hardware component typically includes its unique software, every new function or feature of a car usually means the introduction of an additional control unit. The result is what can be called a highly heterogeneous system architecture; and the resulting challenge of software

---

[1] R. Hegde, G. Mishra and K.S. Gurumurthy, Software and hardware design challenges in automotive embed-ded systems. *International Journal of VLSI Design and Communication Systems* 2 (3), 2011, 164–174.

integration and development has been overwhelming, with tens of millions of lines of code, and thousands of signals travel between different ECUs.[2]

As cost, complexity, and redundancy increase and ECUs reach the limit of their processing power, interest increases in developing a more integrated control architecture for vehicles with organized interconnected networks and serial sharing of data and information.[3] The up-integration of control functions, use of more powerful microcontrollers that can manage multiple functions, multicore processors that can allow one control unit to take on a greater number of operations, integration of functions located near each into single controllers, and other innovations promise to cut the inefficiency, complexity, and cost of current automotive digital systems significantly.[4] In the same vein, an international effort to develop a universal software architecture for vehicle systems, called Automotive Open System Architecture Consortium (AUTOSAR), could allow for improvements and additions of functions to a vehicle without the high initial time and cost investments.[5] The appeal of a standardized modular system with standardized interconnects and scalability is clear.

Still, while there are some challenges to work out, the digitalization of the automobile also offers unprecedented opportunity. Just about every aspect of the modern automobile has become 'smarter', including not just the engine, but also the transmission, brakes, suspension, steering, safety features, and the increasing sophisticated infotainment system. Most notably, the **engine control module** (ECM) of a modern car receives signals from dozens of sensors to actively adjust ignition timing, intake parameters, fuel mixture, injection pulses, and other variables to more ideally tune the engine for enhanced performance, fuel economy, and emissions reduction. This requires ongoing communication between a processor capable of running millions of processes a second with an array of sensors that measure the mass of incoming air, the fuel mixture, engine knocking, system voltage, engine temperature, oil pressure, atmospheric pressure, vehicle speed, exhaust quality, crankshaft position, camshaft position, and many other factors. It gets complicated for sure, but the result is a car that is better, cleaner, and more efficient than ever thought possible.

Sensor Technology

Understanding the new digital landscape requires that we first understand what information we're working with and what sensors are in place. And there are many. Either optical or magnetic sensors are used as **crankshaft position sensors** (CKP) and **camshaft position sensors** (CMP) to allow tracking of rotation within a very few degrees. Updated versions of **engine coolant temp** sensors (ECT), **throttle position sensors** (TPS),

[2] R. Coppola and M. Morisio, Connected car: Technologies, issues, future trends. *ACM Computing Surveys* 49 (3), 2016, 1-36; and A. Sangiovanni-Vincentelli and M. Di Natale, Embedded system design for automotive applications. *IEEE Computer Society* 40 (10), 2007, 42–51.

[3] "Automotive Technology: Greener Vehicles, Changing Skills." Electronics, Software and Controls Report, Center for Automotive Research (CAR), May 2011.

[4] Ibid; and R. Hegde, G. Mishra, and K.S. Gurumurthy, Software and hardware design challenges in automotive embedded systems. *International Journal of VLSI design & Communication Systems* 2 (3), 2011, 165–174.

[5] J. Schlegel, "Technical foundations for the development of automotive embedded systems." Available at: hpi.de/fileadmin/user_upload/fachgebiete/giese/Ausarbeitungen_AUTOSAR0809/Technical_Foundations_schlegel.pdf

vehicle speed sensors (VSS), **fuel pressure sensors** (FPS), and others provided a more precise, responsive, digitized version of the mechanical sensors we once relied on. It is no longer sufficient to simply measure coolant temperature; modern vehicles now continuously monitor intake air temperature, outside air temperature, engine coolant temperature, oil temperature, transmission fluid temperature, and exhaust temperature. Other sensors are more distinctly new and innovative, measuring manifold air pressure, engine knocking, and emissions quality.

A key variable in engine control is the mass of incoming air. A **manifold air pressure** (MAP) sensor functions like a strain gauge, sensing the slightest deformations of a sensing element to measure air pressure. Older systems used a combination of MAP and the engine speed to calculate the approximate amount of air entering the engine. More recently, **mass airflow** (MAF) sensors allow for a direct and more accurate measure. Because air changes its density with temperature, simply measuring the volume of air entering an engine would not be sufficient. It would miss the thinning of the air with rising temperature or altitude and so be inexact. An MAF sensor suspends a heated wire in the path of incoming air. The electrical resistance of the wire changes with temperature, so by precisely measuring the change in resistance of this heated wire, we are able to assess the cooling effect of the mass of air passing the wire. The integrated circuited converts the resistance reading to a useful signal and the ECM is able to calculate the mass of incoming air quickly and with precision. Cool, right? (Image 2.1).

As discussed in the previous chapter, the ability to identify detonation or preignition is vital to engine control, and accomplished with a **knock sensor** (KS). Typically located in the engine block or head, this sensor allows the ECM to adjust engine timing to resolve knocking. Piezo materials make this possible. You'll remember from Chapter 1 that piezoelectric materials alter their shape when exposed to a charge. This also works in reverse, when the crystal experiences a change in its shape, it generates an electric charge. When squeezed or shaken (or knocked), the dipoles are disturbed, and rearrange, resulting in a small electric charge across the crystal. That charge sends a signal to the ECM, which then knows to adjust timing.

IMAGE 2.1
Measuring air.

This mass airflow sensor (MAF) uses a 'hot wire', a heated wire with a resistance that changes predictably with temperature, to measure the cooling effect of incoming airflow. This then allows us to know the air's mass.

A key in defining a cleaner car, and correctly managing the fuel mixture, is the ability to assess the emissions from the engine, and thus enable responsive management of the fuel mixture. The **oxygen sensor**, also called a **lambda sensor**, does this by evaluating oxygen content in the exhaust. The sensor usually consists of a thin, thimble-shaped component of ceramic zirconium dioxide (ZrO_2) coated with platinum on either side. The thimble is placed in the exhaust pipe, so one side is exposed to the exhaust gas and one side is exposed to outside air. The difference in exposure to oxygen causes a chemical reaction that produces charged particles of oxygen (ions) traveling through the ceramic membrane. As the difference in oxygen levels on the two sides varies, so does the flow of ions. This produces a voltage difference, not unlike a tiny battery, but in this case called a Nernst cell. You might also think of this as a difference in the pressure of oxygen on either side of the sensor, leading to a flow of oxygen particles across the zirconium dioxide. The resulting voltage difference won't tell you the actual percentage of oxygen, but it does allow you to identify changes in oxygen concentration compared to the outside air, and that's good enough to see if the engine is burning lean or rich. When the mixture is rich, there is very little oxygen left in the exhaust and the voltage increases. When the mixture is lean, more oxygen is left in the exhaust and a lower voltage results. Because the membrane is only conductive at high temperatures (600°F), a heating element is contained in the sensor housing. The exhaust would eventually heat the sensor, but a heater allows for reliable readings sooner.

Of course, this description is partial, and highly simplified, highlighting a few characteristics of the ECM architecture and some key sensors. Actual operation includes multiple ongoing diagnostic routines providing continuous analysis, data, and communications potential through the **on-board diagnostic** (OBD) system. It includes multiple modes of communication between the powertrain, chassis, suspension, differential, and other vehicle sensors and mechanisms. And, as we'll see in Chapter 9, automotive control systems increasingly incorporate rapid expansion in the integration of echo detection sensors such as radar, lidar, and sonar, as well as navigation and communication technology from advanced GPS to automated vehicle-to-vehicle coordination, and countless additional features. We will explore the possibilities and implications of these innovations later, but for now, let's look at how enhanced digital control has redefined the operation of the internal combustion engine.

Engine Control

The automotive advances made possible through digital control are nowhere more impressive than in the performance of the internal combustion engine. Let's begin with three primary aims to digital engine control: managing the fuel mixture, ignition timing, and exhaust gas recirculation in engines so equipped. (We'll see in a bit that other functions, such as varying valve operation, are also in play.) Digital control allows for the identification of distinct operational modes, based on driving conditions and driver demands, with targeted engine parameters within each mode. Once again, this allows us to avoid the compromises of previous generations, and define distinct targeted priorities and corresponding control logic for each mode. While manufactures differ in the precise differentiation, most ECMs include some variation of seven basic modes: engine start, engine warm-up, normal cruising with both an open-loop and closed-loop variation, high load (acceleration), deceleration (often with a high and low variant), and idle.

Upon engine initial start, the ECM crank control logic is automatically selected and delivers the rich mixture required to start the engine, pulsing the fuel injectors to promote a quick start up. The target air–fuel ratio will be determined by engine temperature, and may vary anywhere from an extreme rich mixture of 2:1 to a modestly rich 12:1, as defined by data stored in a **programmable read-only memory** (PROM). Engine characteristics will change this greatly, of course; with a GDI engine potentially requiring a much less rich mixture thanks to more precise fuel metering, for example. The ECM may also check battery voltage, to determine any needed operating compensation, as fuel injector timing and response speed can be significantly affected by low voltage.

Once idle rpm is achieved, the ECM switches to idle mode, maintaining an air–fuel ratio that ensures the lowest stable engine idle while avoiding stall. The thermostat will ensure that the coolant is automatically recirculated through a short loop that bypasses the radiator to assist in quick engine warm-up, as has been done for decades, but now an electrically assisted thermostat can also manage the ideal temperature for optimal engine operation dependent on initial driving demands. Called a MAP-controlled thermostat since the Manifold Air Pressure sensor provides an indication of engine load to the ECM, this feature allows the thermostat to alter warm-up parameters to suit engine load while addressing combustion and engine temperature, NO_x reduction, and fuel economy. While this is done, the oxygen sensor is warmed, and idle speed, injector pulse duration, and mixture are progressively adjusted to suit engine temperature.

Once the engine has warmed up, the ECM will enter **open-loop** operation, meaning that the ECM will rely on limited sensor data and continue to use PROM fuel ratio data, sometimes called a lookup table, to manage fuel mixture until the O_2 sensors can provide useful information. This is not ideal, as the ECM principally uses engine speed to define ignition timing without compensating for the important impact load has on proper timing. Once the O_2 sensors are warm and providing reliable information, signals from the oxygen sensors in the exhaust can be used to modify mixture for ideal performance and clean operation. So, the ECM can move from a feed-forward control to **closed-loop** operation and begin to reduce emissions levels quickly. Because virtually every car manufacturer in the US utilizes a **three-way catalyst** (TWC) for emissions management, which works best when combustion is stoichiometric, keeping the fuel mixture at or near the stoichiometric ratio is a key concern (Image 2.2). Using information from the MAF, the controller is able to determine the airflow into the combustion chamber to manage the amount of fuel through the injectors.

Unlike open-loop operation, closed-loop control allows for automatic adaptation to mechanical conditions. The ECM is managing more than just mixture; it may be defining fuel injector pulse duration, spark timing, and multiple additional variables based on engine temperature, MAF, barometric pressure, engine speed, knocking, and other variables. Again, a key concern in this process is knocking. The ECM will automatically advance spark timing to achieve improved performance. However, the desirability of spark advance is limited by the threat of denotation. Once the KS identifies excessive knocking, spark timing is moved slightly later, or retarded. Different manufactures do this differently; but generally, the ECM will drop the advance until knocking is eliminated, and then progressively increase it until the knock point is identified, allowing high-performance operation at the cusp of the knock limit. The ECM may even make micro adjustments during operation based on learned past knocking events, a practice Subaru called Fine Knock Learning. Like many ECM functions, managing ignition timing while avoiding knocking requires precise and fast-processing capacity, as the engine rotates a full 18° in a single millisecond at 3,000 rpm.

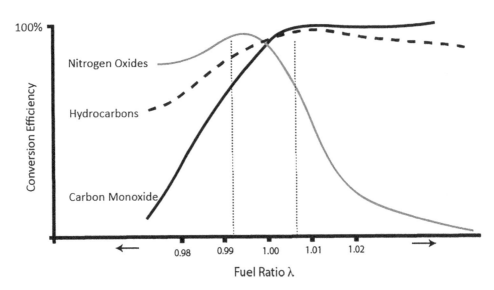

IMAGE 2.2
Three-way catalytic converter operation.

Although NO$_x$ emission can be higher at a stoichiometric fuel ratio, the catalytic converter works most effectively to reduce unburned hydrocarbons, carbon monoxide, and NO$_x$ at stoichiometry. So, in order to ensure a clean burning car, the ECM targets stoichiometry in closed-loop operation.

The ECM's monitoring and responsive control is not limited to mixture and spark timing of course. Multiple variables are monitored and the ECM may initiate additional adjustments at this point. The ECM will initialize EGR, allowing a percentage of exhaust gas to be cooled and recirculated into the combustion chamber in an effort to cool combustion temperatures and reduce the production of NO$_x$. The ECM may trigger the delivery of secondary air to the catalytic converter to ensure complete combustion of remaining emission pollutants if it has a secondary air injection (SAI) system. System voltage may require adjustment to fuel injector signals. More generally, a range of powertrain variations may be initiated. For example, if the vehicle is equipped with a lock-up torque converter (we'll look at this in the next chapter), it may signal a solenoid to lock the clutch mechanism and therefore eliminate efficiency-draining slip. In sum, closed-loop operation allows for sophisticated control logic that ensures engine performance as operating conditions change, both in the management of combustion and the vehicle more generally.

During high-load demand, signaled by the TPS or possibly an additional sensor that identifies wide-open throttle, the ECM will again switch modes to provide a rich mixture that can meet the demand for hard acceleration. This rich mixture will temporarily degrade emissions control and fuel economy; but engineers and regulators view this as an acceptable compromise to provide demanded acceleration, as acceleration capacity is at times needed for safety.

Upon deceleration, the ECM will lean the fuel mixture to allow for reduced emission and improved fuel economy. This also helps avoid an exhaust backfire caused by excessive rich mixture. If the deceleration is severe, the lean-out may be extended to a complete fuel cut-off, allowing for more effective engine braking and reduced emissions. Signals from the MAP and TPS will bring the ECM back into closed-loop mode quickly to avoid stalling, or the ECM will return to idle control mode when the vehicle is stopped.

This might seem complicated, but it's actually a highly simplified rendition of the ECM operation. We're not even considering the integration with the transmission operation, fuel system evaporative emissions control, cabin air conditioning and heating demand, and a host of other variables and functions. But the principle point is clear: ECMs are defining a move beyond conventional control practices. These are no longer single-input/single-output systems (SISO). They define increasingly complex, multivariate operations with sophisticated algorithms and multiple control strategies.[6] Data from disparate sensors are combined to create performance information that exceeds the capacity of any one sensor, a process called **sensor fusion**. The resulting control structure moves toward a form of artificial intelligence, with the ability to monitor, adapt, and modify control logic by 'learning' from past outcomes. Future ECMs may determine what ignition timing, fuel management, or response to knocking work best for your engine, or even your driving, and adapt accordingly. We'll see more on these sorts of systems in Chapter 9.

Variable Valve Actuation

All of this control technology has allowed for a level of management of the engine's processes that was unthinkable a generation ago. We've already seen the results of this in mixture and ignition timing management, though it is perhaps even more impressive in the control of valve movement, a variable that was once mechanically fixed with no possible variation during operation. More varied and precise control of the valves allows us to once again make an end run around the compromises that were once unavoidable. Where once we had to make a tough choice in defining valve movement, now we can have our cake and eat it too. We can design variable valve events that allow our cars to feel like a sports car at high revs without sacrificing efficiency when cruising.

As discussed in the previous chapter, in the classic internal combustion engine, the intake valve opens and closes during the intake stroke; and the exhaust valve opens and closes during the exhaust stroke. The problem with this scenario is that the velocity of incoming air does not vary to suit the engine speed. As the engine speeds up, like the flame front, the speed of the incoming charge often cannot keep up. With a conventional cam, the valve lifts from its seat pretty slowly, so while the valve may be 'open' it takes some time for air to begin to flow freely through the opening. In fact, flow through an open valve is significantly restricted for perhaps two-thirds of the total valve opening time.[7] The solution may be to keep the valves open longer, but it's not that easy. The ideal valve operation changes with engine speed or load. In a perfect world, the opening and closing of each valve could be timed to the precise needs of the engine at a given load.

A key variation in the operation of valves at high engine speeds and loads is to open the intake valve earlier and close the exhaust valve later, defining **valve overlap** (Image 2.3). In addition to allowing more time for air movement, as discussed previously, it can enhance scavenging, allowing the incoming charge to help push out the outgoing exhaust. Typically, the opening of the intake valve (IVO) is timed to begin about 10° BTDC, to allow the valve to fully open for the intake stroke. And, the resulting aggressive pulsation of pressure in the exhaust manifold as all cylinders exhibit powerful cross flow can create pressure waves in the exhaust manifold that can help actively draw the exhaust out of the cylinder.

---

[6] H. Morris, Control systems in automobiles. In J. Haappian-Smith (ed) *Modern Vehicle Design*. Butterworth-Heinemann, Oxford, 2001.

[7] R. Stone and J.K. Ball, *Automotive Engineering Fundamentals*. SAE International, Warrendale, PA, 2004.

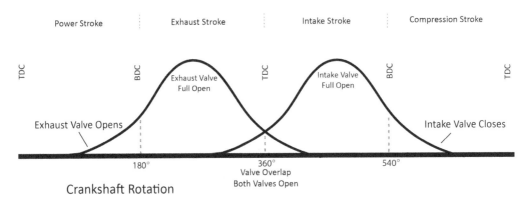

IMAGE 2.3
Valve overlap.

Delaying the closing of the exhaust valve and advancing the opening of the intake valve create an overlap that will allow for more effective scavenging of the combustion chamber by allowing cross flow of incoming air.

At high engine speeds, this is desirable, but the advantages of a large overlap diminish at low engine speeds. In fact, at low loads, keeping some of the exhaust in the combustion chamber is actually a good thing. It means less incoming air is needed when the throttle is more closed, decreasing pumping losses. And, like the effect of the EGR system, more exhaust gas in the chamber at low loads can mean lower NO_x emissions. In effect, this is a bit like making the combustion chamber smaller at lower engine speeds.

The advantage of longer valve opening isn't limited to the benefits of overlap. There is also an advantage to an early exhaust valve opening. Ideally, we'd want to extract all the power we can out of the expanding combustion gas before we open the exhaust valve. However, pushing all the exhaust gas out of the combustion chamber also takes energy. And, opening the exhaust valve before reaching the end of the power stroke can help us save some of the energy required to push out the still-pressurized exhaust gas at the start of the exhaust stroke, called **blowdown loss**. At high load and high engine speed, an early opening of the exhaust valve will allow more time for the pressure inside the chamber to drop before the piston begins to move up. But, in low-load conditions, the early exhaust opening means lost energy and torque, since there's plenty of time for the pressure to drop before the exhaust stroke.

On the other end of the stroke, we can consider the closing of the intake valve. For greatest power at high load, we would want to time the closing of the intake valve to ensure the largest mass of incoming charge. You might expect this to be at the bottom of the stroke; however, the incoming air through the intake manifold develops a momentum that has the effect of pushing a wave of air into the cylinder, thus favoring a late closing of the intake valve at high engine speeds. The intake pressure wave tends to maximize late in the stroke. So, taking full advantage of this effect, and maximizing the mass of incoming air, requires we close the intake valve a bit after BDC. At high rpm, this might be significantly longer than at low power where a late valve closing might be detrimental as it diminishes the compression stroke.

Similar challenges present themselves with defining valve lift. We might be tempted to presume that maximum valve lift is always desirable, since this would allow for the least restrictive flow when the valve is open and so faster intake flow and improved exhaust exit. This is true at high rpm, but, at low rpm, there is a cost to high valve lift. High valve lift means air can flow into the chamber slowly and calmly at low rpm, losing turbulence and risking wetting.[8] Additionally, at partial loads, a smaller valve opening reduces the need for a more restrictive throttle, reducing pumping losses. So, while a smaller valve opening is severely detrimental to the engine's high-load performance, a high lift opening can deteriorate low-load performance.[9] In the past, we conventionally identified a lift pattern that defined a compromise, but that's no longer necessary.

Over the past two decades, control of valve timing and lift has become increasingly more sophisticated. Systems that allow a basic **phase change** of the camshaft for discrete high and low speed cam timing are now common. With a single cam, a valve phase change can retard valve timing by roughly 10°, allowing for significantly improved high-load operation with an earlier intake opening. If the engine has separate intake and exhaust cams, each might be adjusted independently, allowing for greater allowance for engine load and speed. However, while allowing for two or more discreet valve timing settings is an improvement, the possibility of a progressive variation in valve timing to suit changing conditions would be better (Image 2.4).

IMAGE 2.4
Toyota dynamic force engine.

Toyota's 2.0-liter, direct-injection, in-line, 4-cylinder engine uses advanced combustion control technology and variable valve timing to achieve an impressive 40% thermal efficiency.

Image: Toyota Motor Company

---

[8] T. Wang, D. Liu, G. Wang, B. Tan and Z. Peng, Effects of variable valve lift on in-cylinder air motion. *Energies* 8, 2015, 13778–13795.
[9] F. Uysal, The effects of a pneumatic-driven variable valve timing mechanism on the performance of an otto engine. *Journal of Mechanical Engineering* 61 (11), 2015, 632–640.

In fact, greater capacity for variation and control is possible. **Cam profile switching** (CPS) typically uses two sets of distinct cam lobes, or otherwise shift between two distinct cam profiles. As a result, both timing and lift can be changed. But once again, only as a choice between two options, typically a close-to-conventional cam for normal operation, and a cam designed for performance operation at high rpm. Manufacturers have defined multiple sophisticated systems that can allow for control of both lift and timing, and can offer incremental or continuous **variable valve timing** (VVT), **variable valve lift** (VVL), or both. BMW, for example, uses a sprocket on a helical spline of the camshaft to progressively adjust valve timing; and their Valvetronic variable valve system utilizes an electronically adjusted cam and rocker arm to vary the relationship between the camshaft and the valve stem allowing variation in valve lift from 0.3 to 9.7 mm (Image 2.5). This means valve lift variation can be used to control engine speed, and once again eliminate throttling loss from a conventional throttle.

There are nearly as many ways to accomplish VVT as there are manufacturers. Honda was an early adopter of variable valve profiles with its VTEC engine. The engine switched between two cam profiles, allowing smooth operation and efficiency at low rpm, with greater performance possibilities when the engine switched to the more aggressive cam at 4,500 rpm. Honda's recent engines include a three-stage VTEC with separate cams for low,

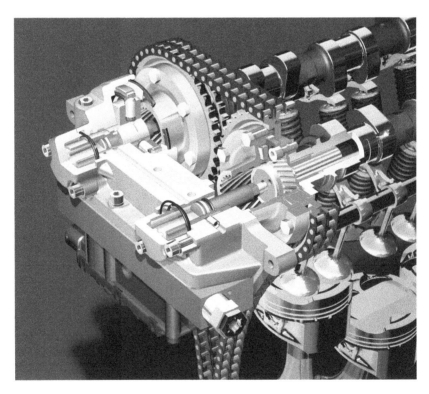

IMAGE 2.5
BMW's VANOS variable valve timing system.

BMW's VANOS variable valve timing system places a sprocket on a helical spline at the end of the camshaft. The sprocket moves in and out with oil pressure, causing a progressive shift in the relative alignment of the camshaft and thus a change in timing.

Source: BMW

medium, and high engine speeds. GM's Intake Valve Lift Control uses an adjustable cam to shift the pivot point of the rocker arms and thus vary valve lift. FIAT's MultiAir system utilizes an electrohydraulic system, using pressurized engine oil to actuate the intake valves, allowing for control of both lift and timing.

As mentioned, more complete control of valve events can allow for a change in the fundamental parameters of the engine's operation, even redefining the basic four-stroke process. Atkinson cycle engine operation offers a great example. The idea of an Atkinson cycle has long defined a variation from the classic Otto cycle four-stroke engine. Dating back to the mid-1800s, the basic idea is this: since there is more combustion power to be extracted at the end of the power stroke, the Atkinson Cycle allows for a longer power stroke rather than the equal length strokes of the Otto Cycle (Image 2.6). The extra piston travel allows the engine to usefully extract more energy from combustion. The Toyota Prius was notably the first to mimic this cycle in a production car by delaying the close of the intake valve. This allows some of the intake charge to return to the manifold, effectively shortening the intake stroke, and by relation, making the power stroke comparatively longer. The result is a significant improvement in efficiency, but a loss of power. Compressing the incoming air with a super or turbo charger can help compensate for this shorter compression stroke and recover lost power, defining a variant called the Miller Cycle. Multiple car manufacturers now use VVT to produce a variant of this Miller/Atkinson Cycle. So, this isn't just about fine tuning engine operation; these systems are basically changing the fundamental operation of the four-stroke engine on the fly as conditions warrant. Is it just me, or is that really, really cool?

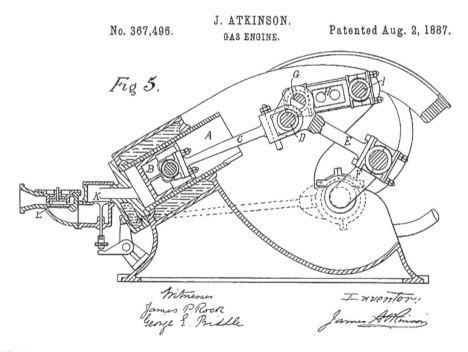

IMAGE 2.6
Atkinson cycle engine.

The Atkinson cycle originally dates to an idea initiated by John Atkinson in the mid-1800s. His design used a multilink connection to allow for a significantly longer power stroke and a shorter intake stroke, allowing the piston to harvest a greater proportion of the combustion energy than is possible when the two strokes are equal.

Equally amazing, advanced valve control can effectively change the size of the engine as needed, on the fly. The ability to effectively shut down cylinders by closing both valves allows for **cylinder deactivation**, letting an engine perform as though some of its cylinders are gone, so acting as a form of **variable displacement**. Fuel delivery is cut off, and both intake and exhaust valves remain closed on the deactivated cylinders (Image 2.7). The sealed cylinder then operates as an air spring, with the energy of the upward piston increasing cylinder pressure and redelivering the energy to the piston on the down stroke. Well over 90% of the energy is recovered, allowing the pistons to simply go along for the ride, neither helping nor seriously hindering the engine when not needed. When more power is demanded, the cylinders are seamlessly reactivated. The result is a significant improvement in efficiency at low load while maintaining power capacity when needed, in effect allowing a V8 to perform like a four cylinder at cruise but like a full V8 when needed.[10]

These technologies are impressive, but the full potential of variable valve control is marked by the promise of individualized and precise control for each valve independently. The idea is to have each individual valve operated discretely, perhaps with an electromagnet as is done on some large ship engines or with a pneumatic actuator, as is done with a prototype engine made by the Swedish firm Freevalve. With computer control, timing and lift of each valve may be altered to suit operating conditions. A key advantage is the immediate opening and closing of the valves. With a camshaft, opening and closing takes time, defined by the elliptical cam lobe, and this dictates valve timing. With a solenoid-controlled pneumatic system, moving a valve from closed to fully open is much quicker, defining a nearly square valve lift profile, and once again requiring less compromise (Image 2.8).

In addition, the weigh and space savings is significant. No camshaft, no pushrods, no rocker arms, no rollers. The resulting friction reduction is also worth noting. The leading

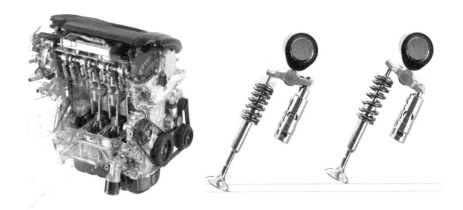

IMAGE 2.7
Mazda cylinder deactivation.

Hydraulic lash adjusters on the first and fourth cylinders are used to shift the pivot point of the rockers and define variable displacement. Normally, the rocker arm pivots around the center point; but when cylinders are deactivated, the pivot point becomes the valve side so the valves remain closed. Because a four-cylinder engine is more likely to experience heightened vibration when only two cylinders are firing, a centrifugal pendulum damper in the automatic transmission is used to help reduce vibration.

Image: Mazda

---

[10] F. Millo, M. Mirzaeian, S. Luisi, V. Doria and A. Stroppiana, Engine displacement modularity for enhancing automotive s.i. engines efficiency at part load. *Fuel* 180 (September), 2016, 645–652.

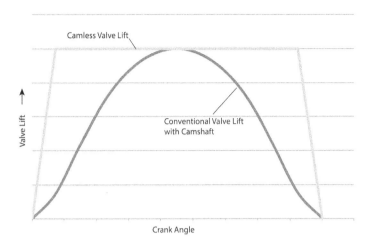

Camless Valve Lift

Valve Lift

Conventional Valve Lift
with Camshaft

Crank Angle

IMAGE 2.8
Camless valve lift.

Controlling valves with independent solenoids allows for faster valve movement than can be achieved with rounded cam lobes, defining a more square lift profile and thus the potential for more precise valve timing.

manufacturer, Freevalve, has reported a 40% peak power increase from an in-line four-cylinder turbocharged test engine, with improved fuel economy and lower cold start emissions, in a package that is significantly lighter, smaller, and no more expensive than a comparable high-end engine (Image 2.9).[11]

IMAGE 2.9
Freevalve.

The Swedish firm Freevalve has defined a radical departure from conventional engine design with its electrically controlled pneumatic valve actuators.

Source: Freevalve

---

[11] Author communications with Andreas Möller at Freevalve.

The possibilities are intriguing, to say the least. This sort of control capacity could allow for more sophisticated and precise variation in engine operation. Exact valve control can be used to define a part-time HCCI or Miller-cycle engine. It could allow for multiple cylinder deactivation modes. It might enable an internal EGR mode. Or, it may well do all of the above.

Induction

If digital control allows for more exact delivery of fuel and spark, it can also allow for improved delivery of air to the combustion chamber. As described in the previous chapter, an internal combustion engine can be thought of as a large air pump, with a performance engine at high speed requiring up to 40 pounds of air, or about 500 ft^3, every minute. The challenge of delivering this air at a pressure and velocity that can readily scavenge and fill a combustion chamber in the few milliseconds the intake valve is open can determine the capacity of an engine.

The aim is to provide air to the combustion chamber as effectively as possible to allow complete combustion, an effort measured as **volumetric efficiency** (VE). VE is defined as a ratio of the incoming air volume to the volume of the chamber. So, since each cylinder has one intake stroke with every two rotations of the crankshaft, the VE of an engine would be 100% if the air intake were half the engine displacement with every revolution. This assumes that the displacement volume is the same as the total volume of the cylinder, which ignores the clearance volume. But, the clearance volume is very small, so for simplicity we use displacement volume. So, an engine operating at a 100% VE is taking in its total displacement with every two crankshaft revolutions. Of course, it is also important that this air be evenly allocated with equal distribution of incoming air to each cylinder. While a typical VE for a conventional, **naturally aspirated** engine might be 75%–90%, higher is possible with precise tuning and control.

Pressure is the strongest determinant in defining VE. And pressure in the induction system is not constant. During an intake stroke, the engine draws in air-reducing induction pressure. This defines a low-pressure wave that travels back through the runner to the plenum. In response, the relatively higher pressure in the plenum gets sucked into the runner and defines a return pulse of high pressure that provides flow into the combustion chamber. At the end of this stroke, the high-pressure air now flowing into the chamber meets a pressure backflow as the piston begins its upward travel and the valve closes, creating a high-pressure **shockwave**. This wave travels back through the intake path and can be reflected and travel back toward the valve. The resulting oscillation of pressure has a frequency defined by the length and volume of the intake chamber and is akin to something called wind throb or **Helmholtz resonation**.

These dynamics can benefit or hinder engine performance. For example, higher engine speeds would benefit from a larger plenum providing dense air to the runners, but this will produce slow throttle response and reduced low-end torque. At a certain engine speed, the high-pressure air pulses returning through the runner can arrive with the next opening of the intake valve helping push the charge into the chamber and possibly increasing VE to over 100%.[12] So, if the intake runner length is right, a synchronization with the engine

[12] J.G. Bayas and N.P. Jadhav, Effect of variable length intake manifold on performance of ic engine. *International Journal of Current Engineering and Technology* Special Issue 5 (June), 2016, 47–52.

speed is possible and the high-pressure wave arrives just as the intake valve opens, providing a **supercharging effect** or ram effect.[13] But this only happens in a narrow rpm range, when the travel time of the air pulses and the engine speed are synchronized. So, once again, in designing an engine a compromise is often required; **acoustic tuning** can provide a beneficial inducting pressure boost and torque increase at a targeted rpm point through proper intake runner sizing, but at other engine speeds, this benefit deteriorates.

So, as engine operating conditions change, the ideal induction system also changes. Runners need to be sized short enough to allow the low-pressure wave to travel down the runner and draw a high-pressure return pulse within the short intake stroke, and long enough to allow the high-pressure pulse caused by the valve closing to travel back through the runner and be reflected back in time to meet the valve as it reopens for the next intake stroke. At higher engine speed, since the pulse wave travels at a relatively fixed speed (the speed of sound), a shorter length would be required to allow the wave to return in time (Image 2.10). At a lower rpm, the length would need to be longer, since there is more time between valve openings. But, long runners would increase flow restriction at high engine speeds and decrease VE, and hence, the need for compromise.

Once again, more advanced mechanical designs and digital controls lessen the need for compromise. **Variable intake systems** (VIS) can adjust the length of the induction run to suit engine speed and load. Short runners provide minimal restriction of flow at high engine speed. With engine speed comes increased induction velocity, again favoring an

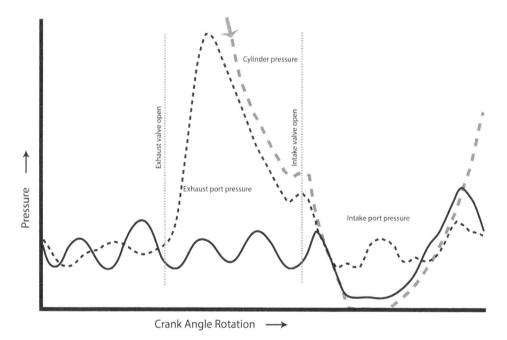

IMAGE 2.10
Port pressure pulsing.

Inlet port pressure varies with the opening and closing of each of the engine's valves. When the intake valve opens, a significant port pressure drop ensues and is followed by a rapid upward pressure bump.

---

[13] J. Bayas, A. Wankar and N.P. Jadhav, A review paper on effect of intake manifold geometry on performance of ic engine. *International Journal of Advance Research and Innovative Ideas in Education* 2 (2), 2016, 101–106.

open short run to provide maximum dynamic pressure pushing air into the cylinder. But long runners can be advantageous at slower speeds, taking advantage of the Helmholtz resonance and the resulting positive pressure wave into the chamber, synchronizing the pressure wave oscillation with engine speed.

Variable length intake runners can be defined in stages, with two or more set length possibilities, or they might allow continuous variation of the runner length. BMW was the first carmaker to offer a continuously variable intake manifold, back in 2001. The Differentiated Variable Air Intake (DIVA) system was defined by a rotating inner rotor that could vary the entry point of the incoming air into a toroidal induction runner. The effective length the incoming air had to travel could therefore vary from over 26 inches to as little as 9 inches. The maximum length was used up to 3,500 rpm, after which the run was gradually reduced. While effective, the necessarily large size of the multiple toroidal runners serving each cylinder limited its application. Since this early introduction in the late 1990s, staged and continuous variable length intake manifolds have become common (Image 2.11).

By ensuring high-velocity airflow into the chamber, a VIS induction can also improve swirl pattern, promote a more complete and fast combustion, and help avoid engine knock. The runner may be tapered to encourage acceleration and proper swirl, particularly if targeting lean burn. The mechanics is straightforward: actuators open and close induction pathways to lengthen or shorten the path of incoming air while referencing MAP and throttle position. So induction pressure pulses are effectively tuned to more closely suit engine speed. This allows positive pressure pulses to contribute positively to engine performance through a wider range of engine speeds. The result is to replace the previous peaked torque curve with a 'thicker' curve defined by distinct performance peaks for each runner length (Image 2.12). Of course, continuously variable induction, rather than two or three length options, can provide a more precise improvement in performance, with the cost of added mechanical complexity. By varying the induction length continuously to

IMAGE 2.11
Mazda variable induction.

This 2.6-liter, 700-horsepower rotary engine from Mazda is fitted with continuously variable intake stacks, allowing the runner length to adjust with engine speed for optimal performance.

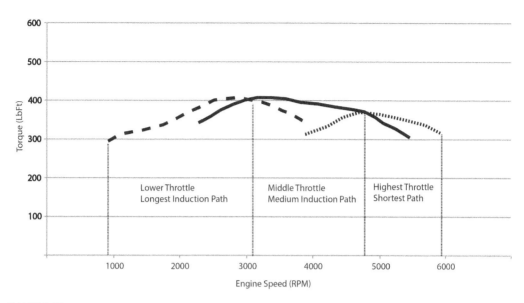

IMAGE 2.12
Torque with three-stage variable induction.

Each runner length in this idealized variable induction system develops a specific peak torque curve. As the engine switches from one to the other, it can utilize the most favorable curve for a given engine speed, effectively widening the engine's peak torque production capacity over a much larger rpm range.

suit engine speed and load, a continuous VE of over 100% can be achieved in a naturally aspirated engine.

Forced Induction

Of course, another way to deliver high-pressure air to the combustion chamber is to actively pressurize it with a pump, called a **supercharger** or more casually a blower. Designs vary, but vane and centrifugal designs are common. A chain or gear could be used to drive the supercharger off the crankshaft, but most commonly a belt is used. Such systems can increase torque by roughly a third, and horsepower by somewhat more.

   This increased power does not come without a cost. A supercharger can increase the pressure in the chamber and therefore the pressure during the compression stroke, increasing knocking. The simplest solution is to use an engine with a lower compression ratio. And, because the supercharger increases the MAF to the engine, it also requires a proportional increase in fuel flow to maintain stoichiometric conditions, and this could mean decreased fuel economy. However, there is a silver lining; forced induction is so effective at increasing power that it can offer the opportunity for engine downsizing, which of course can mean improved fuel economy.

   Of course, even though superchargers can increase power, they also require some of the engine's energy to drive them. This increases the load on the engine, so the benefit is limited by a trade-off. An alternative is to tap into the tremendous waste energy in the engine's exhaust to drive a compressor, defining a **turbocharger** (Image 2.13). This energy

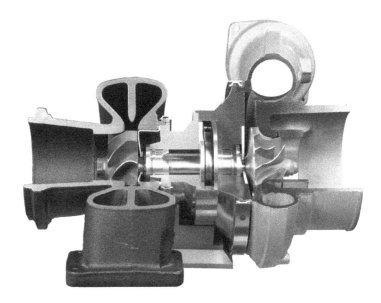

IMAGE 2.13
Turbocharger.

A turbocharger includes a turbine that is driven by escaping hot exhaust gasses (left) connected to a compressor that boosts the pressure of incoming air (right). The resulting boost can increase horsepower significantly.

is significant; roughly a third of the energy from combustion can be lost through the exhaust.[14] A turbo allows us to harvest some of this waste energy to usefully increase power. The basic idea is to place a driven turbine in the exhaust system that can be spun by the escaping hot exhaust gas, and use this energy to drive a coupled compressor in the induction system. The result is an extremely high-speed turbine/compressor system reaching speeds upwards of 150,000 rpm and, because the system is driven by hot exhaust gasses, extremely high temperature. Turbochargers therefore require careful balancing, robust bearings, and precise engineering. Because the induction air rises in temperature, a cooling exchanger called an **intercooler** is used to allow the pressurized induction air to cool and thus increase its density before entering the engine.

Capable of extreme speeds and heat, turbochargers require careful control as they have the potential to define an uncontrolled positive feedback loop: The turbo results in dramatic increase in engine power, which increases exhaust energy, which in turn feeds the turbo. Such a system could easily overstress the engine and lead to catastrophe. To avoid this, a poppet valve, called a **wastegate**, allows some of the exhaust to bypass the turbo and therefore limit the boost. Similarly, to allow the boost to diminish quickly, say for deceleration or gear changes, a recirculation valve, or **dump valve**, can rapidly drop the pressure in the intake system and allow engine speed reduction.

When designing a turbocharger for engine performance, we are once again faced with trade-offs. Turbos take some time to spool up, so there's a delay between hitting the

---
[14] P. Punov, Numerical study of the waste heat recovery potential of the exhaust gases from a tractor engine. *Proceedings of the Institution of Mechanical Engineers, Part D: Journal of Automobile Engineering* 230 (1), 2015, 37–48; and P. Fuc, J. Merkisz, P. Lijewski, A. Merkisz-Guranowska and A. Ziolkowski, The assessment of exhaust system energy losses based on the measurements performer under actual conditions. In E.R. Magaril and C.A. Brebbia (eds) *Energy Production and Management in 21st Century—The Quest for Sustainable Energy*. WIT Press, Ashurst, 2014, 369–378.

accelerator and the generation of boost, called **turbo lag**. Smaller turbines take less time to get spinning due to a lower moment of inertia, so a smaller turbine may be used, sacrificing total boost for the sake of reduced lag. Alternatively two small turbines are often used instead of one large turbine. These can be parallel twin units or alternatively sequential systems can allow one turbine to be used at lower engine speeds and a larger second turbine, or both, to be used at higher speeds. This has the benefit of managing turbo lag while making higher boost available.

A more elegant solution comes from the possibility of a variable geometry turbocharger. Variable geometry turbos (VGT) have been used to improve low-end engine torque in diesel engines for a couple of decade, but this technology was not available to gasoline engines because of the significantly higher exhaust temperatures of gas engines. Manufacturing a variable mechanism that could withstand the high temperatures of a gasoline exhaust system simply wasn't possible. But recent advances in material technology have changed this. A VGT can vary the effective geometry of the turbine, typically with movable vanes, and thus adjust the turbine for quick spool up or high boost throughout the engine speed range. The result is a high total boost capability, with imperceptible turbo lag (Image 2.14).

The potential to integrate new turbocharger technology with other performance features is tremendous. For example, the Hyboost project vehicle by British engineering firm Ricardo achieves radical engine downsizing using an electric supercharger coupled with a turbocharger and energy capture and storage technology.[15] This extremely cost-effective

IMAGE 2.14
Variable geometry turbo.

The Porsche VGT uses guide vanes on the outside of the turbine. At low engine speeds, the electronically controlled vanes tilt to produce a narrow gap that speeds exhaust gas to spin up the turbine more quickly and avoid lag; but at high engine speeds, the vanes open allowing a full turbo boost and avoiding the need for a bypass valve.

---

[15] S. Rajoo, A. Romagnoli, R. Martinez-Botas, A. Pesiridis, C. Copeland and A.M.I. Bin Mamat, Automotive exhaust power and waste heat recovery technologies. In A. Pesiridis (ed) *Automotive Exhaust Emissions and Energy Recovery*. Nova Science Publishers, New York, 2014.

design makes use of technology already on the market, adopting exhaust energy recapture and regenerative braking to drive an electric supercharger on a turbocharged GDI engine. The result is a three-cylinder, 1.0-liter engine that matches the performance of its 2.0-liter counterpart with a dramatic decrease in carbon emissions.[16] Similarly, but with more dramatic effect, Formula 1 cars and Audi's diesel R18 LeMans racer have coupled electric motors with turbochargers. The motor assists with low-speed boost, allowing the use of a larger turbine without lag and enabling a more significant high-end boost. Volvo's production T6 Drive-E engine incorporates a turbocharger and supercharger in a 2.0-liter package that produces 316 horsepower and an impressive 35 highway mpg.[17] Mercedes is taking this one step further by incorporating Formula 1 Motor Generator Unit (MGU) into its hypercar designs. The idea is to place a motor/generator onto the turbocharger shaft, allowing the generator to recover excess energy from the turbo as it spins down, store that in a battery, and use that energy to get the compressor spinning and eliminate lag.[18]

Compression Ratio

In the end, the challenge of forced induction, tuned runners, variable valve action, and just about everything else discussed thus far often comes down to the same issue: avoiding knock. Basically, we want to squeeze as much power as we can out of the fuel, without squeezing it so much that it ignites on its own. Tricky business. But instead of tweaking all the peripheral variables to manage this, why not get to the heart of the matter and adjust the amount the engine squeezes the charge? Advanced engineering and sophisticated digital control offers us the possibility of doing just that.

Defining the CR of an engine has long been the very epitome of automotive engineering trade-offs. An engine's compression ratio is not a variable that can typically be altered, like timing or fuel mixture; it is fundamental to the engine design. So, it needs to be defined carefully at the start. At lighter loads and lower engine speeds, a high compression ratio can provide improved fuel consumption, since higher compression means higher efficiency. An increase in the CR from 8 to 12 might produce a 20% or more improvement in an engine's ability to turn combustion into mechanical work, called thermal efficiency (Image 2.15).[19] While you might expect higher pressure to mean increased knocking, the risk of preignition is low given the low load. However, at high loads, as combustion chamber temperatures rise, the same compression ratio would likely lead to autoignition. And of course, a lower CR could allow for greater boost at high loads.

So, the resulting CR for a production engine tends to represent a compromise. A higher compression ratio offers greater thermal efficiency and thus more power and fuel efficiency, but it also invites knocking. So, in the past, designers defined a middling compression ratio that best suited the intended use of the engine. Compression ratios for gasoline

---

[16] https://ricardo.com/news-and-media/press-releases/hyboost-demonstrates-new-powertrain-architecture.

[17] G. Witzenburg, "Volvo's T6 Engine Part of Bold Powertrain Strategy." *Wards Auto* June 15, 2016. Available at www.wardsauto.com/engines/volvo-s-t6-engine-part-bold-powertrain-strategy

[18] A. MacKenzie, "Revealed: 2019 Mercedes-AMG Project One Powertrain." *Motor Trend* May 30, 2017. Available at www.motortrend.com/news/revealed-2019-mercedes-amg-project-one-powertrain/

[19] K. Satyanarayana, R.T. Naik and S.V. Uma-Maheswara Rao, Performance and emissions characteristics of variable compression ignition engine. *Advances in Automobile Engineering* 5 (2), 2016, 1–5.

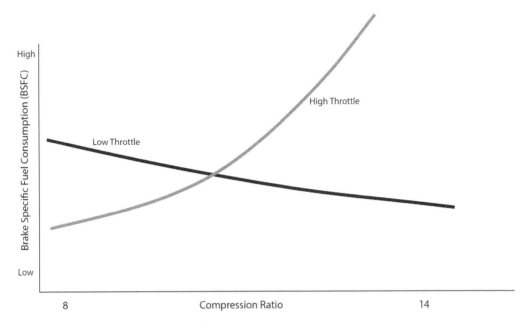

IMAGE 2.15
Fuel efficiency and compression ratio.

At high torque with a wide-open throttle, the engine's fuel consumption per unit of power generated, or Brake Specific Fuel Consumption (BSFC) is lowest at a low compression ratio but increases somewhat exponentially as CR rises. At a lower load demand, fuel consumption decreases as compression ratio rises. So, it makes sense to try to vary compression ratio as the load on the engine changes.

Source: D. Tomazic, H. Kleeberg, S. Bowyer, J. Dohmen, K. Wittek, and B. Haake (FEV), *Two-Stage Variable Compression Ratio (VCR) System to Increase Efficiency in Gasoline Powertrains DEER Conference 2012,* **Dearborn, October 16th, 2012.**

engines can be as low as 6:1, though 10:1 is more typical, with Mazda offering the highest production car CR of 14:1.

Defining an engine that can change its compression ratio mid-operation could end this compromise (Image 2.16). In addition, since the primary limitation to the usage of many alternative fuels is getting the compression ratio right, a variable CR could allow multiple fuel usage including everything from straight vegetable oil or diesel, to hydrogen and methane. With the capacity to offer fuel efficiency and still provide power when needed, a variable compression ratio (VCR) could also allow for more compelling forced induction and 'extreme downsizing' to very small, light, and powerful engines. But all this requires a mechanical configuration that can change an engine's core operation without adding excessive complexity or weight. Conceptually, there are two approaches that could allow the CR to be changed: You can change the clearance volume, or you can change the piston travel.

A simple solution would be to change the crown of the piston and therefore change the clearance volume. Performance engine builders have often machined or swapped piston heads as a way to vary the compression ratio of an engine. In the same vein, Ford and Daimler Benz have experimented with a piston configuration that has a sliding head and skirt unit that allowed for variation in the piston crown height. Hydraulic cylinders in the inner piston slide the outer head up or down relative to the inner piston body, changing the deck height and thus the compression ratio. The design approach has the advantage of

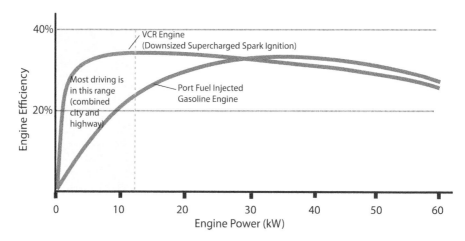

IMAGE 2.16
Improved efficiency of VCR engines.

While a VCR engine typically does not provide improved efficiency across its complete operating range, it does provide notably improved efficiency in the most frequent driving conditions.

Data: A. Baheri and M. Tajvidi, Simulation and analysis of an internal combustion engine with variable compression ratio. *International Journal of Engineering Sciences & Research Technology* 4(12), 2015, 757–763.

not requiring alteration of the basic engine architecture, but it has the significant disadvantage of adding complexity and weight to the engine's oscillating mass.

Moving the cylinder head could offer the same effect without changing the piston. A few years ago, SAAB engineers designed an articulated cylinder head that did just that. The head and cylinder liner were combined into an integrated housing and hinged to the lower crankcase. By moving the upper cylinder housing only slightly the compression ratio could be varied from 8:1 to 14:1. With integrated forced induction, the system promised efficiency, fuel flexibility, and notable power, reaping 225 horsepower from a 1.5-liter engine.[20] Unfortunately, production of the engine was determined to be prohibitively costly.

A similar result can be achieved more simply by adding an adjustable piston to the head. When this piston moves in and out of the combustion chamber, it changes the clearance volume and thus the CR. This approach is used by Lotus' so-called Omnivore research engine. Among other features, the head and block of this two-stroke design would be cast as one piece. A piston, or 'puck', would slide in and out of the combustion chamber to effectively vary the clearance volume and allow the engine to use multiple fuels, earning the engine its name.[21]

The alternative approach is to change the piston stroke. Perhaps the most direct way to achieve this is to change the connecting rod's length. By extending the connecting rod slightly, the clearance volume can be reduced significantly, offering a definitive change in CR. German engineering firm FEV uses an eccentric piston pin connection that is actuated by two pistons integrated into the connecting rod body (Image 2.17). They are able to adjust the CR from 8.8:1 to 12:1 in response to inertia forces and combustion pressure,

---

[20] M.P. Joshi and A.V. Kulkarni, Variable compression ratio (vcr) engine: A review of future power plant for automobile. *International Journal of Mechanical Engineering Research and Development* 2 (1), 2012, 9–16.
[21] Case Study: Omnivore Research Engine. Available at www.lotuscars.com/engineering/case-study-omnivore-research-engine

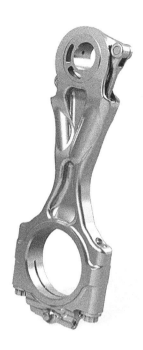

IMAGE 2.17
Variable connecting rod.

This innovative design by FEV uses an eccentric piston pin connection that is actuated by two pistons integrated into the connecting rod body. This allows a VCR engine with minimal engine modifications.

Image: FEV

requiring no active control.[22] Porsche is looking into a similar design produced by engineering firm Hilite International.

Multiple VCR designs offer mechanical variations in the rod-crank linkage. This seems to have become the most popular approach among carmakers, perhaps because the actuation mechanism is stationary, adding little to the engine's oscillating mass and providing easier mechanical control. Some use eccentric gearing to alter the crankshaft rotation, while others use actuators to adjust the crankshaft relative position. Some have been designed with a tip-in/tip-out strategy that adopts a defined lower CR only above a preset throttle position. Others integrate more continuous variation of CR. Nissan, Peugeot, Acura, and others have been playing with such designs for several years. The most promising approach seems to be the use of a multilink connection between the rod and crankshaft that can allow for an effective variation of the shaft throw. This could be controlled by an eccentric rotary actuator, a hydraulic piston, or some other control linkage.

Perhaps the most advanced version of this approach comes from Infiniti (Image 2.18). The Japanese automaker is planning to provide the first production vehicle with a VCR engine in the 2019 QX50. The turbocharged, four-cylinder, 2.0-liter engine is innovative.[23] The rod is not directly connected to the crankshaft, but to a multilink, with the piston rod connected on one side, an actuator arm connected on the other, and the crankshaft throw in between. An electric motor drives an actuator linkage that moves one side of the multilink up and down. The effect of course is to move the other side, where the piston rod is connected, relatively up and down in response and thus alter piston height. The engine can operate in Otto or Atkinson cycle, allowing a longer piston stroke at a high compression ratio. Because the rods are nearly vertical in the combustion stroke, and the multilink is designed to provide improved reciprocating motion, the engine is designed to run smoother and quieter than a conventional counterpart.

[22] S. Asthana, S. Bansal, S. Jaggi and N. Kumar, A comparative study of recent advancements in the field of variable compression ratio engine technology. SAE Technical Paper 2016-01-0669, 2016.

[23] "Infinity Develops a Variable Compression Ratio Engine". Available at www.ADandP.media

INFINITI VC-TURBO ENGINE
COMPARISON OF VC-TURBO TECHNOLOGY IN HIGH AND LOW COMPRESSION RATIOS

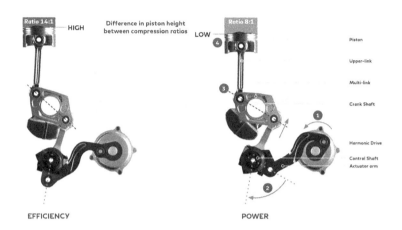

IMAGE 2.18
Infiniti VCR engine.

Infiniti has produced the first variable compression ratio engine in a production car. The harmonic drive moves the actuator arm, which in turn rotates the control shaft, modifies the piston travel, and so varies the compression ratio.

Source: INFINITI

This design is not entirely dislike the French MCE-5 Intelligent Variable Compression Ratio, or VCRi, four-cylinder engine (Image 2.19). Initially partnered with Peugeot and now working with Dongfeng, MCE-5 hopes to have the first variable-CR Miller-cycle engine ready for production within a few years. A lower linkage is moved with a hydraulic piston adjacent to the combustion cylinder to vary piston travel.[24] So, the MCE design allows for independent variation of each cylinder. While an impressive feat of engineering, whether the added value of piston-specific control is worth the additional complexity is not yet clear. The manufacturer points out that this will allow for faster and more accurate CR control, helping knock mitigation and facilitating advanced combustion modes like HCCI. And by reducing the impact of the accumulated small variations in component tolerances of the cylinder system, called tolerance stack, it permits larger manufacturing tolerances and so potentially lower production costs.[25] The CR can vary continuously along a much larger range, and the potential efficiency benefit is major.

---

[24] *Cost, Effectiveness and Deployment of Fuel Economy Technologies for Light-Duty Vehicles.* Committee on the Assessment of Technologies for Improving Fuel Economy of Light-Duty Vehicles, Phase 2, Board on Energy and Environmental Systems Division on Engineering and Physical Sciences National Academy of Sciences, Washington, DC, 2015.

[25] Author communication with Frédéric Dubois, MCE-5 Development S.A.

IMAGE 2.19
Intelligent variable compression ratio (VCRi).

French MCE-5 VCRi could vary CR continuously from 8:1 to as high as 18:1. Using the inertia of the engine to drive the adjustment mechanism allows changes in CR to take place in less than a tenth of a second, and could allow for compression ignition operational mode.

Source: MCE-5 DEVELOPMENT

The design of these mechanisms, and the general design of any modern engine, is a juggling act. Change one thing and three others are affected. Increase the boost and you have to decrease the compression ratio. Increase the valve lift for improved power and suffer deteriorated efficiency. Lengthen the induction runner for better cruising and lose high-end power. The list goes on. But increasingly, the juggling is getting easier and the trade-offs less troublesome. Advanced vehicle electronic management and innovative engineering have eliminated many of the compromises that once defined automotive design. The result is engines that are more powerful, more efficient, and cleaner than ever thought possible. Pretty cool.

3

Getting Power to the Pavement

Recent advances in engine design and control are impressive by anyone's measure. But it's not just about the engine; it can't be. The power coming from the engine only maters if we can effectively transfer it to the wheels. So, the design and capacity of the mechanism that connects the engine to the wheels, called the **drivetrain**, has to be every bit as sophisticated and efficient as the engine itself; and it is. As is true for engines, innovations in drivetrain components and control have revolutionized the industry, and profoundly change the capacity of the automobile and the nature of driving.

The tasks we expect from a drivetrain are not simple. We need a **coupler** that can allow us to connect and disconnect the engine from the rest of the drivetrain. The need for this should be pretty obvious; without it, the car would have to be moving anytime the engine is turning. Though, as we'll see, the coupler does a lot more than just provide stopping ability. We also need gearing. Again, the reason should be fairly clear. A typical internal combustion engine can produce useful power between about 1,500 and, let's say, 6,000 rotations per minute (rpm). Now consider that at 5 miles per hour (mph), an axle on a typical car is turning at about 80 rpm and at 70 mph, that axle needs to turn at about 1,200 rpm, or about 15 times faster. Our engine can't turn 15 times faster than its lowest speed. Much more importantly, while the engine's operable speed range is fairly wide, the range at which it produces desired power and efficiency is much smaller. So, even if we could connect the engine to the axle with no gearing and settle for a car with a 30 mph top speed, the resulting power, torque, and efficiency would be terrible. Lastly, steering a car, or at least turning it, requires that the axles be able to rotate at different speeds, as the outside wheels will strike a longer arc than the inside wheels. If we can't manage this, the tires will slip on any turn, leading not only to rapid tire wear but terrible handling. And all of this is just the tip of the iceberg; we expect much more of a modern drivetrain than these very basic functions. As a result, there has been plenty of need for innovation.

Once again, it's about new ideas and materials, but it is also about precise digital control and the possibilities this offers. We have largely replaced hydraulic control with digital electronic control in both automatic and manual transmissions. Advanced sensors and complex control logic, including artificial intelligence, fuzzy logic, and other advanced algorithms, have changed our understanding of what is possible. Advanced systems now in production or in the works provide predictive distribution of torque and power in any driving conditions, instantly identify slip and compensate with a redirection of torque to any wheel, and even offer the possibility of intelligent tires that can measure, evaluate, and adapt to road conditions. In short, the drivetrain of the future is here.

What Do We Need a Drivetrain to Do?

The drivetrain links the engine to the vehicle's motion. So, we might start by considering what is needed to produce this motion. To propel the car forward we need to overcome the combined **rolling resistance** of the tires and bearing friction, as well as the effect of any incline. We also need to overcome the **aerodynamic drag** of the car. We'll talk about this in Chapter 8; but for now, it's worth noting that drag increases exponentially with speed, while rolling resistance increases linearly. So, although rolling resistance is the primary resistive force at low speed, at high speed drag becomes the definitive force (Image 3.1).

The torque produced by the engine has to be able to match the torque required at the wheels to overcome the forces resisting the car's movement. So, let's look at the torque and power produced by the internal combustion engine. The torque produced by the engine at any given speed is directly related to the force provided by the piston during the power stroke. More particularly, it is the product of the force delivered perpendicular to the crankshaft throw and the throw length. So, we measure it in units of length and force, such as pound-feet (lb-ft) or newton meters (N-m). An engineer might point out that this unit might be more precisely expressed as pound force (lbf) to identify it as a unit of force and distinguish it from a pound of mass. As we've already seen, an engine design is generally tuned to produce performance at a defined speed range; on either side of this range, the efficiency of the combustion process can be compromised. So, it's not surprising that when we graph torque production, we can see that it typically dips at low and high engine speeds, with maximum production somewhere in the middle. As we discussed in the previous chapters, innovative technologies are being deployed to widen the range at which this torque can be produced. So, the torque curve is getting wider, or fatter, and more consistent, or flatter.

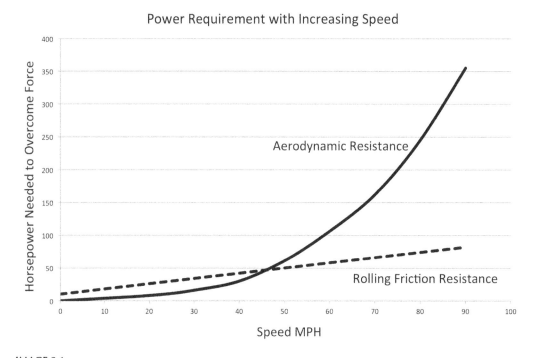

IMAGE 3.1
Power requirement with increasing speed.

As the speed of a vehicle increases, both friction resistance and aerodynamic resistance increase. However, aerodynamic resistance increases as a function of the square of the speed, making it the dominant resistive force at higher speeds.

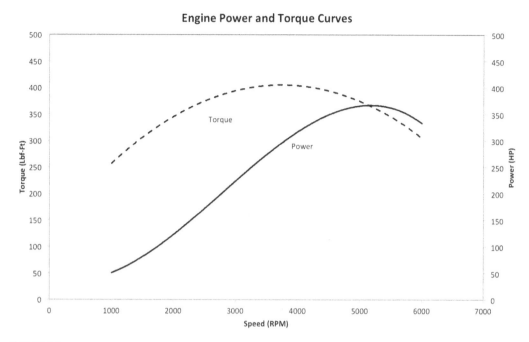

IMAGE 3.2
Engine power and torque.

In this typical torque and power curve, notice how the horsepower continues to climb after the torque has plateaued. As engine speed increases, power production will grow up to a point when the engine is no longer capable of generating more power.

Torque is notably different than power (Image 3.2). Torque is an expression of rotational force at a given point in time. On the other hand, as you might recall from basic physics, work is defined by an application of force over distance. And power is the capacity to do work over time. Consider an example: If you push on your stalled car with 50 pounds of force but can't get it to move, you've accomplished no work. (Though technically, your muscles moved quite a bit in the effort, so they did some work. It just didn't get you anywhere.) If you then get smart, release the parking brake and try again, you're able to move your car 20 ft. So, you have now accomplished 1,000 ft-lbf of work (50 pounds of force times 20 ft). Notice, the unit is foot-pounds for work, to distinguish it from lb-ft of rotational torque; and once more we use the unit pounds-force to distinguish it from pounds of mass or weight. So much for work, but we don't know how much power you were able to produce until we know how long it took you to accomplish this herculean task, since power is work over time. If it took you an hour to get the car to move 20 ft, we can now assess the power you produced. In this case, you've provided 50 lbf×20 ft/60 min, so about 17 ft-lbf/ min of power. Since one **horsepower** is defined as the ability to move 550 pounds 1 ft in 1 min (550 ft-lbf/min), you've produced a whopping 0.03 HP. Nice try. However, if you were able to move the car that 20 ft in 5 s, then you would have produced 50 lbf×20 ft/0.083 min, or about 12,048 ft-lbf/min, that's nearly 22 horsepower. Much better.

The key here is that power is an expression not only of how much force you can produce but how quickly you can produce it. Getting back to engines, torque is generated only in the power stroke, and roughly maximized when the connecting rod is perpendicular to the throw of the shaft. The engine's power, on the other hand, is the result of both the force generated by these strokes, and the number of strokes that the engine can produce in a given amount of time. The faster the engine rotates, the more power strokes in a given period of time, and so the more power that is being produced. Double the engine speed, and all things being equal, the power output doubles. The torque, on the other hand, doesn't change in a fixed way with engine speed, as we saw in the graph above.

Of course the increase in power cannot continue forever. As we saw in Chapter 2, as the engine speeds up, its ability to take in air and produce useful combustion can deteriorate as the engine approaches its maximum speed. So, at some point, we will see a dip in power production, even though speed continues to increase. All this makes the typical tendency to compare engines based on their maximum horsepower a bit foolish. Maximum power is just that, maximum, determined at a particular engine speed. It doesn't always tell you very much about the engine's operation at any other speed (Image 3.3).

Another reason we shouldn't get too carried away with engine horsepower is because what really matters isn't the power of the engine, but the power that can be delivered through the drivetrain to the pavement. Transforming torque and speed through the gearing of the drivetrain allows us to adjust the power and torque available at the drive wheels. Gearing allows us to effectively transform torque to speed and vice versa by applying a form of rotational leverage. As a small gear rotates against a larger gear, it must provide greater speed, but the result in the larger gear is greater torque. If the larger gear, on the other hand, drives the small gear, it can move slowly and generate greater speed in the smaller gear, but the price is torque. In either condition, power at each shaft remains equal; gears do not create power. This means the actual power and torque we can produce at the

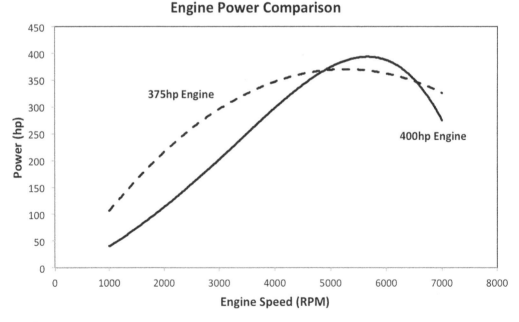

IMAGE 3.3
Comparing engine power.

The common practice of comparing engine capacity by comparing top horsepower production is problematic at best. Notice how in this idealized case the lower horsepower engine actually generates more horsepower for a much greater range of operation, even though the other engine would be rated for a higher peak horsepower. So, looking only at peak horsepower could lead to misleading conclusions about engine performance.

wheels can vary; and while we only have one power curve for our engine, we can produce multiple torque curves for our drive axle depending on the gearing we select.

Remember the power needed to accelerate our car increases with speed. In addition, the more our available power exceeds the resistive force of drag and friction, the more power available for acceleration. So, if we shifted into fifth gear at 10 mph, we would not be able to accelerate, since our available tractive force at the drive axle does not exceed the total resistance to the movement of our car. Dropping the car into a lower gear increases the tractive force at the wheels and allows acceleration. But if we stay in that low gear, we'll reach our maximum speed in that gear pretty quickly. The magic of gear changing allows us to accelerate quickly at low speeds, and change gears so we can continue to accelerate at higher speeds.

Understanding all this allows us to appreciate key characteristics of vehicle performance. For example, at some point the power the engine is able to produce and the power required for continued forward motion are equal. This defines our maximum vehicle speed (all other variables like stability and mechanical integrity allowing). Below that speed, a lower gear can allow for increased surplus power (Image 3.4). So, dropping down a gear can allow for increased acceleration during overtaking. Conversely, a gear beyond the maximum speed gear might allow for lower engine speed, and so quieter and more efficient operation at cruise, even though it would reduce available power somewhat. That is the principle behind **overdrive**. Since early performance transmissions often had the fastest

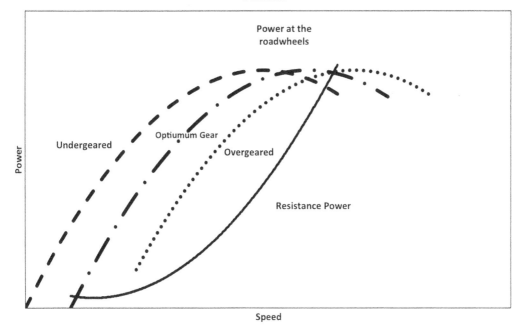

Performance Curve for Undergeared and Overgeared Vehicles

Power at the roadwheels

Undergeared

Optiumum Gear

Overgeared

Resistance Power

Power

Speed

IMAGE 3.4
Performance and gearing.

Accelerative capacity is determined by the power that can be delivered to the wheels in excess of the combined resistance force or friction and drag. Under-gearing can limit top speed, and over-gearing can limit acceleration capacity.

gearset at **direct drive**, or a gear ratio of 1:1, overdrive is usually considered to be any ratio larger than this.

Acceleration typically defines a series of torque peaks and drops with each gear change. Between each peak as engine speed continues to increase there is a marked decline in available torque, called **torque interruption**. Minimizing this interruption and producing a continuous and smooth transition of torque for power throughout a long acceleration is a key to performance. The effect can be felt casually in a typical car as you accelerate; engine speed increases and then drops to accommodate shifts, so acceleration is applied in steps (Image 3.5). The driver experiences a **shift shock** with each transition. While this was once a normal characteristic of driving, it is not ideal, and has decreased significantly with advanced transmissions.

More available gear ratios can help minimize torque interruption by filling the gaps between the curves. It also allows the engine to operate more closely to its most efficient speed, offering improved fuel economy. An engine consumes the least fuel per unit of power produced, called **brake specific fuel consumption** (BSFC), on the lower end of it speed range (Image 3.6). So, a gearbox that allows the engine to stay closer to this efficiency region as the car's speed varies will produce greater fuel economy. In fact, the desire for efficiency has resulted in a general increase in the gearing of the typical production vehicle, allowing the engine to stay in a more efficient operating range and avoid fluctuations in engine speed that can kill efficiency. As a result, the four- or five-speed automatic transmissions that were typical just a decade ago have been replaced with at least six to eight gears, and increasingly nine, ten or more.

At any given point in the acceleration process, there is often more than one possible gear choice, making variations in the timing of the shift an option. As the graph indicates,

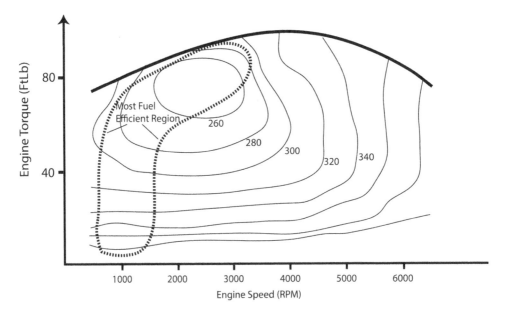

IMAGE 3.5
Fuel map.

This typical fuel map denotes BSFC (g/kWh) within differing regions of engine speed and torque. Fuel consumption dips on the left side of the graph. So, the most efficient operating region for any given torque requirement is along this left side where engine speed is low.

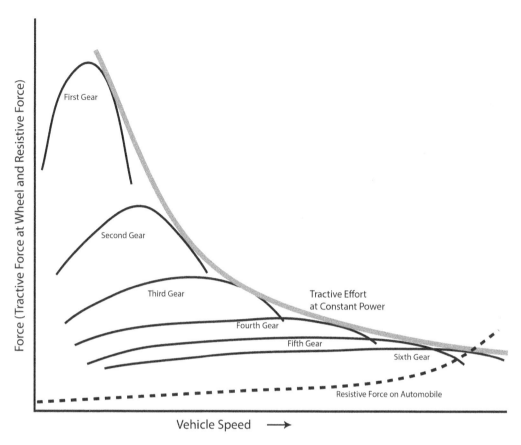

IMAGE 3.6
Tractive effort.

The tractive force available at the wheels varies with each gear. As vehicle speed increases, gear changes are necessary to continue to generate forward force. However, at some speed, the tractive effort available is exceeded by the resistive force of air and friction, and no further acceleration is possible. The more gears available, the less severe the dips in tractive force between gears, and the better the engine is able to remain at a desirable operating speed for a given vehicle speed.

delaying each shift until the maximum tractive force has been reached in that gear ratio can help keep tractive power high, and acceleration quick. However, the cost of this strategy—there is always a cost—is higher fuel consumption. On the other hand, early shifting can significantly increase fuel economy, but the price is accelerative capacity. As we will see, the gear change timing of automated transmissions depends on the design priorities of that vehicle, driving conditions, as well as selections made by the driver if the car is equipped with selectable drive modes.

The basic shift logic that must be managed by any transmission can be understood graphically.[1] The two lines on the graph below represent upshift and downshift points. Normal driving in any given gear is defined between these two lines. The space between the lines provides a lag, or hysteresis, in gear changes so constant shifting is avoided. When driving at a given operating point between the lines, a large throttle position increase can lead to downshift to provide increased power for acceleration. Though the same throttle position

[1] M.G. Gabriel, *Innovations in Automotive Transmission Engineering*. SAE International, Warrendale, PA, 2003.

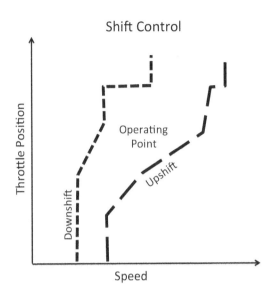

IMAGE 3.7
Shift control.

Shifting speeds vary with driving conditions and throttle position. With a hard acceleration, up shifting is delayed to ensure maximum available torque for overtaking. At low throttle position, upshift happens at a lower engine speed, to help ensure improved fuel economy.

Source: Adapted from M.G. Gabriel, *Innovations in Automotive Transmission Engineering.* **SAE International, Warrendale, PA, 2003.**

increase with an accompanied acceleration might not lead to a gear change, an increase in speed with only a minor increase in throttle position could lead to downshift for improved fuel economy (Image 3.7). A drop in throttle position could lead to a downshift if speed is maintained, or no shift if some speed is lost. And a drop in speed while maintaining throttle position, say due to an incline, could lead to a downshift.

The gearing used for any particular vehicle depends on the characteristics of that vehicle. Greater rolling resistance, say due to high vehicle weight, may require a drop in gearing at the low end of the range. A large frontal area may dictate lower gearing on the high end where drag will define a predominant force. A desire for smooth fast cruising may mean an additional overdrive gear is needed. The performance expectations and priorities of the vehicle will matter a lot, of course. In addition, expected towing capacity, targeted top speed, engine characteristics, and a wide array of other concerns will define the optimal gearing combination. All of that having been said, in general, the essential parameters are the needed upper and lower gear ratios, with the middle gears largely defined to provide a reasonably even distribution of gearing between the two.

Manual Transmission Coupler

With these concepts in mind, we can start to think about how it all comes to life in an actual gearing mechanism. We can start with the most straightforward version, the manual transmission; and a good starting point for that is the basic clutch. A conventional manual transmission, or stick, relies on the driver to physically change gear ratios to suit driving conditions. To make this possible, we have to enable a manual decoupling of the engine and transmission. This not only is needed to allow the car to stop without turning off the engine but to smoothly decouple and then recouple the engine and the driveline to allow gear changes. The principle way this is accomplished in a manual car is a mechanical friction coupling or **clutch**. A key challenge is to avoid a strong jerk, or shift shock, both for initial acceleration and during gear changes, which usually requires an ability to **slip** the friction coupling for gradual engagement.

A friction coupler is a pretty simple device. In its most basic form, imagine two rotating disks placed with their planar surfaces near each other. If the two are pressed together, then the friction contact means the rotation of one will cause the rotation of the other. If we release the pressure, the two can rotate independently. An automotive friction clutch is not that much more complicated. It's composed of three primary components: a flywheel, a pressure plate, and a disk. A cover is bolted to the engine flywheel on one side and connected to the pressure plate on the other. The disk, also called a **clutch plate**, is placed between the pressure plate and the flywheel (Image 3.8). It has friction surfaces on both sides, and is splined so it rotates with the transmission input shaft. We can use a collection of coil springs or a diaphragm spring to squeeze the sandwich together. When we do, the friction between the flywheel, disk, and pressure plate causes them to all rotate together. When we release the pressure, the disk can rotate independently of the cover and pressure plate, decoupling the engine and transmission. When pressure is reapplied, friction serves to lock all the parts back together and the two shafts rotate at the same speed. Importantly, if the force is applied gradually, allowing an initial slip, the torque from the driving shaft will increase proportionally providing a smooth application of torque to the drivetrain.

IMAGE 3.8
The clutch assembly.

The basic components of a clutch assembly are fairly straightforward. This SACHS performance clutch includes a flywheel, a clutch plate or disk, and a pressure plate with a diaphragm spring. A clutch release bearing rides on the diaphragm spring to engage or disengage the grip on the disk, and is actuated by a release mechanism.

Image: ZF Friedrichshafen AG

The basic system is pretty straightforward, but some enhancements are possible. A spring hub is generally used on the friction plate to allow for smoother transitions. A small braking mechanism can be added to stop the clutch plate from spinning once the clutch is disengaged, called a **clutch stop**. And the driven plate can be lightened to reduce its moment of inertia toward the same end. This enables quicker gear changes, since gear changes are limited until the transmission input shaft slows. The friction surface is probably the greatest variable, and can be designed to allow improved wear and smoother engagement.

Ideally, the friction material on the clutch plate will provide a durable and uniform friction surface that wears evenly, is able to conduct heat effectively, and demonstrates friction and strength that is unaffected by high heat. Generally asbestos fiber interwoven with metal wire has been used, though newer materials are changing this. Kevlar, a polymer with five times the strength of steel, improved strength, much-improved heat tolerance, and good friction characteristics that enable a smooth application of braking. Ceramic facings composed of copper, iron, tin, bronze, and silicon dioxide, or graphite are harder and can offer excellent heat transfer for high-performance applications, but with a tendency for sudden engagement that can make them a poor choice for casual driving. Organic composites of advanced resins, rubber, and metals can offer smooth engagement, durability, and good heat management. Groove patterns on the clutch material can be designed to allow improved wear and smoother application without catching. The options are much better than they were even a decade ago.

Similarly, there are some modifications that are not generally applicable, but make sense in specific applications. While a single-plate clutch is common, a multiplate clutch is possible. There are a few reasons this might be desirable. First, the greater the friction surface area of contact, the more effectively the torque is transmitted. And, adding a number of plates can allow for a smaller diameter unit while maintaining adequate surface contact. In addition, to enhance smoothness of interaction as well as cooling, the clutch can be submersed in a fluid, defining a so-called **wet clutch**. Wet clutches are not generally used in conventional manual transmissions where a single-plate dry clutch remains dominant. However, we will see that recent innovations in transmission designs are making this option more desirable for certain applications.

Manual Transmission

The basic design of a manual transmission itself is defined by three shafts: an **input shaft** connected to the clutch; an **output shaft**, also called a main shaft; and what is called a **layshaft** or countershaft (since it counter rotates). Gears are mounted to the layshaft and rotate as one unit (Image 3.9). The gears on the output shaft rotate independently but are always meshed with the layshaft gears. The output shaft is splined, and between each gear is a ring, called a **collar**, that incorporates a cone-shaped clutch mechanism, called a **synchronizer** and dogteeth. As the collar slides toward a selected gear, initial engagement of the synchronizer allows a smooth friction contact so both gear and shaft are rotating at the same speed prior to engagement of the teeth that lock the gear to the shaft. This gear now defines the driving gear ratio, as other gears on the output shaft spin freely. Placing an additional gear, called an idler gear, between a layshaft gear and reverse gear on the output shaft, enables reverse.

IMAGE 3.9
Manual transmission.

A basic manual transmission includes a layshaft (lower), an input shaft connected to the clutch (left) an output shaft connected to the drivetrain (upper), collars between the gears on the output shaft, and forks that move the collars as well as a shift rail (upper) that moves the forks.

Allowing for multiple gear ratios is then simple, as various gear combinations on the layshaft and output shaft provide differing gearing combinations (Image 3.10). A basic actuator mechanism controls linkages that enable the movement of each collar, and so the engagement of any gear on the output shaft. Because each collar incorporates a synchronizer, the gnashing of gears that was common long ago is avoided nearly completely. The driver can disengage the clutch, and so release tension in the drive train, move the shift lever which actuates forks that slide a collar disengaging one gear combination and engaging another, and reengage the clutch by releasing foot pressure from the actuator. Because the gears are always engaged, there is no need for double clutching, and no damage to gear teeth.

If you live in the US, you might be thinking that this is all tired technology. Manual transmissions have lost popularity in the US almost entirely, and now represent less than 5% of new vehicles. They are declining in Asia as well. However, they remain popular in Europe, representing about 75% of European auto sales.[2] Despite declining interest, there are clear advantages that make manual transmission worth another look. The solid mechanical connection of manual gearboxes means low gear loss, defining the most efficient way to transfer engine power to the driveshaft (despite the recent improvements in automatic transmission we'll look at soon). A typical manual transmission can provide an overall efficiency percentage in the mid to high 90s, a notch above the overall mid to high 80s efficiency of a typical automatic transmission.[3] They also offer lower cost, particularly in Europe where existing facilities can be used without retooling for a new transmission type. Moreover, with the

---

[2] *European Vehicle Market Statistics 2016/17.* International Council for Clean Transportation. Available at eupocketbook.theicct.org; A. Isenstadt, J. German, M. Burd and E. Greif, *Transmissions* International Council for Clean Transportation, Working Paper, August, 2016.

[3] M.S. Kumbhar and D.R. Panchagade, A literature review on automated manual transmission (AMT). *IJSRD—International Journal for Scientific Research & Development* 2(3), 2014, 1236–1239; and M. Kulkarni, T. Shim and Y.S. Zhang, Dynamics and control of dual-clutch transmissions. *Mechanism and Machine Theory* 42, 2007, 168–182.

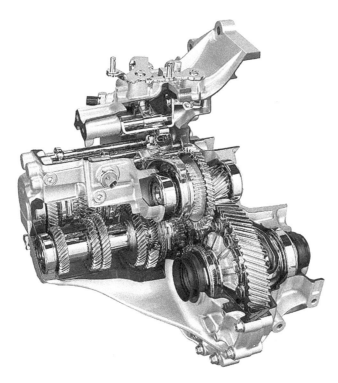

IMAGE 3.10
Toyota six-speed manual transmission.

Toyota has developed an advanced manual transmission primarily for the European market. The design is notably lightweight and compact, with a total weight of about 88 pounds or 40 kg. Toyota's iMT (Intelligent Manual Transmission) controls automatically adjust engine speed when shifting to ensure smooth gear transitions and reduced torque interruption.

Image: Toyota Motor Company

performance and automation upgrades now possible with digital control and advanced manufacturing, a manual transmission no longer means more work behind the wheel, but it can still mean a more engaged and dynamic driving experience.

Nevertheless, increasing engine speeds and increasing vehicle top speeds are challenging the capacity of the old manual transmission. Automatic transmissions, as we will see, have responded with an ever-increasing number of gear ratios, enabling improved fuel economy and performance. Manuals are somewhat limited here; while seven-speed manual transmissions are available in numerous performance cars, an eight- or nine-speed manual transmission would be large and offer questionable drivability.

Automated Manual

As is true for so much we have seen so far, modern digital control has enabled a new phase in the manual transmission's evolution. A variety of transmission control strategies have been applied to what are called semi-automatic transmissions, or the seemingly oxymoronic, **automated manual transmission** (AMT). By adding hydraulic or electrome-chanical actuators that allow for the clutch as well as shifting to be controlled digitally, manufacturers have dispensed with the clutch pedal and developed a transmission that can incorporate varying degrees of automated function. The digital logic can enable fea-tures that would normally not be possible with a manual transmission, such as fuel saving strategies, start-stop technology (to be discussed in Chapter 6), or idling or stopping the engine when coasting in high gear, called **sailing**.

The basic idea is to combine the efficiency, simplicity, and affordability of a manual transmission with the drivability of an automatic. Though it should be said that fuel econ-omy and manufacturing cost, not necessarily driver preference, have probably been the strongest motivating factor in the development of the AMT. Early versions of this technol-ogy simply replaced the clutch pedal with an automated electrohydraulic clutch that was activated upon shifting. Such systems can still provide an economical alternative that can make a low- cost manual transmission more attractive, such as Schaeffler's recent develop-ment of the eClutch affordable automated clutching mechanism.

Typically, an AMT system will include a fully automatic mode, with no driver input needed. So, an AMT can drive a lot like an automatic transmission, although they will tend to provide a sharper responsiveness because of the direct mechanical engagement of the drivetrain. The aim is to combine the efficiency and control of a manual transmission with the drivability of an automatic, in a package that can appeal to a driver's desire for a more engaged experience when wanted and simplified driving when not. So, shifting can take place automatically, with a conventional lever or with paddles on the steering column. Adopted from Formula One, shift paddles offer a sportier driving experience, typically with a paddle on the right of the steering wheel shifting to a higher gear and one on the left dropping to a lower gear.

Clearly, digital control is indispensible to the AMT. A **transmission control unit** (TCU) combines torque, vehicle speed, engine speed, and throttle data from the ECU with trans-mission specific information from gear, clutch, and brake position sensors, as well as transmission output speed. Moving the shifter sends an electronic signal, not a mechani-cal transfer, a so-called shift-by-wire system. When a shift is called for by the driver or indicated by the algorithm, optimal torque, engine speed and shift timing are defined for a fast and smooth gear transition. Clutch disengagement, shift, and reengagement then all occur automatically. When downshifting, the control module revs the engine upon disen-gagement to achieve rev matching (matching the engine speed to the transmission shaft speed) and so allow for smooth reengagement. In fact, while not done, it could be possible to operate such as system with no use of a clutch at all given the precise control possible.[4] In most systems, gears can be skipped to allow the most effective ratio, while precisely matching engine and transmission speeds to avoid lurching or lugging the engine.

The control logic can be changed to suit driving needs. In normal drive mode, for exam-ple, the shifting happens earlier to ensure optimal fuel efficiency and smooth driving. In sports mode, on the other hand, shifting will happen later to allow the full advantage of high power operation, and downshifting will become more aggressive to enhance respon-siveness. More than this, the TCU can include a measure of artificial intelligence, utiliz-ing throttle position, vehicle speed, engine speed, and other parameters to estimate the driver's intention and manage shifting to provide performance characteristics that match the driver's intent without even needing to change the drive mode. A speed, throttle, and

---

[4] Z. Zhong, G. Kong, Z. Yu, X. Xin and X. Chen, Shifting control of an automated mechanical transmission without using the clutch. *International Journal of Automotive Technology* 13(3), 2012, 487–496.

torque combination that indicate uphill driving, for example, might prevent an upshift to ensure desired acceleration. Downhill driving, on the other hand, might automatically trigger a downshift to avoid unwanted acceleration.

Even though automation is impressive, in itself it does not solve one of the fundamental challenges of manual transmissions: the time required to shift gears. Any gear shift has to entail four steps: a disengagement of the clutch, a move out of one gear ratio, a move into another gear ratio, and clutch reengagement. That takes time, even if done automatically. The result is an undesirable torque interruption. But, as always, there is another option: This challenge can be met by the integration of two clutches, a **dual-clutch transmission (DCT)** (Image 3.11). If we use two clutches, the disengagement of one gear and the engagement of another don't have to happen sequentially, and that means a faster transition. Originally branded by Volkswagen as the Direct Shift Gearbox (DSG) and Audi as the S-Tronic transmission, versions of dual-clutch units are provided by just about every major automaker.

The trick to the DCT is to superimpose two independent main shafts in a single-transmission housing. Odd numbered gears are placed on one of the coaxial shafts and even numbered gears on the other. Each shaft is served by a separate clutch mechanism, so

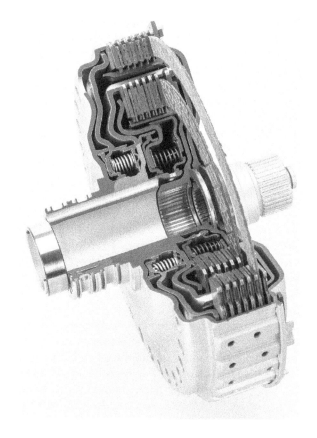

IMAGE 3.11
Dual-clutch module.

Initially designed for European sports cars, this dual clutch module produced by BorgWarner is ideal for high-rev, high-torque applications and allows performance shifting with minimal torque interruption.

Image: BorgWarner

a gear on one shaft can be disengaged at the same time that a gear on another is engaged. This can make shifting much faster and smoother, and nearly eliminate torque interruption by allowing the subsequent gear to already be selected before a shift takes place. For example, in shifting from second to third gear, the even-geared clutch is disengaged to allow exit from second gear while third gear is already selected on the separate odd-geared clutch. The effect is impressive; in an automated system a DCT can provide shift speeds in microseconds with minimal torque interruption.[5]

In order to take advantage of a dual clutch, the digital control logic requires that the control module predict the direction of the next gear change. So, based on speed, throttle position, and driver behavior, the control unit must correctly preselect the likely next gear for the idle main shaft. When the driver shifts to fifth gear, for example, the control unit algorithm must decide whether the idle shaft should be preselected in fourth or sixth gear. In normal conditions, the logic is not particularly challenging, with acceleration, the system preselects the next higher gear in a matter of milliseconds. With deceleration, a lower gear is selected. An unusual condition, however, can present more of a challenge. For example, slowing in traffic might trigger a lower gear preselection. As a result, if the traffic suddenly moves forward, the quick change in throttle position for acceleration can provoke a jerky response. However, even in this unusual scenario, the response will still be considerably faster and smoother than a driver-controlled manual transmission.

Clearly, electronic control of the transmission is more complicated than simply identifying the ideal gear ratio. Remember that neither the shifter nor the accelerator pedal is a direct mechanical actuator. Both simply convey driver preferences electronically to a microprocessor. The control logic then synthesizes these demands with engine speed control, vehicle speed, and other factors to identify shifting points based on identified driving strategy. For example, a call for hard acceleration needs a delayed shift to maintain high power. An indication of leisurely acceleration requires early shift to maintain high fuel economy. Shift points can be informed by a drive mode selection. But they can also be informed by driver behaviors, driving settings, and even road conditions. An ability to sense high load, due to climbing or towing for example, or forward acceleration due to a decline, is indispensible to correct automated control. The system must also interact with other electronically controlled vehicle systems, engine control, active chassis control, steering, traction control, electronic stability control, and others. In short, the TCU must be able to directly or indirectly sense every variable that the most proficient driver would sense and accommodate while driving. At the same time, the control system needs to protect against driver error, a call for an inappropriate gear for example. Advanced algorithms, Kalman filters, and fuzzy logic can allow for a useful synthesis of the data and the identification of an appropriate control scheme (we'll cover a bit more on these algorithms in Chapter 9).

So, while obviously very cool, DCT systems also face a common difficulty, providing high-torque operation without excess heat. Launch, for example, needs to take place while engaging a single clutch. We will see later than an automatic transmission relies on a torque converter to multiply low-speed torque. But a DCT does not typically have that option, and particularly in a high-torque condition, say a turbo charged engine or a high-load diesel, this can lead to problems. The extended clutch slip needed to allow smooth

[5] F. Vacca, S. De Pinto, A.E.H. Karci, P. Gruber, F. Viotto, C. Cavallino, J. Rossi and A. Sorniotti, On the energy efficiency of dual clutch transmissions and automated manual transmissions. *Energies* 10, 2017, 1562; and G. Shi, P. Dong, H.Q. Sun, Y.Liu, Y.J. Cheng and Y. Xu, Adaptive control of the shifting process in automatic transmissions. *International Journal of Automotive Technology* 18(1), 2017, 179–194.

engagement at high engine speed can lead to excess heating, performance deterioration, and mechanical damage. This brings us back to that idea of submerging the clutch in oil. A wet clutch can provide smoother engagement and cooling but with some lost efficiency due to the need for a pump and the losses caused by the resistance of the clutch pack turning in oil, called **churning**.[6] An innovative option is to just add a torque converter to a DCT as Honda did with the 2.4 L Acura TLX. As we will see soon, a torque converter can provide torque multiplication during launch, though lost efficiency and higher costs are still challenges.

So, the manual transmission has truly come of age. Twin-disk clutches, more closely spaced gears, improved synchronization, and enhanced automation make manual transmissions a promising option for performance, without even considering the improved fuel efficiency and lower cost. As a result, they are continually relevant in Europe, even if they have yet to impress American buyers. It is possible that since most US drivers have never driven a manual transmission they are less favorably disposed to the more abrupt torque shift of a manual gearbox and less inclined to use the manual feature, making the AMT feel like it is just a rougher version of an automatic. In addition, their relative complexity and fragility mean DCTs are increasingly challenged to handle the output of high-power engines. So, American automakers are largely looking past automated manuals toward improved automatic transmissions for the US market. The breakthrough may come with the growth in hybrid drives since these dual-disk systems could be well suited for hybrid systems, where the additional torque of the electric drive can more easily handle the launch load. More on this in Chapter 5.

Automatic Transmission Coupler

If we want to understand the automatic transmission systems, we have to start with the coupler. The friction clutch is efficient, reliable, robust, and inexpensive. But, until recently it required driver input to operate, and so was not an option for an automated system. An alternative coupling mechanism called a torque converter provides a fluid rather than mechanical coupling, and can effectively allow coupling and uncoupling with no driver input. The basic idea is pretty simple. Imagine a fan blowing on another fan. The rotation of one causes a rotation of the other. But, if you stopped the rotation of the second fan, it wouldn't bother the driving fan. Got that? Okay, now contain the whole thing in a fluid filled container that can direct and concentrate the flow and provide a viscosity that offers a much-improved transfer of force. This is your basic fluid coupling.

In practice, the essential mechanism includes three key components: a toroidal impeller that serves as a pump and is fixed to a housing that connects to the engine flywheel; a turbine that is driven by the pumped fluid, has no mechanical contact with the pump, and is mounted to the transmission input; and a set of stator blades between the two that serves to redirect the fluid flow from the turbine to the pump. This last component is key because it allow for torque multiplication. The impeller is driven by the engine and defines a rotary flow that produces a centrifugal pressure inside the housing. The resulting pressure causes an outward movement of fluid defining a vortex flow that impacts and rotates

---

[6] F. Vacca, S. De Pinto, A.E.H. Karci, P. Gruber, F. Viotto, C. Cavallino, J. Rossi and A. Sorniotti, On the energy efficiency of dual clutch transmissions and automated manual transmissions. *Energies* 10, 2017, 1562.

the driven turbine. The fluid is then redirected through the stator and back to the impeller. The stator's capacity to redirect the flow means the return flow from the turbine can be reversed so it does not work against the rotation of the impeller. This means torque is increased, making this a more useful device than a simple fluid coupling and earning the name **torque converter** (Image 3.12).

Understanding how this works requires a bit more detail. Let's begin at idle. Because there is no mechanical connection between the driving and driven members, decoupling is not a problem and the car sits with the pump turning and the turbine still. As the engine accelerates from idle, the pump's speed increases as does the resulting rotational flow and pressure (indicated below by the arrows). As this pumped fluid impacts the turbine with growing force, it defines an increasing torque that rotates the turbine and output shaft. The fluid then continues its helical flow and moves back toward the center of the impeller with a movement that is opposite the pump rotation. Normally, you would expect this flow to act against the impeller's rotation. However, the flow is channeled through the stator and because a one-way clutch is used to allow the stator to rotate only in the direction of the impeller, the fluid flowing from the turbine and through the stator is redirected back to the rotational direction of the pump. The one-way clutch is the key here. By linking the impeller to the transmission casing, a counter rotational force on the impeller is avoided and this allows for torque multiplication since the turbine is now being rotated by both the

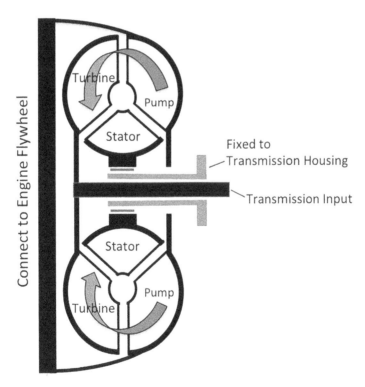

IMAGE 3.12
The basic torque converter.

A basic torque converter includes a toroidal impeller connected to the housing that serves as a centrifugal pump, a turbine that is driven by the pumped fluid (indicated by the arrows), and stator blades between the two. The stator allows for torque multiplication since the high-pressure fluid that returns to the fast-moving pump can be redetected, ensuring it does not work against the pump.

force pushing it due to the pressure of the incoming fluid as well as the benefit of ejecting the exiting fluid against the fixed stator.

As turbine speed increases, the relative difference in speed of the pump and turbine decreases, and so the torque multiplication decreases. In this sense, it works a bit like gearing: the less the speed decreases, the less the torque increases. At higher speed, when the turbine is turning nearly as fast as the pump, the fluid leaves the turbine with much less velocity, and the relatively geometry of the turbine and stator change. Effectively, the turbine is now rotating so quickly that the fluid's relative transaxial motion changes and the fluid leaving the turbine begins to hit the backside of the stator. So, the stator no longer reverses the flow, it now freewheels with the impeller. At this point, called the **coupling point**, there is no longer torque multiplication, and the torque converter operates as a basic fluid coupling. The ability of the stator to freewheel avoids the drop in torque that would result from the changed geometry at high speeds if the stator were fixed, and it allows a continued rise in efficiency to perhaps 92% or so.

Fundamentally, the capacity for torque multiplication is possible because the pump is rotating faster than the turbine. The greater the difference in the rotation speeds, the greater the torque. This has the happy effect of providing the greatest torque multiplication, typically about 2:1, when the car just begins to move. As the speed picks up, the relative speed difference of the impeller and turbine decrease and the multiplying effect drops off. The difference between the two speeds is called **slip**. Fortunately, at higher speed when no torque multiplication is needed, the slip is reduced. But there is always some slip; otherwise no torque could be transmitted through flow. At lower speeds, as slip increases, the efficiency of a torque converter drops notably.

Some of this inefficiency can be addressed by adding a clutch mechanism that can lock the assembly together at cruise speed called a **lock-up clutch**. This is particularly advantageous in overdrive, since the negative relative slip could cause fluid cavitation and overheating. While this technology has been around for a while, and is nearly universal in cars manufactured in the last two decades, more precise digital control now allows for earlier and more frequent engagement of the lock-up mechanism by identifying conditions for lock-up engagement with low load and high speeds in every gear ratio but the first. The result is improved fuel efficiency as well as reduced heat accumulation.

Automatic Transmissions

While we now know that a layshaft transmission can be made to operate automatically, this was not always the case. So, when engineers set out to build an automatic transmission more than half a century ago, they began with an entirely different starting point. As a result, the heart of the conventional automatic transmission is not a series of gears lined up on dual shafts; it is an epicyclic or **planetary gearset**. The most basic example is defined by four components: A center gear, called a **sun gear**, multiple pinion gears that mesh with the sun gear and can rotate around it, called **planet gears**, an outer ring with teeth on the inside, called a **ring gear** or annulus, and a carrier that connects the planet gears together and allows them to rotate as a unit (Image 3.13). Each component can be fixed in position or allowed to spin though the application of clutching mechanisms or band brakes. Changes in the drive ratio are then simply a matter of changing the driving and driven components and fixing or releasing others to define various combinations of

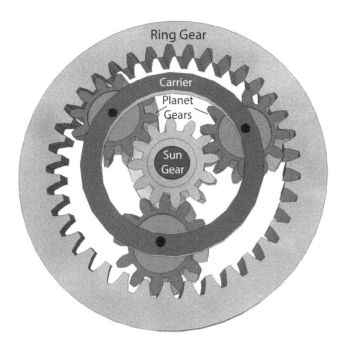

IMAGE 3.13
Basic planetary gearset.

gearing. The key characteristic that made this indispensable to automated operation is that because all gears in the gearset are in constant mesh, torque flow need not be interrupted to allow changes in gear ratios.

So, a single planetary gearset provides multiple drive ratios by holding one of the three components in place with a clutching mechanism and engaging the other two to define a torque transfer. Maximum reduction, or underdrive, can be achieved with the sun gear as the input and the carrier as the output. The ring gear is held in place and the planet gears slowly walk around the sun gear as it rotates, defining a reduced rate of rotation. Maintaining the sun gear as the input, if the carrier is held in place with a clutch pack, the ring gear can define the output, offering a less reduced underdrive. Alternatively, the ring gear could be made the input, and while holding the carrier, the planet gears will define a reverse rotation of the sun gear, providing a reverse output. Variations in increased speed, or overdrive, can occur if the carrier is used as input and either the sun or ring is output, with the other held in place. Of course, if all the components are lock together, the gearset can provide direct drive.

The planetary gearset has a few advantages over the layshaft transmission. Most importantly, they are compact and because all elements rotate around a single axis they easily can be placed in series or nested together in differing combinations to define a **compound gearset**. The most well-established compound gearset is the **Simpson gearset**, with applications in production cars dating back to the 1960s. Effectively, this is simply two planetary gearsets with their sun gears linked. The front and rear planetary components can have different sized gears and so can define differing gearing options. This basic configuration allowed a significant improvement on the two-speed automatic transmission that preceded it; and this configuration is still used as a low-cost unit in contemporary production cars.

The commonly used Ravigneaux offers all the basic features of the Simpson gearset, but in a somewhat more compact package and with increased tooth contact and increased torque capacity. Unlike the Simpson system with a shared sun gear, two different sized and independent sun gears are used with a common ring gear (Image 3.14). A set of short planet pinions mesh with the small sun gear, and a set of longer planet gears connect with the large sun gear, the three short pinions and the ring gear. The two sets of planet gears are on a common carrier. Because the carrier is one of the larger and more expensive components of a gearset, this typically makes the gearset smaller, lighter and less expensive than the Simpson gearset.

Once again, four speeds are possible: The lowest gearing ratio can be achieved with the small sun gear connected to the engine and the planet carrier prevented from rotating. This results in the small planetary pinions counter rotating against the sun gear and so the long planetary gears and ring gear turn in the direction of the initial input but at a reduced speed. Using the same input connection but locking the large sun gear requires the long pinion gears to walk around the large sun gear producing less of a reduction. Of course, direct drive is achieved by locking the sun and carrier. An overdrive ratio is attained by linking the engine to the carrier and allowing the long planetary pinions to walk around the large sun gear, driving the ring gear output at an increased speed.

A breakthrough in transmission innovation came with the development of the Lepelletier gearing mechanism, making new configurations and expanded gearing ratios possible. First proposed in the early 1990s, this system combines a simple planetary gearset with a Ravigneaux unit on a common central shaft. The input is linked to the simple planetary ring gear. However, a key feature is the possibility of two inputs, as input can also be concurrently connected to either the carrier or large sun gear of the Ravigneaux unit or both. The carrier of the simple gearset is connected by clutches to the large or small sun gear in the Ravigneaux set, providing a reduced input in underdrive gears. And, typically, the sun gear of the simple gearset is connected to the housing and does not rotate. The sun gear of the Ravigneaux gearset is driven by the output of the single planetary gearset. And the output of the compound gearset is defined by the Ravigneaux ring gear. The key

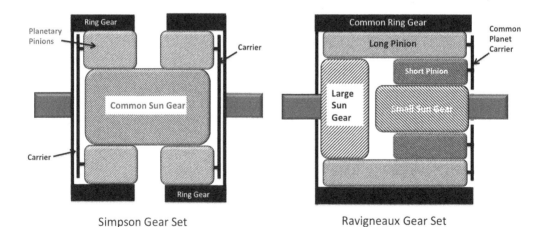

Simpson Gear Set   Ravigneaux Gear Set

IMAGE 3.14
Simpson and Ravigneaux gearsets.

The Simpson gearset offers a simple configuration that has made it is the most historically common planetary gearing configuration. The Ravigneaux offers a more compact unit with improved torque capacity.

characteristic to note is the sun gears and carrier of the Ravigneaux unit can be driven at different speeds, allowing for a larger number of possible overall gearing ratios since the resulting gearing of the compound unit is defined by the combination of the two input ratios. We'll talk more about this in Chapter 5.

While the mechanics of the Lepelletier system have always been possible, it is only with digital control that the complexities of the multiple clutch pack combinations could be practically managed. The resulting unit is impressive, relatively simple to control, lightweight, compact, and robust. The engineering firm ZF used this configuration to produce the first production six-speed automatic (6 HP 26) in 2001, and since then the number of speeds in an automatic transmission has increased dramatically (Image 3.15). The Lepelletier gearset makes 11 speed ratios possible.[7]

This capacity to increase available drive ratios can define a notable advantage over the manual transmission. With the ability to nest gearsets and increase gearing combinations without increasing the number of clutches, more gears does not always mean more weight or size for an automatic transmission as it generally does for a manual. As a result, choosing a manual transmission often means choosing limited performance. So, for example,

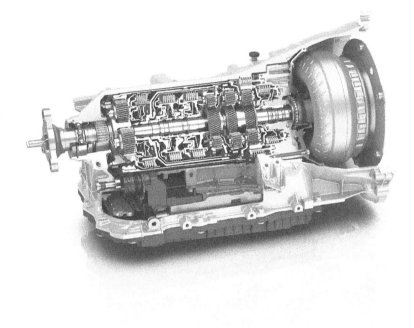

IMAGE 3.15
Automatic transmission.

The ability to nest gearsets means increased gearing does not always mean significantly increased size or weight in an automatic transmission. This eight-speed ZF unit makes efficient use of space, fitting clutch packs and gearsets neatly in a compact design.

Source: ZF Friedrichshafen AG

---

[7] E.L. Esmail, Configuration design of ten-speed automatic transmissions with twelve-link three-DOF lepelletier gear mechanism. *Journal of Mechanical Science and Technology* 30 (1), 2016, 211–220.

while the 2018 Ford Mustang comes with an impressive seven speed manual as stock, the optional automatic is able to provide a full ten gear ratios, gear changes that a professional driver would have trouble matching, and undeniably improved performance. The downside is perhaps in driving experience and certainly in cost, in the Mustang's case adding $1500 to the sticker price while barely nudging the fuel economy.

Transmission Control

The result of this ever-growing number of gearing ratios can be a mixed bag. Adding more gear ratios allows the engine and transmission to operate at lower BSFC; and so with rising efficiency expectations, nine- and ten-speed transmissions are becoming common. Moreover, more gear options can help achieve specific design goals, such as improved low-end torque or high-speed cruising. However, more gears can also mean more shifting, which can effect driver perception of ride quality. So, control systems need to be carefully designed to exploit the possibilities entailed in more gears without compromising ride quality or performance. It's worth noting that it is not only the number of gears but more importantly the spread of the available gear ratios that can determine ride quality and performance. With a greater spread of gears, the smaller engines being used now can provide desired launch acceleration with a lower launch gear, while still managing high-speed ride quality and performance with a higher top end overdrive.

Whatever gearset configuration is used, it is clear that sophisticated computer control and innovation have dramatically changed the possibilities. Initially, the planetary gearset was adopted for automatic transmissions because the goal of automated operation required it. At the time, it was easier to devise an automated hydraulic mechanism for engaging and disengaging hydraulic clutches than it was to automate the actuators in a layshaft gearbox. Until recently, these transmissions were controlled with a network of hydraulic fluid channels in a complex **valve body** that actuated the various clutching mechanisms. However much of these functions have now been integrated into a digital TCU that uses electronically controlled hydraulic solenoids, expanding design possibilities greatly. Computer control enables enhanced **clutch-to-clutch** actuation, with a clutch to one gear disengaged the instant a clutch for the other is engaged. At idle, the transmission can be automatically placed into neutral, helping manage temperature and improve fuel efficiency. Multiple variable displacement pumps can be used to produce just the hydraulic pressure needed, and significantly reduce associated loss. With pump loss accounting for nearly two-thirds of transmission power loss, this can improve efficiency significantly.[8]

The new range of possibilities was evident when Ford and GM introduced the result of their joint effort to produce an advanced 10-speed automatic in 2017. Using sophisticated programing they are able to achieve a high-performance transmission for trucks and sports cars. The mechanics was kept relatively simple, using as much existing componentry as possible, the new gearbox links four simple gearsets and six clutches. The real innovation is in enabling each automaker to develop specific control software for their applications. Ford can therefore define a transmission that prioritizes towing capacity and fuel economy in its F-150. Closely set overdrive gearing helps improve towing

---

[8] M. Gabriel, *Innovations in Automotive Transmission Engineering*. SAE International, Warrendale, PA, 2004.

performance, allowing imperceptible shifting to accommodate higher overdrive torque needs. At the same time, GM can use the same mechanics to define a sports transmission for its Camaro. Low-friction components, a bypass that allows for a faster rise to operating temperature, a capacity to shift without requiring the torque converter to unlock, and the use of high-performance, low-viscosity automatic transmission fluid (ATF) make this one of the most efficient transmission ever designed. All in a package that is only an inch longer and 4 pounds heavier than the six speed it replaced.[9]

This search for greater transmission efficiency can be helped with careful consideration of the control logic, but this can also present challenges. Once again, there are trade-offs. For example, the enhanced efficiency that can be achieved by shifting at an earlier stage during acceleration, often called shift optimization or **aggressive shift logic** (ASL), tends to be tied to reduced acceleration performance. So, often combined with early torque converter lockup, ASL can provide notable efficiency improvements at acceleration and cruise. And, as mentioned previously, lockup can occur whenever torque multiplication is not available, in every gear but first. Disengagement occurs when torque demand drops, to allow coasting with low engine rpm. However, locking the torque converter at a lower speed can cause shuddering that can impact the driving experience and shape perceptions of vehicle quality. Improved engineering has resolved some of this, but driver acceptance of noise is still a concern. And because the effect of downsizing and higher turbocharging have meant very high torque in the low-speed range, the challenge of addressing **noise, vibration and harshness** (NVH) has grown more complex. Features like cylinder deactivation, particularly as it is increasingly being applied to smaller engines, have made this even more challenging.

To help address the challenge of increased low-speed torque and early shifting in both manual and automatic cars, the humble flywheel and modest torque converter are getting a makeover (Image 3.16). Dual-mass flywheels are an established technology that has significantly reduced vibration caused by the irregular torque of the crankshaft. However, a centrifugal pendulum-type absorber goes a step further. By integrating a pendulum mass that can rotate outward and significantly increase the flywheel's moment of inertia temporarily, the flywheel can help cope with drivetrain vibration. At low speed, the deflection of the mass can counteract torsional vibration and dramatically reduce appreciable shudder and noise in a manual or an automatic application. This makes early shifting possible while protecting ride quality. So far, this has not been used in dual clutch units, perhaps because the larger mass needed to provide thermal stability in a dry DCT provides enough dampening on its own. This works for both AMTs and automatics, though in automatic transmissions the centrifugal absorber is immersed in oil within the torque converter.

In fact, changes to the torque converter have defined what is sometimes called a **multifunction torque converter** (MFTC). A torsional dampener mechanically insulates the outer shell of the torque converter from the inner components with a sprung dampener. This can allow for a more responsive TC while minimizing slippage and dampening the vibration of early shifting.[10] In addition, overall reduced weight from conventional torque converters can further improve efficiency and responsiveness. The addition of a clutch on the impeller can be used to allow the engine to spin at a higher speed than the pump,

---

[9] B. Chabot, "The Need for 10-Speeds." *Motor* June, 2017. Available at www.motor.com/magazine-summary/need-10-speeds/

[10] A. Isenstadt, J. German, M. Burd and E. Greif, Transmissions. Working Paper, The International Council on Clean Transportation, August, 2016.

IMAGE 3.16
Torque converter with centrifugal pendulum.

This advanced torque converter by LuK includes a centrifugal pendulum absorber that is able to absorb disruptive bumps in rotation and return the energy to the system more smoothly later. This can markedly improve the operation of engines with fewer cylinders and lower operating speeds that offer improved efficiency.

Image: LuK USA

enabling higher torque faster, and is particularly useful when trying to address turbo lag.[11] So, even the humble torque converter has come of age.

Continuously Variable Transmissions

Conceptually, a more desirable transmission would allow for ideal engine speed and torque at all times, without compromising for the gap between gears and without disruption for gear changes. In short, ideally, we would want to select from an infinite possibility of gear ratios, to choose just the right one for a given condition, and then seamlessly and continuously adjust to any other drive ratio as needed. Sound impossible? Actually, the idea of a transmission that can do this isn't fantasy; it's not even new. Leonardo DaVinci sketched one in the fifteenth century, and they have appeared in a variety of automobiles here and there since the early twentieth century. The 1934 Austin 18 offered an admirable example. Recently, the application of a **continuously variable transmission** (CVT) has enjoyed a revival; but it's future is less than certain.

The most established CVT type is a variable pulley and belt system called a **push belt.** Like any other pulley, these pulleys are defined by two opposing tapered sides. The difference is that these sides move, opening and closing the width of the pulley, defined by one

[11] K. Buchholz, "New-gen Torque Converter Aims at 2017 Vehicle Intro." *Automotive Engineering* SAE October, 2014.

fixed and one adjustable pulley half. Called a **variator**, the sides can be drawn together or pulled apart to widen or narrow the width of the riding surface for the belt and increase or decrease the effective radius of the pulley (Image 3.17). Clearly, this system cannot use a simple V belt. In fact, the belt is defined by several steel bands attached to multiple radial plates to provide a ridged cross section that can ride on the edges of the variator. This steel belt is commonly called a **Van Doorne belt** and is used by nearly every production push belt CVT. As the variator moves in, the belt is wedged between the narrowing sides of the pulley and so rides higher up the tapered walls, effectively defining a larger pulley radius. With the belt having a fixed length, the walls of the other pulley are forced to separate, allowing the belt to ride lower between the sides and define a smaller radius. This also creates compression on one side of the belt and slack on the other, hence the name push belt. The result is seamless adjustment of relative pulley size and drive ratio on the fly. With smooth acceleration from a stop possible in a low ratio, and modest and reliable low-speed movement, or **creep**, a torque converter can be eliminated in favor of a simple starting clutch. With fuel economy as the target, the CVT typically operates to keep the engine near a constant speed at low BSFC, shifting drive ratio to accommodate throttle position.

IMAGE 3.17
Toyota direct shift CVT.

Toyota's new CVT utilizes a launch gear to start from a stop, leaving the belt drive for higher gear ratios. This enables reduced belt load, a wider gear range and a resulting 6% improvement in efficiency. A reduced belt angle and smaller pulley allow faster shift speeds.

Image: Toyota Motor Company

In the past, the push belt CVT was only an option for low power applications. The limits of the belt's traction restricted the amount of torque that could be provided. However, improved chain drive belts developed by Schaeffler Automotive have defied this restriction. Most notably, Audi chose a link plate steel belt approach in its Multitronic CVT. The result can be as flexible as a V belt and able to transmit force through hard pin ends. The rolling-pin type pulley contact means less friction, though it can also mean more slip. Nevertheless, while newer belts such as this offer greater strength and improved traction and so efficiency, the friction force required to provide belt traction unavoidably results in energy loss. And performance is heavily dependent on precise lubrication and accurate control of the clamping force of the pulleys. The result is efficiency somewhere between 85% and 92%, generally somewhat lower than an automatic transmission and notably lower than a manual transmission.[12] These results have been disappointing to some manufactures. However, while CVTs used to be limited to smaller cars because of the torque limitations of belt slip, belt improvement have changed the calculus. The Nissan Murano, for example, a midsize crossover SUV with a 3.5-liter engine relies on a CVT.

However, let's not throw the baby out with the bathwater. If belts are the problem, it is possible to avoid the use of belts altogether. A notable example is offered by the toroidal traction drive CVT. Defined by two opposing toroidal disks and two or three oscillating rollers between them, this system offers variable drive ratios without high friction contact. If the rollers are made to turn toward the driving disk, the outside of the roller is driven by the outside of that disk, and the roller then drives the inside of the opposing disk, so a high ratio is defined. As the rollers are turned toward the other disk, the ratio is increasingly reduced. Because the friction and wear of a system with such high friction contact between hard surfaces would be extreme, a viscous fluid with high shear strength and friction characteristics well above ATF is used to transmit the torque between surfaces. Called **elastohydrodynamic lubrication** (EHL), this fluid allows the sheer resistance of the film between the surfaces to transmit force, with the surfaces themselves never touching. Used by Infinity's Q35 and Nissan's Cedric, this drive offered efficiency and a smooth ride with absolutely no torque interruption. The inherent slip at startup of the toroidal drive was addressed by adding a torque converter. However, challenges with reliability and consumer acceptance continue to plague this system and have led Nissan to discontinue its use.

Hope for the future of the CVT is not lost, but it's plenty shaky. There's innovative work being done to improve on these first efforts. The Torotrak system offers an interesting example. While existing toroidal drive systems utilize a half toroidal geometry, the Torotrak system used a full-toroidal mechanism, by placing two sets of three orbiting rollers between mated toroidal disks. The result was a variator with a much wider spectrum of potential drive ratios. However, the system presents some design challenges, particularly given the large range of roller angles; and before these were resolved, the maker succumbed to a heavy debt burden and ceased operations.[13]

A more recent variation on this theme is defined by the engineers at Dana Inc. Their VariGlide planetary variator could offer the heart of a compact and relatively simple CVT. The company points out that their system represents a significant departure from conventional CVTs.[14] The key is rolling spheres set between input and output traction

[12] H. Heisler, *Advanced Vehicle Technology*, 2nd edition. SAE International, Warrendale, PA, 2002.

[13] M. Gabriel, *Innovations in Automotive Transmission Engineering*. SAE International, Warrendale, PA, 2004.

[14] Communications with Jeff Cole, Senior Director, Corporate Communications Dana Incorporated

IMAGE 3.18
VariGlide system.

The VariGlide planetary variator offers a new option for CVTs. By collectively altering the axis of rotation of spheres set between traction rings, the input and output rings are made to run on smaller or larger rotational diameters of the spheres.

Image: Dana Incorporated

rings (Image 3.18). The spheres can collectively shift their axis of rotation; so, the input and output rings can be made to run on larger or smaller diameters of the planetary balls' rotation. The result may be more compact and flexible than the toroidal and more durable and powerful than the belt drive. With a full ratio sweep taking just two rotations of the unit, the system can allow a wide range of vehicle calibration options and drive modes. Like the toroidal drive, an EHL fluid is used, avoiding excess wear and offering efficiencies that are potentially in the high 90s. While not yet in production, the company hopes to have it ready for the road by 2022.[15]

Still, the shift from simple gears to spheres or toroids is anything but certain. When compared to the four-speed automatics available upon their initial development, CVTs looked promising. After all, an improvement of as much as 8% could be achieved by replacing a four-speed automatic with a CVT.[16] This is particularly true for low-speed

[15] Communications with Jeff Cole, Senior Director, Corporate Communications Dana Incorporated; and T. Murphy, "Planets Aligning for Dana's VariGlide Beltless CVT." *WardsAuto* August 22, 2017.
[16] A. Isenstadt, J. German, M. Burd and E. Greif, *Transmissions*. Working Paper, The International Council on Clean Transportation, August, 2016.

city driving, when automatic transmission inefficiency can be very high. However, when now compared to the high-geared transmissions available, the CVT loses a bit of luster. Moreover, they face a significant challenge of driver acceptance. Used to the experience of shift shock, and having learned to associate that with performance and acceleration, drivers can find the experience of CVT cars wanting, particular by those looking for performance over efficiency. As a result, while still maintaining popularity in Japan, CVTs are losing appeal in the US and Europe. For many, advances in DCTs and high-geared automatics have made the compromises of the CVT unnecessary. Nevertheless, their relative simplicity can make a lot of sense for a modest cruiser looking to gain some fuel economy. It's likely that it is just that goal that convinced Chevrolet to replace the standard six-speed automatic with a CVT its 2019 Malibu, the car makers only full size non-hybrid with a CVT.

Differentials, AWD, and Torque Vectoring

Once shaft speed and torque is established through the transmission, it needs to be distributed to the wheels. This might seem like a simple task; it is not. There are some tricky issues. First, the basic point: cars turn. As a result, all wheels on a car strike a distinct arc during a turn and so need to be able to turn independently. Second, a bit more tricky, proper handing and safety can be significantly enhanced by the possibility of apply differing torque to each driven wheel. Third, this is made difficult because each wheel experiences differing downward and lateral forces and differing friction. Lastly, it's worth noting that all of this gets harder when we're using all-wheel drive.

Managing the wheel speed variation needed for turning can be pretty simple. A basic open differential can allow for differential rotation of the wheels on a single axle, allowing turns with no necessary tire slip. This is accomplished with a basic pinion gear and ring gear, coupled with a set of beveled gears to allow differential rotation at each end. The challenge with this solution is that torque will be lost if either wheel fails to produce traction. If a wheel slips on ice, for example, it will spin freely, allowing the opposing wheel to remain stationary while the full driveshaft rotation goes to the wheel with no traction, the opposite of what we want. Alternatively, a **limited slip differential** (LSD) can offer the ability to either completely or partially lock the axles to avoid dumping torque at the least tractive wheel. While there are various designs out there, the typical system uses clutches to lock the axle whenever the differential's ring gear and driveshaft torque are uneven, that is to say when there is acceleration. A so-called one-way LSD responds only to forward acceleration, and a two way unit also responds to deceleration.

Electronic stability control (ESC) offers an improvement on this technology. Essentially an extension of **antilock braking** (ABS), ESC monitors the vehicle's motion and slip and applies individual wheel braking to correct slip and maintain a hold on the road. Basically, asymmetric braking creates a rotational moment around the car's vertical axis, called a **yaw** moment, and can correct unwanted turning due to **slip**.

Of course, for this system to work, we need to know when the car is slipping. More precisely, the yaw rotation of the vehicle relative to its intended direction of travel needs to be sensed. This can be tricky. A steering angle sensor can be used to determine the intended direction of travel. So, that's simple enough. And wheel speed sensors are used to identify slip at each wheel. Measuring sideways slide, or lateral acceleration, is a bit more complicated, but not much. A **lateral acceleration sensor** uses a bending element comprised of two piezoelectric layers. When a car starts sliding sideways the lateral acceleration produces a bending force on the element resulting in compression on one side and tension on the other, and the crystalline sensors produce resulting voltage differences that are proportional to the rate of acceleration.

We also need to know the yaw rotation of the car; this gets a little thornier. Again, a piezoelectric element is key; and in this case, it is incorporated into an interesting example of a **quartz rate sensor** (QRS). A yaw sensor is composed of a microminature double-ended tuning fork. Put simply, the tinny two-ended fork is placed vertically, with two tines pointed upward and two pointed downward. The fork is made of a single crystalline cell of quartz, so it can be made to vibrate with the application of an alternating current. The upper two tines of the quartz fork oscillate toward and away from each other, and since the tines are identical, their vibrations cancel each other out, and there is no energy transfer to the center of the fork. But when the fork is rotated the symmetry is disturbed, the rotational effect combines with the linear movement of the car to create a lateral force, called a Coriolis force, which is proportional to the rate of rotation. The magnitude of the force is sensed by piezoelectric sensors on the fork, defining a useful yaw signal.

The ESC control unit monitors yaw acceleration, vehicle speed, and steering angle continuously, and activates ESC correction when a loss of steering control is determined. At this point it actuates asymmetrical braking on wheels to counter the car's slide. In an **oversteer** condition, when the vehicle is rotating too much into the turn, ESC can apply braking to the outside front wheel to create a counter rotational force. In an **understeer** condition, when the car's rotation is less than needed and the car's turn is wide, braking can be applied to the inner rear wheel to increase the rotational force into the turn. The system can also reduce engine power or even shift the transmission into a low gear to maintain traction and control.

However, a more effective option comes once again with the possibility of even more sophisticated digital control. The challenge with ESC is that it relies on the application of braking. This is fine in a critical cornering, deceleration or crash-avoidance situation. However, the loss of energy entailed in braking means it is not an ideal way to improve general cornering performance. If you are trying to take a turn fast, automated braking is not always your best friend. Moreover, as might be expected, ESC increases fuel consumption.[17] A preferable system would actively distribute differential torque to each wheel to produce the same effect without braking. That's where **torque vectoring** comes in. As the name implies, torque vectoring can deliver discrete torque to each drive wheel using an active differential, or **torque-vectoring differential** (TVD). The system compensates for variations in traction and applies yaw correction to enhance cornering. So, rather than differential braking, differential torque is applied, pushing the car forward while enhancing road holding and cornering. In the simplest example, when slipping is indicated, torque can be reduced at that wheel and increased at the other. And, in cornering, a TVD can send power to the outside wheel, avoiding the braking effect of excess torque at the inside wheel. Using two wheels rather than one, a TVD can mark a noted improvement over

[17] D. Piyabongkarn, J.Y. Lew, R. Rajamani and J.A. Grogg, "Active Driveline Torque-Management Systems." *Control Systems Magazine*, IEEE 2010, 30, 86–102.

basic ESC. In fact, with the exception of the most critical loss of control situations, torque vectoring has demonstrated far improved handling over ESC.[18] And the efficiency losses associated with ESC are avoided.[19]

What's new in advanced torque vectoring is the capacity for active control. Rather than rely on engine torque or friction to allocate torque to the wheels, computer control uses sensor input to provide a wider range of regulation, with the capacity to possibly allocate all torque to a single wheel, or remove torque from a wheel entirely. Predictive algorithms can respond to road conditions, driving behavior and tire performance to predictively adjust torque allocation before slip even occurs.

A typical TVD is composed of differential gearing in the center, very much like the old open differential (Image 3.19). On either side is a multiplate wet clutch pack, like the LSD. An electric motor or hydraulic actuator can engage and disengage the clutches and allow reduced torque to either side as needed. So, when a clutch is actuated on a wheel, the balance of the available torque is redirected to the alternative wheel. No braking required. Advanced systems such as Ricardo's Cross-Axle Torque-Vectoring system used in the Audi A6 place planetary gearsets inboard of each clutch allowing differential speed at

IMAGE 3.19
Active torque vectoring.

Magna's TWIN Rear Drive System offers limited slip all-wheel drive with torque vectoring. Torque can be independently directed to each wheel thanks to twin couplings on the rear axle. Multiplate wet clutches are controlled by an electric motor and managed by an ECU to offer optimal torque distribution under changing driving conditions.

Image: Magna

---

[18] S.M.M. Jaafari and K.H. Shirazi, A comparison on optimal torque vectoring strategies in overall performance enhancement of a passenger car. *Journal of Multi-body Dynamics* 230(4), 2016, 469–488.

[19] M. Hancock, R. Williams, T. Gordon and M.C. Best. A Comparison of Braking and Differential Control of Road Vehicle Yaw-sideslip Dynamics. *Proceedings of the Institution of Mechanical Engineers, Part D: Journal of Automobile Engineering* 219, 2005, 309–327; and J. Deur, V. Ivanovic, M. Hancock and F. Assadian. Modeling and analysis of active differential dynamics. *Journal of Dynamic Systems, Measure, and Control* 132, 2010, 061501–061514.

the clutches. This allows wheel torque to function more independently of the incoming engine torque. When driving straight, the gears in the gearset are fixed, but in turns, torque is distributed to each wheel with a response time of less than a tenth of a second (Image 3.20).

When incorporated in an AWD system, the effect is impressive. For our purposes here, let's make a distinction between AWD and 4WD. Four-wheel drive relies on a transfer case to provide torque to the front axle. However, there is no differential rotation possible between the two axles. This makes it unsuitable for dry road conditions, since the necessary relative slip of the tires on a turn would significantly increase wear and decrease road holding. All-wheel drive, on the other hand, utilizes a central differential, allowing differentiated rotation at each wheel and axle (Image 3.21). So, when combined with torque vectoring this allows for the delivery of a precise torque at each wheel. During hard acceleration or cornering, an AWD system that normally distributes 90% of torque to the front and 10% to the rear, might even the torque to a 50/50 split; and the bulk of that rear torque would be sent to the outside wheel, ensuring a more controlled turn. The challenge in AWD is typically efficiency, as driving both axles entails much more mechanical friction. However, the addition of torque vectoring can address this by permitting disengagement of an axle entirely, effectively allowing the drivetrain efficiency to approach that of a two wheel drive vehicle when the all-wheel drive is not needed.

IMAGE 3.20
Vector drive.

The ZF Vector Drive TVD used by BMW incorporates a planetary gearset that allows wheel torque to function more independently of engine torque. So, corrective torque can be precisely applied to each wheel even if the car is decelerating, for example, while descending a winding mountain road.

Image: ZF Friedrichshafen AG

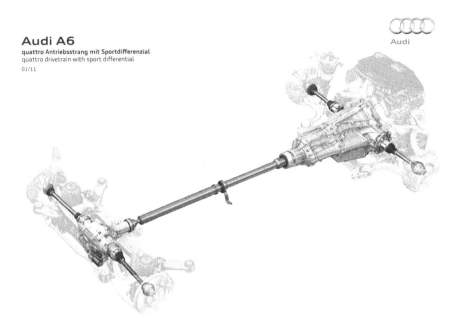

IMAGE 3.21
Audi all-wheel drive.

The Audi quattro drivetrain offers a base power distribution of 40:60, meaning 40% of the power is delivered to the front axle and 60% to the rear. However, this distribution is adjusted when needed, with up to 70% at the front or 85% at the rear to counteract slip. The sports differential further enables torque to be distributed in continuously variable proportions to each rear wheel based on vehicle dynamics to help ensure optimal road holding.

Image: Audi

Advanced Tires and Control

Lastly, if we're going to talk about getting 'power to the pavement' we need to address where the 'rubber meet the road', literally. The last step in forward traction is the connection between the tires and the pavement. We have incorporated smart technology into every step of the powertrain, from generation at the engine to the coupling, transmission, and differential. It would not do to then have the last and arguably most important step in this chain left dumb. While the transformation hasn't happened just yet, tires are on their way to becoming a critical and integral component of an active and intelligent drivetrain.

Significant enhancements in tire design have already occurred to be sure. Contemporary tires are more durable, precise, and efficient than ever. Run flat designs not only enable continued operation with a puncture, but also increasingly provide improved performance, ride and affordability, reflected in their growing popularity. However, we have yet to fully exploit the potential for digital monitoring and control technology in this vital last step in the powertrain. This is, after all, the only component of the automobile that contacts the road. But tire engineering is at a turning point; the integration of advanced technology to define the next generation of intelligent tires is around the corner.

The possibilities for tire monitoring go well beyond the tire pressure monitoring systems (TPMS) now standard in all US production vehicles. But it's worth noting that it was the development of this system, with the associated capacity to transfer data from inside a pneumatic tire, which catalyzed the development of more advanced monitoring systems. Intelligent tire systems are developing the capacity to monitor not only air pressure but strain, contact patch size, temperature, acceleration, slip, tread depth, and load. This data can be provided to a tire control system that integrates with the drivetrain.

Measuring tire performance can be difficult, but there are multiple options being explored. Using tire deformation, vehicle speed, wheel speed, and force on the tire or axle, algorithms can estimate tire friction and other key factors.[20] Data from the yaw sensor, acceleration sensor, and steering angle sensor can enhance the estimation of wheel grip. Even the vibration of the wheels as they rotate at different speeds can be used to improve vehicle control.[21] A more direct measurement capacity is being developed that could utilize a strain gauge embedded in the tire. The challenge is to design a gauge material that is as flexible as the tire itself, so the elongation and flex of the tire is not compromised by the gauge. Even the possible use of surface acoustic wave (SAW) sensors is being considered to monitor tire deformation with road contact. Defined by two metallic interlocking comb shaped electrodes, or **interdigital transducers** (IDT), on the surface of a piezoelectric material, an ultrasonic sensor mounted inside the tire can measure sidewall deformation. Alternatively, an ultra-flexible sensor made though photolithography might provide a thin, flexible sensor that would not interfere with tire function or durability.[22] And there are a variety of other possible microelectomechanical sensors, or MEMS, that are being considered.[23] While not yet ready for mass production, the implementation of smart tie technology seems imminent.

Powering such systems is also a bit tricky. Of course, intelligent tire sensors could be powered by batteries. But, given the high-power requirements of such systems, this option may be limited. A passive wireless system using electromagnetic coupling could be used, and we'll look at such systems the next chapter; but this would be likely to be inefficient. A more interesting option is a passive batteryless system that could harvest energy from tire motion and deformation to energize its sensors. Appropriately placed piezoelectric materials in the inner-tire could be used to convert the mechanical movement of rotation or deformation into a useful electric charge.[24] This may sound crazy, but it's not. The energy of low-frequency vibration could potentially be harvested by using piezoelectric zinc oxide nanowires radially entwined with the tire's fibers. Or, a capacitive generator could be used by charging capacitor plates and then moving the plates apart to generate electrical energy.[25] It might also be possible to use the heat generated in tires as a power source. Work is being done on all these ideas and more.

The ways intelligent tires could be integrated with existing control systems opens a world of possibilities. Smart tires could inform the traction control systems to provide

---

[20] R. Matsuzaki and A. Todoroki, Wireless monitoring of automobile tires for intelligent tires. *Sensors* 8, 2008, 8123–8138.

[21] T. Umeno, Estimation of tire-road friction by tire rotational vibration model. *R&D Review of Toyota CRDL*, 37, 2002, 53–58.

[22] R. Matsuzaki, T. Keating, A. Todoroki and N. Hiraoka, Rubber-based strain sensor fabricated using photolithography for intelligent tires. *Sensors and Actuators A: Physical*, 148, 2008, 1–9; and R. Matsuzaki and A. Todoroki, Wireless monitoring of automobile tires for intelligent tires. *Sensors* 8, 2008, 8123–8138.

[23] R. Matsuzaki and A. Todoroki, Wireless monitoring of automobile tires for intelligent tires. *Sensors*, 8, 2008, 8123–8138.

[24] A.E. Kubba and K. Jiang, A comprehensive study on technologies of tyre monitoring systems and possible energy solutions. *Sensors* 14, 2014, 10306–10345.

[25] S. Meninger, J.O. Mur-Miranda, R. Amirtharajah, A.P. Chandrakasan and J.H. Lang, Vibration-to- electric energy conversion. *IEEE Transactions on Very Large Scale Integration (VLSI) Systems*, 9, 2001, 64–76.

optimal delivery of tractive force to the pavement. They might also communicate with active chassis, suspension, transmission controls, steering, and a host of other systems to adjust to changing road conditions. We'll look at this in Chapter 7. However, more than just monitoring tire traction, smart tires could actually alter traction to suit driving conditions. Active inflation management could adjust tire inflation to suit the surface. Tread compounds could adjust to road conditions, defining a more stiff tread when dry to improve efficiency and handling, and absorbing moisture to become softer when wet, to allow improved handling. As the tires sense the state of the road, they could potentially adjust their physical properties to differing road materials, temperature, wetness, and friction, and report these changes to the rest of the automobile.[26] Who says tires aren't exciting and drivetrains aren't innovative?

---

[26] B. Schoettle and M. Sivak, "The Importance of Active and Intelligent Tires for Autonomous Vehicles." The University of Michigan Sustainable Worldwide Transportation Report No. SWT-2017-2 January, 2017.

4

Electric Machines

Unless you've been asleep for the past 10 years, you know that internal combustion engines are no longer the only way to power an automobile. Increasingly, automakers are turning to partial or full electric drive to achieve improved fuel efficiency, reduced emissions, and even upgraded performance. In fact, the potential for the electrification of the drivetrain is remarkable, leading virtually every major carmaker to declare aggressive targets for the electrification of their fleet. But, before we get into the cars themselves, we should take a look at the particulars of the electric machines that power them. Like the internal combustion engine, we can't appreciate the possibilities for the whole vehicle system until we understand the particulars behind the propulsion mechanism.

You might think that electrifying a drivetrain is an easy task. After all, electric motors are a mature technology, with millions of applications. Slapping one on a car can't be too difficult, right? However, the demands placed on the traction motors of electric vehicles (EV) or hybrid electric vehicle (HEV) are very different from the typical demands placed on most motors. In most cases, whether we are talking about the motor in your washing machine or a large industrial motor, the unit is fixed, mounted to a stationary structure or floor, so the weight and size are not particularly important. In addition, typically these motors operate at a defined speed for a predictable period of time in a relatively predictable and controlled environment. However, none of this is true for cars. The speed and torque demands placed on an automotive motor vary significantly and quickly with no set pattern. The weight and size and even shape of an EV or HEV motor matter greatly. And the environmental conditions vary dramatically, as a car must be able to operate reliably in any weather condition, from subfreezing to desert hot. And, on top of all this, a motor must be able to function easily as a generator to allow us to harvest the kinetic energy of the car during deceleration, called regenerative braking.

So, while there are many differing sorts of electric motors out there, with differing architecture, differing control requirements, and differing configurations, a relatively few of them are suitable for use as automotive traction motors. In general, we need motors with a capacity to produce reliable torque for acceleration and hill climbing as well as an ability to manage periodic overload for overtaking when needed, ideally in as small a package as possible, what we can call **torque density**. And we need a broad speed range that allows low-speed creeping and high-speed cruising, an ability to produce constant power over a wide section of this range, and low maintenance and high reliability in differing environmental conditions. Ideally all of this will come in an affordable package. Needless to say, not all motors fit the bill.

As was true for the internal combustion engine, control is everything. The capacity for advanced precise digital control has allowed for remarkable innovation in electric machine design. In fact, it is no exaggeration to say that innovations in the field of **power electronics** (the combined electronic, electromagnetic, and electrochemical components that control and convert power) have fundamentally reshaped what is viable in electric motors and thus what is achievable in EVs and HEVs. As a result, ideas and concepts that have been known for many decades as conceptually possible but not feasible, have now become technically practicable options.

The Principles of the Electric Motor

As always, let's start at the beginning. Motors function by exploiting electromagnetic force, one of the four fundamental forces of nature. As might be obvious from the name, electromagnetism defines a relationship between magnetism and electricity. In fact, magnetic and electric forces are really fundamentally the same sort of thing; both are defined by the exchange of photons between charged particles, called an **exchange force**. In our case, the particles are electrons. You can think of photons as elementary particles or waves that act as **force carriers**, traveling between electrons and exerting force, and defining both electric and magnetic fields.

Of course, to understand magnetic fields we need to understand magnets. Three elements, iron, nickel, and cobalt, demonstrate the property of **ferromagnetism**, or the ability to be permanently magnetized when placed in a magnetic field. This magnetization happens at the atomic level. The atoms that compose these materials are themselves like tiny magnets with opposing poles, or **magnetic moments**, that produce a magnetic field, interact with other magnetic moments and change their orientation in response to magnetic fields. When these atoms are similarly oriented throughout the material, their fields combine together and define a uniform magnetic domain. So, by exposing a ferromagnetic material to a powerful magnetic field, the orientation of these crystals can be aligned so that all the magnetic axes point in the same direction, thus creating a magnet. Importantly, with these three materials, when the magnetic field is removed the polarization remains, defining a **permanent magnet** (PM). When certain rare earth elements, in particular neodymium and samarium–cobalt, are combined with these elements, they can form magnets that are several orders of magnitude more powerful than the simple ferrite magnets on your refrigerator. In addition, as we will see later, some metals, such as copper and aluminum, do not become PMs themselves, but can exhibit magnetic qualities when an electric current is passed through them. This allows us to define controllable magnets that can be switched on and off, called **electromagnets**.

As you probably discovered sometime in grammar school, similar magnetic poles repel each other and opposing magnetic poles attract each other. The force between these poles can be described as a field shaped by contour lines of equal magnetic force. This magnetic field can't really be seen of course; but drawing out the flux lines of a field offers a useful way of imagining the effect, and will help us understand how motors work (Image 4.1).

Because magnetic flux tends toward the path of least resistance, placing a needle in this field would cause it to align with the lines of flux. This is because the needle has much lower resistance to flux, called **reluctance**, than the surrounding air. The concentration of flux in a material with low reluctance, like our needle, forms strong temporary poles that pull the material toward areas of higher flux. So our needle turns, and when aligned with the magnetic lines of flux, offers a path of lowest reluctance. If this needle were magnetically polarized, this rotational force would be directional, as the north side of the needle would seek the south pole of the magnet, and vice versa since opposing magnetic forces attract each other.

All of this may seem a bit dense and academic. But these two dynamics—the attraction of opposing poles and the force of reluctance in a magnetic circuit—aren't just interesting,

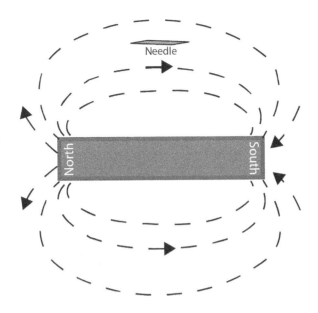

IMAGE 4.1
Magnetic fields.

Lines of flux are (somewhat arbitrarily) said to flow from the north to the south pole of a magnet. As we get closer to the polls, the lines of flux get closer, indicating more flux per unit area, or that the intensity of the field gets greater. A needle place in the field would align with the lines of flux to define a path of least reluctance. Two polarized magnetic fields can interact, defining magnetic attraction and repulsion. The resulting **magnetomotive force** (MMF) is proportional to the strength of the field and the distance between the poles.

they're amazing. They allow us to use magnetic fields to create mechanical motion. As we will see, these two forces are present in any motor or generator and the essential motive core of every electric or hybrid car on the planet.

We can already see that a motor's operation is fundamentally defined in the interaction of movement and magnetism. Now what we need to better shape this interaction is an ability to define and control the magnetic field, and that requires that we add electricity to the mix, bringing us to a phenomenon called electromagnetic **induction**. Induction defines the relationship between **electromotive force** (EMF), or voltage, and magnetic fields. Very simply, if we move an electrical conductor within a magnetic field we will induce a voltage or EMF within that conductor. This voltage in turn will cause current to flow if connected to a circuit. So the relative movement of a conductor, let's say a wire, through a magnetic field will result in an electric current in the wire. This relationship between changing magnetic fields, called **magnetic flux**, and current works the other way around too. As current travels through a conductor, it induces a magnetic field around it (Image 4.2).

These two related observations—that electric current can induce a magnetic field and that magnetic flux can induce current—are critical to understanding electric motors. If we place a current-caring conductor near a magnet, the magnetic field induced by the current will interact with the magnetic field of the magnet and define a physical force. This essential relationship between electricity and force makes all motors possible. After all, that is the central purpose of a motor, to turn electrical energy into movement, or take physical movement and generate electricity as a generator. Since the devices we need for cars have to be able to do both, we can call them electric machines or motor-generators rather than just motors.

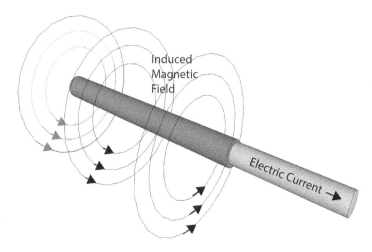

IMAGE 4.2
Current and magnetic fields.

As current travels through a conductor, it induces a magnetic field around it. The direction of the field can be predicted with the right-hand rule. If you place your right hand with curled fingers and your thumb in the direction of the current, your fingers will point the direction of the resulting field.

Making an Electric Machine

With this basic information, we could assemble a primitive machine. If we make a loop of conductive wire, for example, we can place a current through that loop. The current will induce a magnetic field around the wire. Placing that wire loop between two poles of a magnet would cause the magnetic field induced by the current and the magnetic field of the magnets to interact. If the current flow in the loop is oriented properly we could cause the wire to jump a bit as the two magnetic fields interact (Image 4.3). Like I said, primitive. This isn't much of a motor.

We could improve the motor my shaping our wire into a hoop. If our wire hoop can rotate and put the two poles of the magnet on opposing sides of our axis, we could cause that jumping motion to push the hoop around, and our wire loop would revolve. By adding slip connectors to provide the current to the loop, we can arrange to mechanically reverse the flow of current with each half turn of the loop. We call this sort of connection a slip-ring commutator. Small carbon contact points, called brushes, are used to slide against the commutator and allow electric current to travel from the stationary frame to the rotating commutator. The result will be that each time the loop rotates to resolve the force caused by the magnetic fields, the current is reversed and the rotational torque reappears. So, the loop remains in constant rotation as it continuously attempts to resolve the force while the force is continually reestablished with reversal of the current. So, in sum, we've defined a component that rotates, the **rotor**, a component that stays put and provides a magnetic field, sometimes called the **stator**, an axle, and a slip-ring commutator. The term **armature** refers to the power-producing component of the machine, in this case the loop. These are the basic components of any simple direct current motor (Image 4.4).

The same basic principles allow us to understand the function of a generator. However, to get a sense of how things can differ, let's stagger the commutator rings instead of

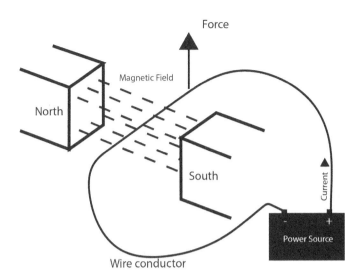

IMAGE 4.3
Motive force.

Running a current through a conductor will induce a magnetic field. If we place that conductor within the magnetic field of a permanent magnet, the two fields will interact and a force on the wire will be the result.

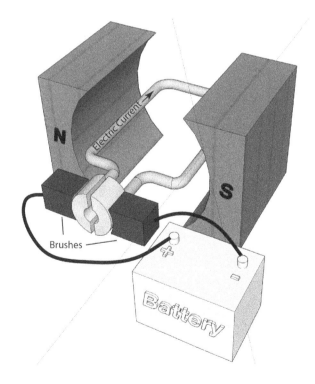

IMAGE 4.4
Basic DC motor.

As the armature rotates due to the interacting magnetic fields of the permanent-magnet stator and the current in the armature, the slip-ring commutator reverses the contact with the brushes, and therefore reverse the current, allowing the armature to continue to rotate.

splitting them. And, in this case, since we want to generate electricity, we can put a handle on our loop, like Image 4.5.

As we rotate the conductive loop, it swings past the alternating magnetic poles, cutting across the magnetic field. As a result, an EMF is induced causing a current flow. This current is quickly reversed as the polarity of the field shifts in the second half of the rotation. The shifting potential back and forth in the loop defines an **alternating current** (AC), reaching maximum values as the loop passes each pole with movement perpendicular to the lines of flux.

With this alternating output, we can now imagine a motor that doesn't require the same sort of split commutator we had previously. If the current is consistently shifting from negative to positive, there's no need for the commutator to reverse the polarity of the loop, that's done automatically by the nature of AC current. In fact, if we took our generator and placed it on the receiving end of our output, it would now behave as an AC motor (Image 4.6). When it is cranked, it's a generator providing current; when current is applied, it is a motor providing rotational force. The resulting motor could be called a **synchronous motor**, because its rotation will be defined by the frequency of the AC current supplied to it. Turn the generator faster and the motor will turn faster.

Don't run out and make an electric motor for your car from a coat hanger and kitchen magnet just yet. You might be disappointed in its performance. To have a useful motor, we need to first produce a strong magnetic field, and second allow it to interact with as much of a conductor as possible. More windings of the conductor to maximize the conductor's cross-sectional exposure to the magnetic field will help. To accomplish this we shape the conductor into a **coil** with multiple loops. With each additional loop, we increase the cross-sectional exposure to the field. The strength of the magnetic field is not shared among the

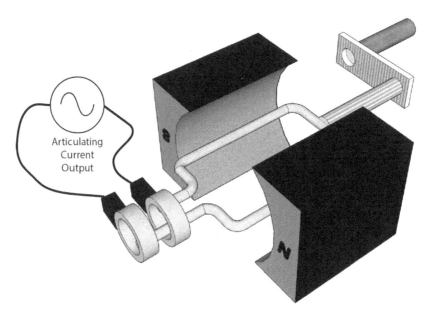

IMAGE 4.5
Basic AC generator.

When the handcrank is turned, the resulting current through the slip rings alternates as the magnetic field reverses the induced current, defining an AC output.

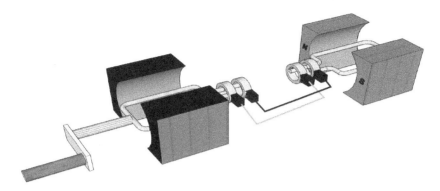

IMAGE 4.6
Motor-generators.

Turning one motor-generator produces an alternating current that will induce an alternating magnetic field in the second motor-generator and cause it to turn synchronously.

conductive loops; the force is multiplied. The more the conductor is exposed to the magnetic field, the stronger the effect.

We might also add additional **poles** to our motor, so the armature doesn't abruptly flip from one orientation to the opposite (Image 4.7). More poles can allow a smoother rotation and reduce the rapid fluctuation in torque that follows changing rotor angle, called **torque ripple** (Image 4.8).

So far, we've relied on magnets to produce our initial field. We called these PMs, not because they are strictly permanent. As we discussed, they can be made and unmade. In fact, a serious constraint in the use of these magnets is that too much heat can cause

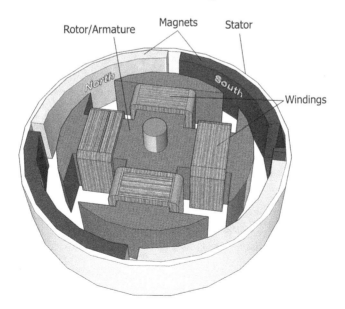

IMAGE 4.7
Four-pole motor.

The components of this basic motor include a rotor with field windings, a stator with mounted permanent magnets.

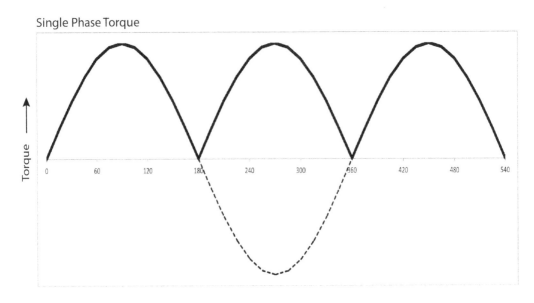

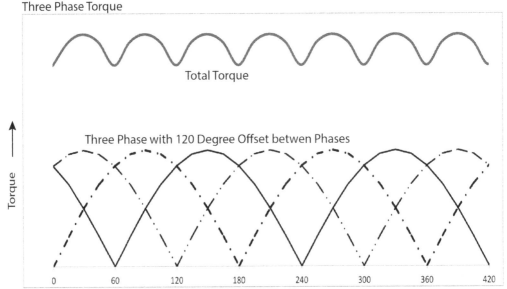

IMAGE 4.8
Torque ripple.

Motor torque varies sinusoidally. Reversing the current in a single-phase machine results in torque bumps called torque ripple. A three-phase motor can help reduce torque ripple by filling in the torque dips with the additional phases. However, torque ripple is never fully removed.

a magnet to lose its magnetism by allowing the molecular polarity to reorganize. Rather they are called permanent because they are magnets all the time, defining a continuous magnetic field. This can sometimes be troublesome, as by their nature they are always on, defining a possible safety issue since this means we can never fully shut down our machine. And, as we will see in a bit, the permanent and fixed nature of PMs can have undesirable effects at high speed. On top of all this, powerful magnets can be expensive.

However, the use of PM is not our only option. As discussed earlier, current flowing through a conductor can define a magnetic field. So, a PM could be replaced by a properly defined conductor. Once again, a single wire is not going to produce an adequate field for a useful magnet. We can instead wrap our conductor with a coil that will multiply the strength of the magnetic field in our core, defining an electromagnet (Image 4.9). Since a coil's magnetomotive force is the product of the number of turns and the current, with each added coil we add to the magnitude of our field.

Placing an iron core in our coil can allow us to shape the magnetic force more effectively. So far, our magnetic fields have traveled entirely through air. But air is not a great conductor of MMF. In fact, it has a very high reluctance. Providing a low-reluctance conductor for magnetic flux allows us to potentially shape or concentrate the magnetic field. In fact, the ability of iron to support a magnetic field, called **permeability**, is roughly 1,000 times better than air. Consequently, far less of it will be lost as **leakage** to the surrounding air. The force produced is proportional to the flux density we can produce. Double the density and you double the force. Of course, there is a limit. Any given magnetic conductor has a maximum amount of magnetic flux it can accommodate until **saturation**. At this point, all the atomic dipoles in the material are lined up and no further increase in the field is possible; after that increasing the magnetic field results only in lost energy. When the core is saturated, the reluctance of the material goes up very quickly. This is typically not a major factor, as a motor's core is typically sized to suit its purpose and operate below saturation; but it is the primary reason a larger power demand generally requires a physically larger motor.

Since the rotor has to be able to turn, we need to maintain some space between it and the stator. We call that an **air gap**. Ideally, this gap will be as small as possible to minimize loss. Flux is generated in this air gap, so the larger it is the less flux density that can be achieved; and remember, flux density is what defines the rotational force we can generate. So, defining this air gap becomes a key step in any motor design.

All of this now allows us to imagine a more sophisticated motor. We can define a coil armature, to maximize the magnetic flux exposure. We can utilize more poles, to offer better control and smooth operation of our motor. We might choose to use electromagnets with iron cores in the stator, to generate a strong and more controlled magnetic field.

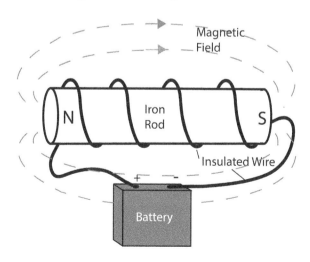

IMAGE 4.9
Electromagnet.

These basic motors, whether using PM or electromagnets, are an affordable, well-established technology. For these reasons permanent-magnet DC (PMDC) motors were once widely used by hobbyists to electrify older cars, so-called de-ICEing (get it?). Motors like this are still frequently used in cars, to move your window up and down for example. However it's not much use as an EV traction motor. The principle problem is that everything we've discussed so far requires a commutator. Commutators can cause torque ripples, they can limit the motor speed, the carbon brushes generate friction that impacts efficiency and also generates radio frequency interference. Most importantly, the brushes are subject to wear that seriously impacts reliability and maintenance requirements. No one wants to swap the brushes on their car's motor every few thousand miles. So, once we've discussed the basic performance characteristics of electric machines, we will see that while current technology is based on these same fundamental principles, the actual machines used in contemporary EVs have moved well beyond these simple motors (Image 4.10).

Motor Performance

As you might expect, the performance of an electric machine depends greatly on the design, and designs vary enormously. However, these machines do share some general qualities worth noting. First, they are far more efficient than internal combustion engines. Most gasoline engines are about 20%–35% efficient in converting the chemical energy in gasoline to useful torque and speed. A high very performing engine might achieve

IMAGE 4.10
Brushed DC EV.

One of the very few production cars to use a brushed DC motor is the Reva G-Wiz. This much-criticized micro-car used a 6.4 hp (4.8 kW) motor powered by eight lead-acid batteries under the front seat.

Image: CC BY-SA 2.0

efficiency in the mid-40s. However, traction motors in EV regularly achieve efficiencies orbiting 90% or higher.[1]

If we're going to achieve maximum efficiency, we're going to have to think carefully about how to control the speed of our motor. For our purposes, this is more challenging than it might be for a washing machine or blender for at least two reasons: First, unlike many motor applications, vehicles require a wide range of precise speed control. And second, we need to do this with as little loss as possible, since any inefficiency reduces our vehicle's range. At first blush, we might think speed control is pretty easily achieved. Varying the voltage at the stator will vary the field magnitude, which should modify the force on the rotor and so the speed. Putting a variable resistor in series with the stator would allow us to vary this voltage easily. In fact, this is the approach that was used successfully for a long time by analog power systems. The problem is that such an approach entails significant inefficiency. The combined rheostat and stator circuit draws power consistently. So even when the motor is turning with minimal load, the energy used would remain high. The excess energy would simply be absorbed by the rheostat and turned to heat. So, a basic linear power source tends to entail a lot of loss and a fair amount of heat.

A better approach is offered by a modern digital control method called **pulse width modulation** (PWM). The core of PWM is essentially a switch that can turn a signal off and on very quickly. By accurately controlling the amount of time the signal is on as a percentage of the total cycle, called **duty cycle**, and the frequency of the cycle, we can create an output that behaves like a precisely defined constant voltage. So, while the signal behaves like low voltage, it is in fact a very rapid succession of full-voltage pulses. As a result, power loss is small, though loss still occurs during the actual switching, called **switching loss**, but this can be reduced with improved control logic. Essentially, the goal of all this is to digitally encode a precisely modulated analog signal. To get a particular speed at a given load, a simple lookup table can be used to define the PWM duty cycle needed to produce the targeted motor speed (Image 4.11).

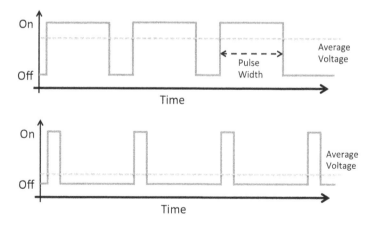

IMAGE 4.11
Pulse width modulation.

By turning the signal off and on quickly, the effective output voltage can be efficiently controlled. The effective voltage value is the average of the on–off voltage over time.

---

[1] A. Hughes, *Electric Motors and Drives: Fundamentals, Types and Application*, 3rd edition. Elsevier, London, 2006.

This does not imply that electric motors can now achieve near perfect efficiency. In fact, inefficiencies in electric motors are a significant concern in the design process. There is, of course, inefficiency due to friction of the bearing surfaces, and to lesser degree air resistance on the rotor. However, a more pressing design concern is the loss of energy due to electrical resistance in a machine's coils. We call this **copper loss**, and it is principally lost through the generation of heat. The loss of energy due to resistance in any conductor is defined by the resistance and the square of the current. This means that even a small change in current can result in a large change in loss, and high-torque operation that requires high current is likely to lead to excessive heat. So, efficiency can change significantly as operating conditions change. The resulting generation of heat is problematic for the obvious reason that inefficiency is bad and results in reduced range or diminished power. But in addition the generation of heat is problematic because PM can suffer demagnetization when heated, resulting in reduced motor capacity. Moreover, because resistance of a conductor is defined not only by the conductor's dimensions and composition but also by the temperature, increasing temperature can in turn increase copper loss and result in more heating. Giving us good reason to size electric machines carefully and be concerned about motor temperature control.

In addition to the loss due to resistance in the coil, there is also a loss due to the magnetic field. When a fluctuating magnetic field travels through a ferromagnetic material, loss occurs. As previously discussed, the polarization that defines opposing north and south charges also occurs at the microscopic level with individual molecules defining dipoles. These dipole moments respond to the alternating magnetic fields with a torque force that repeatedly shifts the alignment of the atomic dipoles. This entails internal friction and requires energy; and it results in the generation of heat, physicists call this **hysteresis loss**.

An additional loss comes from the generation of uncontrolled currents within the material. As the machine's core is exposed to a fluctuating magnetic field, a current is induced that opposes the external field that created it. As the resulting current flows in a plane that is perpendicular to the magnetic field, it defines small loops of current. These micro whirlpools of lost energy are called **eddy currents**, and once again result in heat generation. Combined with hysteresis loss, this effect defines **iron loss** or **core loss**.

The balance of these losses varies with condition. The more the rotor turns in synchronization with the field, the lower these losses, since there is no change in flux at any location in the rotor and so no induced currents. But, of course, no system is perfect, and neither the field flux nor the rotor is flawless, so eddy currents are generated even in a synchronous machine. In general, at higher torque and lower speed, copper losses tend to be the most significant. And, at high speed and low torque, when the flux in the core is potentially at its greatest, iron losses are dominant.

Torque and Power

With all these complications, you might be starting to wonder why we want to use electric machines to power cars at all. However, despite these generally shared challenges, electric machines also share some very desirable characteristics. Most notably, they produce maximum torque upon initial start-up, and maintain that capacity until the speed of maximum power generation, or **base speed**, is attained. As we will see later, this can make an electric drive an ideal companion to an internal combustion engine that has limited torque at lower speeds. Once a machine's base (or rated) speed is achieved, torque will decrease while maximum power is achieved and maintained (Image 4.12). As we will see, variations in the type and design of the machine will change these parameters greatly, but the basic pattern of torque and power generation will hold true for nearly any machine.

Understanding why this occurs is important. As previously mentioned, the most basic mode of adjusting speed is by varying the EMF on the rotor. This alters the resulting current and thus the strength of the magnetic field. However, as the rotor spins, and the flux through the rotor changes, it also creates an induced opposing voltage, called a **back EMF**. In essence, the machine is also operating as a generator when it is motoring. When this back EMF nearly equals the supplied voltage this defines the motor's rated speed. When a load is applied, the motor slows, reducing back EMF and so increasing the supplied current, in effect automatically compensating for the change in load. If the load gets high enough to stop rotation, torque will be at its maximum.

This capacity to provide maximum torque with increasing load is ideal for a traction device, but it makes speed control tricky. Once at rated speed, getting the motor to spin faster cannot be accomplished with more voltage, as this would only serve to increase the back EMF. However, if instead we decrease the magnitude of the stator field by reducing current, called **field weakening**, the back EMF will also decrease, and more current will be able to flow through the rotor resulting in increased speed. The cost, however, will be reduced torque. The motor now enters the **constant power** range of operation. As the torque decreases it is offset with increasing speed to provide constant power. If you keep decreasing the field, the speed will continue to increase and the torque will decrease until mechanical failure or excess heat cause the machine to stop. This once again underscores the importance of precise control technology and thermal management.

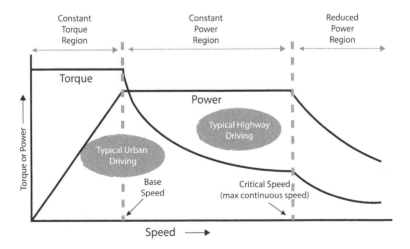

IMAGE 4.12
Ideal power/torque performance for EV electric machine.

Cooling

Even at this early point, it is clear that heating is a key limiting factor on the performance of electric machines (Image 4.13). In fact, the challenge of cooling the machines and the power electronics that provide control without unnecessary weight, bulk or complexity is a major concern for manufacturers. Of course, avoiding the generation of heat with efficient and well-designed machines and power electronics is a good option, and we will see many examples of this. But even the most efficient machine will generate some heat; and as we will see, the magnets used in modern machines can be very sensitive to heat, resulting in significant performance deterioration with inadequate thermal management. As an example, a temperature increase from 20°C to 160°C could result in a loss of nearly half the motor's torque.[2] However, thermal management can be tricky. Any electric machine has multiple components made of multiple materials with differing thermal characteristics, all defining multiple heat transfer pathways. In a hybrid system, where the electric machine is in close proximity to the internal combustion engine, the task is even thornier. And the cooling requirements of the power electronics can be equally difficult to address. So, defining a thermal management system once again presents us with a trade-off between costs, performance, size, and efficiency.

Cooling can generally occur through active or passive methods. Passive cooling simply uses careful material selection and design to allow ready paths for conductive heat transfer away from the machine's core components. In fact, the need to provide thermal conductivity between heat generating components and the housing and external sinks can be a key design goal. Active cooling, on the other hand, can be as simple as a cooling jacket through the stator housing, much like an engine's cooling systems. More advanced cooling systems place the cooling jackets at the coil rather than the housing. Coolant channels can run between each winding bundle and along the stator slots. Forced air cooling can augment the effect. An alternative approach might tap into the cabin AC condenser to cool the fluid or provide cooled air.

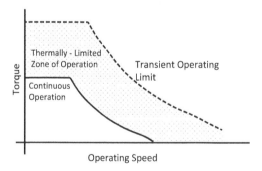

IMAGE 4.13
Thermal management and motor performance.

A chief constraint for electric machines is the generation of heat. A motor has a defined torque capacity at any given speed, as discussed previously. That limit defines a region of safe continuous operation. That torque capacity may be exceeded momentarily, but the excess heat generation limits this to transient burst rather than continuous operation.

2 B. Bilgin and A. Sathyan, Fundamentals of electric machines. In A. Emadi (ed) *Advanced Electric Drive Vehicles*. CRC Press, New York, 2017, 5–27.

Jet impingement cooling is used to spray a coolant either directly on the surface of the coil or an attached cooling plate, allowing the coolant to evaporate and extract heat. When materials change phase as a result of cooling or warming they absorb heat energy without changing temperature. So when a substance changes from liquid to vapor for example, it absorbs a high amount of heat energy without getting hotter, this is called **latent heat**. Impingement takes advantage of this to extract maximum heat energy. As a result, plates less than a few square inches are able to absorb impressive amounts of heat energy. And the addition of fins or microgrooves can increase heat flux several orders of magnitude.[3] However, the effective transfer of heat through the machine and onto the cooling plates can be a challenge.[4]

Oil spray cooling using a fluid like automatic transmission fluid (ATF) is another option. The use of oil offers a clear improvement over air cooling.[5] The motor can be partially filled with oil, and splash cooling can be used as the rotor tosses the oil around. Alternatively, oil can be fed onto the rotor ends, and made to splash across the windings. While not yet used by a major manufacturer, an emersion method can be imagined that takes its lead from the cooling of transformers by immersing components in an electrically non-conductive fluid such as transformer oil. The challenge, of course, is managing the oil's path and protecting the air gap.

Whatever the method, the trick is to design something small, which adds neither excess weight nor size to the motor but provides adequate cooling. There are a lot of innovative ideas out there. A direct winding heat exchanger, developed by Georgia-based DHX, for example, places a micro-heat exchanger between the stator windings, using tiny cooling channels that significantly increase the cooling area and exploiting an effect they call localized turbulence to increase the flow rate. The result is a four-fold increase in current capacity.[6] The developers of a cold plate system that uses multiple miniature nozzles and a plate covered in tiny pin-like copper fins developed for Toyota at Purdue University claim it can extract a very-impressive $1,000\,\text{W/cm}^2$ from power electronics.[7] These and other cooling technologies can allow improved reliability and performance from a smaller machine.

Induction Motor

The problem with the motor we've discussed up to now is in the commutator. As mentioned, it is responsible for inefficiency, wear, limited speed, and other problems. So, why not get rid of it? In fact, we might rethink the need for a conductive connection with the

[3] D. Wadsworth and I. Mudawar, Enhancement of single-phase heat transfer and critical heat flux from an ultra-high-flux simulated microelectronic heat source to a rectangular impinging jet of dielectric liquid. *Journal of Heat Transfer* 114 (3), 1992, 764–768; and H. Sun, C. Ma and Y. Chen, Prandtl number dependence of impingement heat transfer with circular free-surface liquid jets. *International Journal of Heat Mass Transfer* 41 (10), 1998, 1360–1363.

[4] T. Davin, J. Pelle, S. Harmand and R. Yu, Experimental study of oil cooling systems for electric motors. *Applied Thermal Engineering* 75 (1), 2015, 1–13.

[5] T. Davin, J. Pelle, S. Harmand and R. Yu, Experimental study of oil cooling systems for electric motors. *Applied Thermal Engineering* 75 (1), 2015, 1–13.

[6] S. Andrew Semidey and J. Rhett Mayor, Experimentation of an electric machine technology demonstrator incorporating direct winding heat exchangers. *IEEE Transactions on Industrial Electronics* 61(10), 2014, 5771–5778.

[7] E. Venere, Research team develops new cooling technology for hybrid and electric vehicles. PHYS.ORG, 13 September, 2016. Available at: https://phys.org/news/2016-09-team-cooling-technology-hybrid-electric.html

rotor entirely, and instead rely on induction to create an electromagnetic connection with the rotor. Since induction takes place without physical contact, this eliminates the need for commutator and brushes.

The basic principle is straightforward. Remember that a current induces a magnetic field, and as a magnetic field cuts across a conductor, it in turn induces a current. We can then imagine a chain of current and induction, with the stator current defining a magnetic field that cuts across the rotor and thus inducing EMF and current in the rotor. This current flows in the opposite direction of the current in the stator, and induces its own magnetic field, a process called **mutual induction**. These two fields will have opposing polarity. So, we have two magnetic fields that are opposing each other, one on the rotor and one at the stator; the result is a rotational force on the rotor. Because it is the change in magnetic field, or the flux, that induces current, the stator would have to be provided with AC current, as DC would define a steady magnetic field rather than needed changing flux.

The **induction motor** is a well-established and mature technology, and has earned a reputation for reliability. Unlike the PM machine that we have discussed so far, this approach has the advantage of not requiring brushes to transfer current to the rotor. With no commutator and no brushes, this can be a simpler, reliable, and efficient machine for EV.

For EV applications, we need to define a rotor that is mechanically durable, a good conductor, and can offer a maximum exposure to the magnetic field of the stator. While wound armatures can and are used in induction machines of this sort, because of the high speed and reliability expected of automotive traction machines, a more robust design is needed. Coils can have trouble enduring the centrifugal force of a high-speed automotive application. Instead, we start with a set of conductive rods that define a maximum cross-sectional exposure to the stator field. Often aluminum is used in such motors; however, once again the demands of an automotive traction motor are high, so more conductive copper is used despite the added expense in automotive applications. The ends of the bars are connected together by rings to define a closed circuit, allowing the current to operate throughout the copper cage (Image 4.14).

This **squirrel cage** armature is embedded in iron to provide minimum reluctance, but the iron core is assembled of stacked laminations separated by insulation to reduce eddy currents. Ideally the air gap between the rotor and stator is minimal, usually a couple of hundredths of an inch. Variations in the impedance, resistance, and size and shape of the rotor can define the generated torque and speed of the motor.

The stator is defined by stacked laminations wrapped in coils, similar to our previous machines; but the fundamental architecture of the coils is significantly changed to allow the use of three-phase AC. This entailed three sequential phases of AC, each set off by a third of the time it takes the sinusoidal current to complete a cycle, known as the **period**. So, current is provided in three sine waves in sequence, offset by 120°. The sequential fluctuations are used to define motor rotational speed by providing each phase in sequence to successive coils, and so defining a rotating field of flux across the rotor (Image 4.15).

To do this, the coils of an induction machine are defined in paired poles, with each pair in opposite polarity. When the poles are energized, or switched, in sequence with each phase of current, a rotating sinusoidal flux in the stator windings is defined. With the cage bars shorted on either side, the relative motion between this rotating field and rotor cage induces a current in the rotor with the same number of poles as the stator's field. This current in turn induces a field that is opposed to the stator's field. The two fields meet in the air gap to define torque.

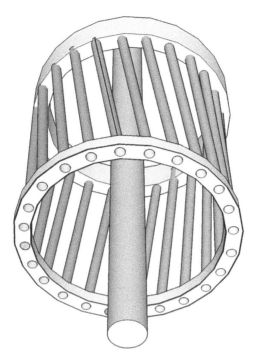

IMAGE 4.14
Basic squirrel cage armature.

It's clear how the basic **squirrel cage** armature configuration gets its name. With the stacked laminations removed, the rotor bars and end rings on either side define a conductive cage. The rotor bars are typically slightly skewed to reduce torque ripples by distributing them over a wider degree of rotation. Rotor losses can be high in an induction machine, leading to possible heating of the rotor, which is far more difficult to cool than the stator. As a result, Tesla Motors, the sole current user of induction motors in EVs, uses copper exclusively.

An induction machine's rotor therefore must always rotate at a slower speed than the stator field since it's the relative motion between the two that defines the production of torque. The difference is known as slip. The greater the slip, the greater the torque. At zero slip, the rotor and stator fields rotate at the same speed, and therefore the stator field does not cut across the rotor bars, and no torque is produced. This turns out to offer an advantage as it allows a motor controller to shift to a zero slip condition quickly to reduce torque, say when tires slip.

On the other hand, when using the machine as a generator, the rotor needs to lead the stator. This defines a negative slip condition, and reverses the direction of energy flow. However, the stator must continue to receive current, since there are no magnets in the rotor and so no way for it to generate flux if it does not receive current induced from the stator windings. But with the rotor leading the stator, the rotor's magnetic field now cuts across the stator and induces a voltage back into the stator coils, enabling the recovery of energy during regenerative braking.

As you might guess, controlling an induction motor entails somewhat more complexity than a simple DC motor, and is typically provided by a voltage-fed pulse width modulated inverter. An **inverter** is the power electronic circuit that can convert the DC from the battery into the multiphase sinusoidal signal needed by the motor. Inverters generally

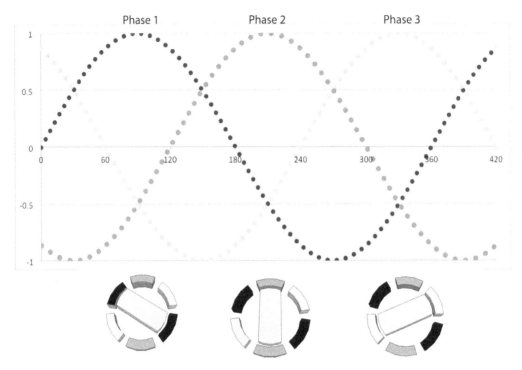

IMAGE 4.15
Three-phase AC.

Three-phase AC provides a more even distribution of torque. Each phase is offset by 120°, and maximizes when the rotor is aligned with that pole on the stator.

hold either current or voltage constant when defining sinusoidal output. To provide control, PWM is typically used to adjust the width of the switching pulse based on feedback voltage. This provides narrow or wide pulses for low or high voltage, respectively, and adjusts for changes in power consumption while providing a smooth and harmonic waveform. Soft switching can be added so switching takes place at points of zero current or zero voltage. This prevents switching during simultaneous high current and high voltage, or high-power points, to minimize switching loss and enable a smaller and quieter inverter.

The embedded control functions required to manage an induction motor are challenging, requiring complex mathematical models and high-performance control algorithms. At first glance, the task might look easy. After all, more current to the stator increases the magnetic field strength at each pole, which increases the resulting magnetic field of the rotor, in turn increasing the rotational force and thus the motor's torque. However, unlike the DC motor, simple PWM or varying the voltage alone will not offer good speed control, since the frequency of the changing magnetic field is the principle determinant of rotational speed. At the same time, changing the frequency of the AC current alone is problematic as it combines with voltage to define the air gap flux, and shifting one without the other would have undesirable effects. So, we can adjust frequency and voltage together, maintaining a constant ratio between the two to alter the speed of the motor, a method called **variable voltage variable frequency** (VVVF). However, as discussed with DC machines, above the rated speed we can no longer usefully increase voltage. So, we increase frequency alone and experience reduced torque with decreased flux. Unfortunately, this method is imperfect, control is imprecise and response can be sluggish. So, it is not good for multi-motor EV configurations that require precise motor speed coordination, and it is similarly not good for hybrids that need to be accurately controlled to work with the ICE. So, in general VVVF is not suitable for advanced vehicle applications.

A preferred method of speed control is called **Field Oriented Control** (FOC), also called **vector control**. To understand FOC, we need to recall that the magnetic field generated by the stator is constantly changing; and at any point in time, the field has a distinct magnitude and orientation. So we can understand the stator field as having a direction and magnitude, what physicists call a vector. We can resolve this vector into two components, one operating perpendicular to the axis of rotation, and so defining rotational force or torque, and the other operating parallel to the axis and defining magnetic flux. Since these two are **orthogonal** components, which means they operate perpendicular to each other in three dimensions, they are also independent of each other; so changing one has no effect on the other. FOC uses this principle as the basis of a control logic that allows us to vary the torque without changing the magnitude of the magnetic field, enabling a more dynamic torque response.

To accomplish this, you might expect that a precise and instantaneous measure of the stator field is needed to adjust winding current and thus define a current vector that can define torque exclusively. But placing a precision sensor in the air gap to measure the stator field is expensive and not easy. So in EVs we largely rely on an **indirect vector control** that uses measured rotor speed and current to calculate a predicted flux angle and magnitude and identify a reference angle of slip. The actual rotor flux interaction does not need to be fully identified or measured; instead we only need to identify the slip speed to deduce rotor flux position.[8] A slip lookup table is used to define the generation of torque. Not unlike the tables used in ICE control, the desired number is based on parameters such as load, motor speed, and throttle position. The result is that, in effect, the torque-producing component of the stator flux is managed separately, allowing us to maintain torque at a near optimal level throughout the operating range. However, FOC is hardly perfect, most notably, dependent on a fixed lookup table, it can have trouble accommodating changes in operating temperature and magnetic saturation, and solving this only compounds an already-complex control operation.

As a result **Direct Torque Control** (DTC) is gaining interest. This advanced control scheme directly controls the stator flux linkage and torque by controlling the switching modes of the constant voltage PWM inverter. Basically, DTC estimates the magnetic flux and torque based on the current and voltage of the motor, and when these estimated values vary too far from set reference values, switching is used to bring them back into the targeted band. The process is quick, simple, and efficient, allowing torque to be directly controlled, offering faster torque response, while simplifying computational requirements. However, this control logic can present challenging torque ripple and sluggish response. Recent work on improved algorithms and the application of fuzzy logic may offer solutions.[9]

---

[8] K.T. Chau, *Electric Vehicle Machines and Drives: Design, Analysis and Application*. Wiley-IEEE Press, Singapore, 2016.

[9] F. Korkmaz, İ. Topaloğlu and H. Mamur, Fuzzy logic based direct torque control of induction motor with space vector modulation. *International Journal on Soft Computing, Artificial Intelligence and Applications (IJSCAI)* 2 (5/6), 2013, 31–40; and Y. Bendaha and M. Benyounes, Fuzzy direct torque control of induction motor with sensorless speed control using parameters machine estimation. *2015 3rd International Conference on Control, Engineering & Information Technology* (CEIT), Tlemcen, Algeria, 2015.

Control complexities aside, induction motors may offer a promising option for automotive applications. While a notable improvement on the previous DC machines, the induction machine's power density and efficiency tend to be somewhat lower than their PM counterparts we will examine next. So, an induction machine is likely to be larger and heavier than a PM machine with the same performance. Typically a high-speed design is preferred, as this allows for more power from a smaller and lighter machine. They generally provide a high starting torque and robust configuration that offers good reliability; while overall efficiency tends to be lower than the PM motors we will discuss next, they still operate at about 85% efficient even at maximum load.

The basic induction machine is a mature and established brushless motor technology, dating back to Nikola Tesla's introduction in 1887. So, it is not likely that we will see dramatic improvements in performance or efficiency in the future. Faced with rapidly developing innovations in other areas of electric machine technology, future applications for induction motors in production vehicles may be limited.

Permanent-Magnet Machines

While induction machines clearly still have a place in EVs, the development of high-performance magnetic material and innovations in machine design are making PM motors much more promising in most electric and hybrid vehicle applications. You might be inclined to think that this will raise that ugly problem we discussed previously, the negative effects of a commutator and brushes. However, the solution is fairly straightforward: instead of placing the magnets on the stator and necessitating commutation to the rotor, we can place them at the rotor. The stator windings define a shifting magnetic field that crosses the fields of the magnets in the rotor, resulting in rotational force and defining the basic operation of a brushless permanent-magnet (BLPM) machine. No commutator required.

The stator windings can either be concentrated or distributed. Distributed winding spreads the armature windings across the stator evenly; this can offer smoother operation and higher efficiency because of the ability to harvest reluctance torque (something we'll discuss soon). In addition, the distribution of the conductor means less heat buildup. Concentrated windings, on the other hand, place only one phase coil at each slot on the stator. The advantage is this can be smaller, lighter, and easier to manufacture. While copper loss can be lower in concentrated windings, this configuration generally requires more poles to avoid torque ripple, and that complicates manufacturing as well as high-speed operation. In addition, these machines tend toward greater heating and vibration.[10] As a result, with the exception of the Hyundai Sonata hybrid, all major production cars use distributed windings.

Two key aims in defining the stator are effective heat shedding and the reduction of resistance. But these two are linked since the generation of heat in the coil is a function of resistance. Resistance is defined by the cross section of the wire, the material used, and the

---

[10] B. Sarlioglu, C.T. Morris, D. Han and S. Li, Benchmarking of Electric and Hybrid Vehicle Electric Machines, Power Electronics, and Batteries. *2015 International Aegean Conference on Electrical Machines & Power Electronics (ACEMP), 2015 International Conference on Optimization of Electrical & Electronic Equipment (OPTIM)*, 2015; and Y.Y. Choe, S.Y. Oh, S.H. Ham, I.S. Jang, S.Y. Cho, J. Lee and K.C. Koa, Comparison of concentrated and distributed winding in an IPMSM for vehicle traction. *Energy Procedia* 14, 2012, 1368–1373.

square of the current. So, to allow high current capacity without excess heating, the cross section can be adjusted. One option is using wire with a rectangular cross section to allow the wire wraps to fit together with less wasted space, enabling more coils in a given cross section. The result can maximize current density, offering greater continuous performance in the same sized package (Image 4.16). Using rectangular wire in a simplified wave-style winding with exposed ends, called a **bar wound** stator, can offer greater surface area exposure and result in improved heat rejection. These exposed turns can be cooled with oil flow to improve heat rejection by 50% or more.[11]

A more advanced alternative may soon be defined by extremely low resistance materials. A true superconductor operates with zero resistance but only at extremely cold temperatures. However, work on so-called high-temperature superconducting materials is improving and could soon redefine the electric machine. For example, a single layer of carbon shaped in an extremely fine tube, called **carbon nanotubes** (CNTs) can be used to define a web of fine woven CNT threads called nanotube yarn, to replace copper windings. This could potentially halve the resistance of a coil and dramatically lower weight.[12] As a bonus, heat shedding would increase notably.

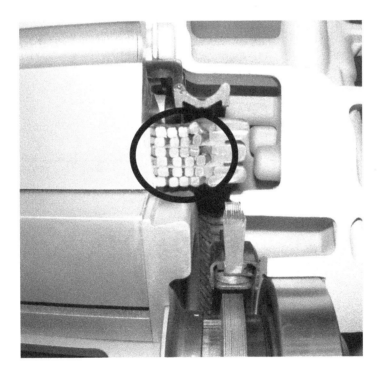

IMAGE 4.16
Square wire.

A square wire profile can allow more efficient windings.

[11] S. Jurkovic, K.M. Rahman, J.C. Morgante and P.J. Savagian, Induction machine design and analysis for general motors e-assist electrification technology. *IEEE Transactions on Industry Applications* 51(1), 2015, 631–639.

[12] D. Johnson, Carbon nanotube yarns could replace copper windings in electric motors. *IEEE Spectrum* 3 October 2014. Available at https://spectrum.ieee.org/nanoclast/semiconductors/nanotechnology/carbon-nanotube-yarns-set-to-replace-copper-windings-in-electric-motors

The fundamental mechanical difference between a PM and induction machine is in the rotor. Rather than a squirrel cage, the rotor is essentially a low-reluctance housing for PMs. A common version of this places the magnets in a V shape within the rotor (Image 4.17). Used in the Prius hybrid and multiple other production vehicles, this configuration defines a more significant space between the magnets and can incorporate a designed air pocket in the core to change saliency and help reduce counter EMF during regenerative braking. An advantage of this approach is the angulation of the magnets allows for a more gradual onset and release of maximum torque, diminishing **torque ripple** which can be prominent at low speed and high current.

An alternative would be to place the magnets at the surface. This can provide higher efficiency because the rotor magnets are closer to the stator coils, so less energy is lost. However, gluing magnets to the surface of the rotor presents a mechanical weakness that can limit the upper speed of the machine. This is a bigger deal than may first be apparent. Automotive electric machines can run at well over 10,000 rpm, with speeds likely to continue to increase to perhaps well over 20,000 rpm. As a result, the durability of the rotor is an increasingly important concern. In addition, because surface-mounted magnets have

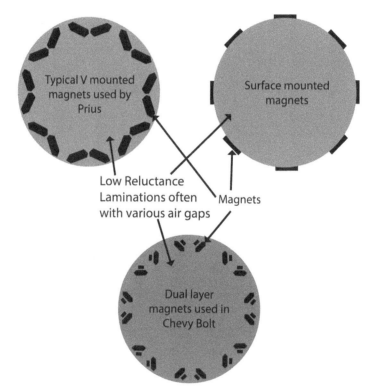

IMAGE 4.17
Rotor magnet configurations.

Reduced torque ripple in PM machines can be achieved by skewing the magnets, effectively offering the same effect as a skew in the induction squirrel cage. However, this requires a compromise in torque. As an alternative, the Chevy Bolt utilizes dual layer V-shaped configuration to offer smoother operation.[13]

[13] F. Momen, K. Rahman, Y. Son and P. Savagian, Electrical propulsion system design of Chevrolet Bolt battery electric vehicle. *Energy Conversion Congress and Exposition (ECCE)*, Milwaukee, WI, 2016.

about the same reluctance as air, the space taken up by the magnets is effectively like an increase in the air gap, and this larger gap can cause a loss of flux density. These may be the reasons that the Hyundai Sonata is the only production car to use surface-mounted magnets.

Magnets

More important than their particular mounting configuration, the composition of the magnets has changed greatly over the past generation. In fact, the development of high-power PM has made an array of motor topologies possible that would not have been achievable previously. In the 1970s and 1980s, rare earth elements were found to be particularly promising in the creation of PMs when combined with metals such as iron, nickel, or cobalt. This is because the atomic structure of these elements contains the unusual condition of multiple orbits containing only a single electron. Unlike the more common paired electrons with opposing spins, these **unpaired electrons** can be made to all spin in the same direction, generating an especially strong magnetic field. In addition, the crystalline structure of these materials defines a high directional property, or **anisotropy**, which makes a magnetization that is aligned with the crystalline axis easier to achieve and more durable.

The resulting rare earth magnetic alloys are nothing short of remarkable. Previously, magnets were made either of a ceramic compound of iron oxide and other metals, called **ferrite**, or an alloy of aluminum, nickel, and cobalt that produced a much stronger magnet called **Alnico**. Yet compared to current rare earth options, both of these have a very poor specific power, and would need to be many times larger to offer the same performance. In fact, the samarium–cobalt alloy (SmCo) that defined the first family of rare earth magnets proved more than twice as powerful as Alnico magnets and many times more powerful than ferrite magnets. An important point of comparison is the magnetism remaining after the external magnetic field that magnetized a material is removed, called **remanence flux density**. Also important is a magnet's resistance to demagnetization, called **coercivity**. As the table below indicates, these factors, and the magnetic energy of the material, are remarkably greater in rare earth magnets. However, the very high cost of these magnets limits their use. An even more impressive and feasible option was defined with the next generate of rare earth magnets, neodymium–iron–boron ($Nd_2Fe_{14}B$ or NIB). With just about twice the power of SmCo magnets, they are also lower cost. With the price of NIB magnets having dropped significantly over the past two decades, they are now the go-to choice for PM machines for everything from cordless tools to automobiles.

However, while impressive, rare earth magnets can be fragile and are susceptible to accidental demagnification from flux and also from heat. NIB magnets in particular are vulnerable to high heat, and must be kept at operating temperatures below 300°F to avoid thermal demagnification, a point defined as the **Curie temperature** of the material. At high heat, the magnets are weakened and power is lost. And, because of their low resistance, these magnets can generate high internal eddy currents intensifying the problem. In fact, frequently the limit to a PM motor's high-speed operation is heat. This has defined a continued need for SmCo magnets in high-temperature applications, as they can withstand temperatures approaching 800°C (Table 4.1).

Besides heat vulnerability, there is one more challenge to these super magnets. They require the use of dense rare earth metals. Metals such as dysprosium and terbium are

TABLE 4.1

Comparing Magnet Characteristics

Sources	Remanence Flux Density B_r (Gauss)	Intrinsic Coercive Force H_{ci} (kA/m)	Maximum Magnetic Energy BH_{max} (kJ/m³)	Curie Temperature (Magnetism Is Lost) T_c (°C)	Cost
Ferrite	200–400	100–300	10–40	570	Very low
$Nd_2Fe_{14}B$	600–1,400	600–1,800	200–440	315	High
SmCo5	900–1,150	450–1,200	60–240	750	Very high
Alnico	600–1,200	50–120	10–80	700–860	Low

Source: S.J. Collocott, J.B. Dunlop, H.C. Lovatt and V.S. Ramsden, *Rare-Earth Permanent Magnets: New Magnet Materials and Application*. School of Electrical Engineering, University of Technology, Sydney, NSW, 2007; J. Liu, "Some Design Considerations Using Permanent Magnets." *Magnetics Magazine* March 13, 2016; J.M.D Coey, *Magnetism and Magnetic Material Cambridge*. Cambridge University Press, Cambridge, UK, 2009; and J.D. Widmer, R. Martin and M. Kimiabeigi, Electric vehicle traction motors without rare earth magnets. *Sustainable Materials and Technologies* 3, 2015, 7–13.

added to improve resistance to heat; if not, the thermal performance of NIBs would be even worse. However, these metals are largely mined in China, raising concerns over reliable global access as well as cost. As a result, Honda has developed a commercial motor that uses no heavy rare earth metals. Instead of the typical process of heating the material and pressing it together, called sintering, a hot deformation technique is used that defines much smaller crystalline grains. These nano-grains are about 100–500 nm across, or about a tenth of the size resulting from sintering, and allow for greater heat resistance without the use of heavy rare earth elements.[14] The cost and uncertainty of rare earth elements has even made ferrite magnets of interest, leading GM to use ferrite magnets in one of the motors in the second generation Chevy Volt. When combined with reluctance torque, the weaker magnets offered adequate performance and good heat tolerance at a lower price. In fact, ferrite magnets increase coercivity with increasing temperature, just the opposite of the neodymium.

More powerful magnets are not always better. As we know, machines for automotive application need to perform over a wide speed range. Because PMs are fixed, and can't be turned up or down, the designed strength of the magnet represents a trade-off. In particular, at high speed and low torque, the PM produces a high flux that cannot be turned off. The back EMF can be accommodated with an opposing stator current, to a point. But this can seriously degrade efficiency, so it can be better to avoid overly powerful magnets for high-speed machines. Using a weaker magnet means it is principally the rotor's saliency that defines torque at low speed and high load. This is a bit more like the reluctance machines we will discuss in a bit.

BLPM Control

Advanced magnets are vital to the performance of modern PM machines, but brushless PM motors would not be possible without precise digital control. Power electronics provide each phase of a rotating magnetic field that defines the armature's rotation. So, although powered by DC electricity from the battery pack, the control of these DC machines is not

---

[14] How Honda Developed the World's First Heavy Rare Earth-Free Hybrid Motor. Available at http://world. honda.com/environment/face/2016/case59/episode/episode01.html

unlike the control of a synchronous AC machine. This can be accomplished by the rectangular switching of DC or the generation of a true sinusoidal current.

Rectangular switching defines a **permanent-magnet brushless direct current motor (PM/BLDC)**. A controller provides pulses of current to the stator windings, shifting the stator phases off and on. Voltage is provided in two of three phases at any one time, and switches between phases every 60°. So, each phase is on for 120° of field rotation. This defines a rotating square wave that produces the rotation of the motor. The resulting geometry of field interaction with a rectangular signal produces extremely high torque.

If we instead produce a true sinusoidal current, we can define a synchronous machine, or **permanent-magnet synchronous motor** (PMSM). The resulting torque and therefore power density is diminished; but the sinusoidal current smooths the torque, reducing torque ripple that can be associated with PM/BLDC machines. This is because the synchronous field control allows the stator and rotor to interact at consistent angles throughout the rotation. A synchronous machine can utilize the same control strategies as induction machines, since they both run on a sinusoidal waveform. In particular, FOC is commonly applied to synchronous machines. The high efficiency, good field weakening capacity, and high torque have made the PMSM the dominant EV machine.

The PM/BLDC motor offers a good fit for HEVs where the machine principally augments the primary internal combustion engine. Strong torque density, particularly at low speeds, and good efficiency makes a good fit with this hybrid architecture. So, for example, Honda saw this as a good choice for the early Honda Integrated Motor Assist (IMA) system. Brushless AC (BLAC) machines, on the other hand, with greater control, smoother operation, and better power performance are more favorable to being the primary traction device in a hybrid system or a modest EV. Efficiencies can reach into the mid- and high 90s.[15] Consequently, the brushless AC motor has been adopted in cars as varied as the Nissan Leaf, Chevy Bolt, BMW i3, Mitsubishi i-MiEV, Toyota Prius, and Citroën C-Zero, for example.

However, brushless sinusoidal machines do require more complicated control. In particular, the need for precision FOC control of sinusoidal flux at the air gap requires that the angle of the rotor be exactly known. This necessitates an extremely high-resolution position sensor.[16] The sort of encoder that can convert the angular position and movement of the shaft to a digital signal with high precision adds significant cost and complexity that were not necessary for the open-loop control system of the induction machine. And position sensing for the trapezoidal flux of the BLDC machine only needs to know rotor position at each commutation point, so each 60°, and can be managed with a less-expensive Hall effect sensor. However, a variety of sensorless coil control methods that measure the back EMF and other parameters to determine armature position are being developed for PMSM applications.

More generally, both PM machines offer good flexibility. Either machine can easily function as a generator or motor. When a motor, the rotor magnets follow the stator field movement. When a generator, the rotor magnets lead the stator field. Also PM machines allow for high adaptability and a variety of geometries. An axial flux machine with the air gap perpendicular to the axis of rotation can be an attractive option, for example. Sometimes called a pancake motor, this configuration allows for a lot of poles and can provide

[15] K.T. Chau and W. Li, Overview of electric machines for electric and hybrid vehicles. *International Journal of Vehicle Design* 64(1), 2014, 1–34; and Patrick Hummel et al., UBS Evidence Lab Electric Car Teardown—Disruption Ahead? UBS Group AG, May 18, 2017.

[16] J.X. Shen, Z.Q. Zhu and D. Howe, PM brushless drives with low-cost and low-resolution position sensors. *The 4th International Power Electronics and Motion Control Conference*, 2004.

improved power density over the traditional radial flux PM motors we've been discussing, and improved heat shedding given the large surface area and radius of the rotor. In addition, the higher moment of inertia can offer smoother operation in certain applications. This flexibility of design can be useful when trying to integrate a machine into an engine bay for a hybrid car. Similarly, PM designs can allow for single- or double-sided stators and multiple rotors. For all these reasons, PM machines are seen as the best option for a possible wheel-hub configuration to be discussed in Chapter 5.

These PM machines also present challenges related to the use of PM. First, with an unregulated PM flux, high-speed back EMF can be problematic. Second, sensitivity to high temperature can be a challenge and significantly reduce performance if not properly managed. Lastly, the rising cost of PM presents a problem for low-cost applications. For these reasons, there is increasing attention being paid to motors that use reluctance as their driving force and so do not need to rely on magnets.

Reluctance Machines

Making a motor that relies on reluctance could potentially define a much simpler and less-expensive option for EVs. Remember that a force is exerted on any low-reluctance ferromagnetic material in a magnetic field. This force tends to align the material with the density of magnetic flux, like the needle we considered at the start of the chapter that turned to define a path of lowest reluctance. This is the basic driving principle of the **switched reluctance machine** (SRM). If we properly define a rotor with a low-reluctance ferromagnetic path, the flux of the stator can be designed to cause the rotor to turn like that needle to define the path of least reluctance and so exert a rotational force.

In fact, the basic design of a functional SR motor is relatively straightforward. We can begin with a rotor with projected poles made of laminations of iron, called **salient poles**. These will define our low-reluctance paths. We couple this with a stator with similar salient poles wrapped in coils to generate the flux (Image 4.18). We can then produce a path of magnetic flux from one pole on a stator to an opposing pole. The rotor will rotate to align its poles with the flow of flux, providing a path of least reluctance to the field. With proper digital control, we can rotate this flux around the stator poles, defining significant and continuous torque on the rotor. So, in sum, current flows through a coil on a salient stator pole and defines a flux that pulls the salient rotor pole to align with the stator pole and so reduce the reluctance of the magnetic field. When the poles are aligned, the reluctance is minimized; and if the pole remained energized, the rotor pole would be held in position, but instead it is switched to another stator pole that attracts another rotor pole and continues the rotation. With multiphase operation, as this pole is aligned an alternative pole is feeling the pull of another stator winding and the rotation continues.

This is not a new idea; in fact, it dates back to early nineteenth century Scotland. However, because smooth and controlled rotation requires that we precisely define the rotating stator field, a reliable performance reluctance motor was not possible before the development of modern advanced power electronics, sensors, and digital control technology.

The torque is defined by the current and speed of the rotating stator field. Unlike PM machines, only a pulling force is exerted on the rotor pole, so torque is independent of current polarity, and an inverter specific to SR machines is needed. Precise switching needs to be provided to each phase in series but independently, allowing for high-speed operation

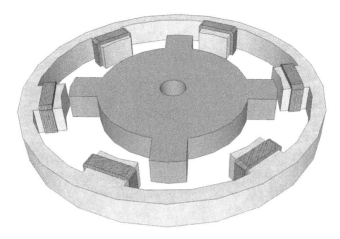

IMAGE 4.18
Switched reluctance machine.

This SR machine has four salient poles on the rotor, defining low-reluctance paths. The stator provides six salient poles each wrapped in coils. As the stator generates rotating paths of magnetic flux, the rotor rotates to allow the path of least reluctance.

with adjacent phase currents overlapping. PWM is used to allow control of field magnitude; but the discrete excitation of phase windings still makes the SR highly vulnerable to torque ripple. To address this a **torque sharing function** (TSF) can be implemented that distributes torque in relation to armature rotation so that the sum of all phase torques is defined as the targeted torque production.

Therefore, particularly for SR machines, synchronization of winding excitation and armature rotation needs to be precise, especially for high-speed operation. A hall effect or optical sensor can be mounted on the shaft to provide rotor position. In addition, sensorless control that estimates rotor position based on known parameters such as current can improve the fault tolerance of the machine.[17]

Even with these challenges in mind, the great advantage of a switch reluctance machine is its simplicity. As demand for electric motors grows and magnets become more expensive, SR offers an inexpensive and robust alternative. The stator has concentrated windings around each pole. The rotor has fewer poles, a simple laminated structure, and no windings. With neither magnets nor coils, the rotor can handle high torque and high rpm with no trouble. Distinct salient poles means the phase windings can operate in isolation from each other, reducing the problem of mutual inductance, and offering high fault tolerance.

So, with the rising cost of magnets, the potential for SR motors is being seriously examined in EV applications and elsewhere (Image 4.19). The machines offer simple control and cooling, reliability, very high speed operation, excellent heat tolerance, good torque, and low cost. Although these motors still generally suffer from lower torque density and high noise, advanced digital control and computational methods are being developed to address this.[18] Increasingly, a well-designed machine with carefully considered pole and

[17] M. Yilmaz, Limitations/capabilities of electric machine technologies and modeling approaches for electric motor design and analysis in plug-in electric vehicle applications. *Renewable and Sustainable Energy Reviews* 52 (December), 2015, 80–99.

[18] M. Cheng, L. Sun, G. Buja and L. Song, Advanced electrical machines and machine-based systems for electric and hybrid vehicles. *Energies*, 8, 2015, 9541–9564.

IMAGE 4.19
Switched reluctance in action.

This Land Rover all-electric test vehicle, seen here at the 2013 Geneva Auto Show, is one of the few uses of an SR machine in an EV. It uses a 70 kW (94 bhp) **switched reluctance** motor, coupled with a 300-volt lithium-ion battery pack with a capacity of 27 kWh.

Image: Norbert Aepli, Switzerland

phase configuration can provide control and reduce torque ripple. While still large, with continued development, increasing power is becoming available from a smaller package. So, look forward to seeing an SR machine in a production vehicle soon.

A variation on this theme that is somewhat less developed is the synchronous reluctance machine (SynRM). The basic idea is to devise a rotor with integrated flux barriers and guides, defining a minimum reluctance in one direction and maximum in the other. The rotor could then be made to turn synchronously with a sinusoidal voltage. The SynRM occupies a space somewhere between a PM and induction machine, and offers a new sort of reluctance drive. With a small number of poles, and the same number of poles on rotor and stator, it bears only a little resemblance to the SR machine.

This machine is decidedly not ready for deployment in production vehicles. It has high torque ripple, low power density, and poor efficiency. But, it is still early, and its progress is interesting. With no heat loss, no eddy currents, and a simple variable frequency drive managing speed, the idea has clear potential. A PM-assisted SynRM machine may offer improved efficiency and torque density thanks to the magnets, and might only need a little assist from modest ferrite magnets.

Advanced Motor Possibilities

The PM motors discussed so far are widely used and demonstrate reliable and impressive performance. However, they are not perfect. In particular, they pose two general challenges. First, they require PM be mounted on a rapidly spinning and vibrating rotor.

The mechanical stress placed on rotors on PM machines due to centripetal force at high speed is significant. This is particularly true for surface-mounted magnets, but presents a problem for all PM rotors. Second, buried in the heart of the machine, these magnets can get easily overheated and are difficult to cool. Heat stressing may be a more significant challenge than mechanical stress, since it can lead to the partial demagnetization of the rotor magnets and can have a devastating impact on power capacity. A possible solution comes from a deceptively simple and radical idea: put the magnets on the stator instead.

An interesting example is based on an ingenious modification of the basic SR machine. By putting the magnets on the stator of an SR machine, the rotor is left with neither windings nor magnets, maintaining a robust and simple architecture, and making the machine easy and inexpensive to manufacture. But the addition of magnets fundamentally shifts the power capacity of the reluctance drive. The rotor incorporates salient poles coupled with PM-equipped salient poles on the stator, earning the name: **doubly salient PM** motor (DSPM). Operation is somewhat like a square waveform BLDC, but with a much-improved **constant power speed range** (CPSR). The result is high efficiency and power density, but at the cost of some control capacity since the PM flux is uncontrolled.[19] And, of course, the DSPM still entails the cost of PM magnets.

An alternative could replace the PMs with electromagnets to define a doubly salient electromagnet machine. This promises improved control, as the magnetic flux can now be regulated. However, the power loss through excitation can tend to be high, so the efficiency drops. And because an electromagnetic coil might require five times the size of an equivalent neodymium magnet, the motor size increases.[20]

A promising example of this idea is **flux switching PM motors** (FSPM). This machine places magnets between U-shaped magnetic cores on the stator. The magnets are circumferentially magnetized, meaning that one side is north and the other south as you move along the circumference; and magnets are placed with alternating polarity. Concentrated windings are then wrapped around the two adjacent segments and the magnet, defining a single tooth (Image 4.20).

As the rotor coil aligns with the next stator tooth of the same phase, the polarity of the flux linkage is reversed. By providing opposing direction current at each flux linkage, torque is generated in both flux locations. With coil excitation, the field on one side is reduced and on the other side increased, and the rotor moves to the stronger field. As the rotor travels across the stator field, adjacent rotor poles align with successive stator teeth and there is a reversal in the PM flux linkage, earning it the name 'flux switching'. In essence, this machine operates a bit like a brushless AC motor.[21] By using the high-energy PM excitation and the flux concentration effect, the FSPM provides very good power and torque density. In addition to the general cooling advantage provided simply by putting the magnets and coils on the stator, the need for more limited windings means lower copper loss and so less heating. An added advantage of this machine and others like it is the inherent redundancy, and thus fault tolerance, offered by multiple coils; if a circuit fails, the motor control can reconfigure and maintain traction capacity.

---

[19] J.T. Shi, Z.Q. Zhu, D. Wu and X. Liu, Comparative study of biased flux permanent magnet machines with doubly salient permanent magnet machines considering with influence of flux focusing. *Electric Power Systems Research* 141, 2016, 281–289.

[20] J.D. Widmer, R. Martin and M. Kimiabeigi, Electric vehicle traction motors without rare earth magnets. *Sustainable Materials and Technologies* 3, 2015, 7–13.

[21] M. Cheng, L. Sun, G. Buja and L. Song, Advanced electrical machines and machine-based systems for electric and hybrid vehicles. *Energies*, 8, 2015, 9541–9564.

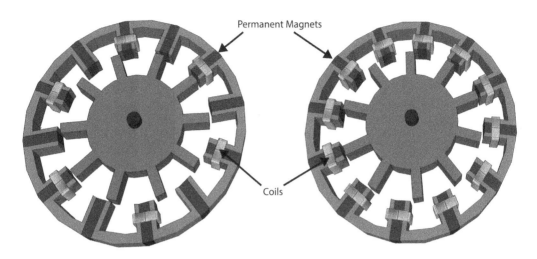

IMAGE 4.20
Flux switching permanent-magnet (FSPM) machine.

There are multiple configurations of poles and windings for FSPM machines. For example, a single-phase coil can be wound around each stator pole with four coils defining one phase. This would define a double-layer coil, since a coil slot would have two coil phases. Alternatively, each slot could have only one coil, with alternating poles left without a coil. The torque capacity could be maintained, but this could result in increased torque ripple.

But this motor is not perfect. Accidental partial demagnetization can be a concern, as the PMs are mounted adjacent to the coils. The uncontrolled PM field still presents a challenging control problem, and it can make the constant power operation range small as the back EMF is difficult to accommodate. This is a particular challenge for EVs that require a wide address this CPSR.

This can be addressed by a variant of the doubly salient machine that can be called **hybrid excitation doubly salient machines** (HEDSPM). The name may be a mouthful, but it really says it all. While maintaining the general doubly salient design, this machine uses a hybrid combination of both DC field windings and PM to provide improved flux control. The additional DC winding offers desirable and straightforward controllability, avoiding the cost and complexity of vector control while widening the CPSR.[22] The rotor remains without coils or magnets, offering motor robustness and improved thermal stability for the magnets. A positive DC current strengthens the PM flux during high torque and a negative current can weaken the flux for constant power in high-speed operation. So the DC coil acts as both an electromagnet and a mechanism of flux control. This allows for improved motor efficiency over a wider speed range, offering the advantage of a DSPM while providing magnetic flux control with field windings. But the trade-off is in power density and efficiency resulting from the added DC coils, as well as added complexity. In essence, we trade power density for improved controllability and a wider constant power range.

This hybrid approach has advantages. The flux density at the air gap becomes easily controllable. This enables flux strengthening for much-improved short periods of high torque for acceleration or starting. The constant power range is widened greatly by enabling

[22] Q. Wang and S. Niu, Overview of flux-controllable machines: Electrically excited machines, hybrid excited machines and memory machines. *Renewable and Sustainable Energy Reviews* 68, 2017, 475–491.

simple flux weakening. Adjusting the flux density also allows for stable voltage output while operating as a generator, simplifying battery charging.[23] It is unfortunate that the DC field windings of this hybrid system can increase coil loss and reduce power density; but as you might expect, there is work being done on yet another option.

The option that can preserve some of these advantages while also providing lower copper loss is a specially defined flux memory motor. The key to this design is the use of a magnet that can be readily re-magnetized, exhibiting what is called a low **coercive force**, but is also able to maintain magnetism once polarized, called high **remanence**. The dominant neodymium magnets do not fit the bill, so Alnico magnets are used. Current pulses can be used to reset the polarity of these so-called **memory magnets** during motor operation. This allows control of the magnetic flux, without requiring the copper losses entailed in a DC electromagnetic coil, earning the name **Flux Modulated Permanent Magnet**, or sometimes Flux Mnemonic Permanent Magnet (FMPM).[24]

This innovative FMPM offers significant advantages. Since the needed pulse is very brief, there is minimal coil loss. The controllable magnet also offers a simpler and more efficient way to provide flux weakening. The additional windings mean reduced power density, but the added control means torque can be boosted for short periods such as start-up and overtaking. The only real disadvantage is the complexity of the system, incorporating two distinct sets of coils, and thus increasing manufacturing difficulty and costs.

If that's not impressive enough, a variation on this design uses two magnet types to provide improved power. Since the remanence of Alnico magnets is less than traditional NIB magnets, power density can suffer in a memory motor. However, by placing two magnets, one Alnico and one NIB on the rotor, we can have the best of both. The higher remanence NIB provides needed power density, while the Alnico magnet enables the benefit of flux memory current pulsation (Table 4.2).[25]

Before we finish, we should briefly look at one more variation on this theme. Removing both coils and magnets from the rotor on these advanced machines has resolved the mechanical issues related to mechanical failure of the rotor at high speed and eased the problem of cooling. However placing both coils and magnets on the stator can present new challenges. First, mechanically it can be very difficult to fit multiple coils and PM on the stator. The stator needs to be made larger and this means a larger motor, which can be a

TABLE 4.2

Comparing PM Machine Characteristics

	Power	Torque	Efficiency	Controllability	Robustness
DSPM	Moderate	Moderate	Good	Moderate	High
FRPM	Good	Good	Good	Moderate	Moderate
FSPM	High	High	Good	Moderate	Moderate
HEPM	Moderate	High	High	Excellent	Moderate
FMPM	Moderate	High	High	Excellent	Moderate

Source: K.T. Chau, *Electric Vehicle Machines and Drives: Design, Analysis and Application.* Wiley-IEEE Press, Singapore, 2016.

---

[23] A. Emadi, *Advanced Electric Drive Vehicles.* CRC Press, Boca Raton, 2017.

[24] C. Yu, K.T. Chau, X. Liu and J.Z. Jiang, A flux-mnemonic permanent magnet brushless motor for electric vehicles. *Journal of Applied Physics* 103(7), 2008, 07F103–07F103-3; and Y. Fan, L. Gu, Y. Luo, X. Han and M. Cheng, Investigation of a new flux-modulated permanent magnet brushless motor for EVs. *The Scientific World Journal* 2014, 2014, 1–9.

[25] M. Cheng, L. Sun, G. Buja and L. Song, Advanced electrical machines and machine-based systems for electric and hybrid vehicles. *Energies*, 8, 2015, 9541–9564.

major problem in some automotive applications. Second, having the coils so close to the outer periphery of the motor invites increased flux leakage.

However, if we swap the location of the rotor and stator, making the outside ring the rotor and the inside cylinder the stator, these problems can be addressed. Making the inner cylinder the stator means the inner-core of this component can be used to accommodate DC coils and magnets, allowing more efficient use of space and no need to enlarge the overall motor dimensions. Essentially, this defines two layers to the stator, an outer layer with armature windings and an inner layer with magnets and possibly DC coils. With both magnets and DC coils fully incased, flux leakage is minimized. And, because the armature windings and PMs are located in different layers of the stator, the threat of accidental demagnetization is reduced. The outer ring can then just provide salient poles, with no magnets or coils, defining a more robust component with no associated flux leakage. Moreover, with an increased radius the moment of inertia of the rotor is now usefully increased without increasing overall machine size. The higher inertia offers smoother, quieter operation.

Maybe you have already guessed the principle drawback of this design. With the stator now tucked into the core of the motor, it will be difficult to dissipate heat, making this motor susceptible to magnet degradation. In fact, the outer magnets can act as an insulator, making cooling even more challenging. Yet since the inner stator does not rotate, providing cooling to the core though a fluid or oil system could be a relatively simple task.

In fact, continuing in this same vein, for both PM and induction configurations, there are any number of possible motor topologies that we could imagine. For example, a dual-stator–dual-rotor motor can be defined by applying the same techniques as a conventional motor. The basic idea is to place two concentric stators on either side of a rotor, or vice versa. Each flux interaction generates torque. The total torque is the sum of the torques from each stator. While this could add additional weight and complexity, the added reliability and torque capacity are notable, and excellent for high-power applications.

Nissan's so-called super motor is an interesting example of such innovation taken to an impressive finish. Intended to replace the use of two conventional motors, the super motor has dual rotors, one on the inside and one the outside of the stator. Each rotor is attached to a separate shaft, and a compound current to the stator is used to control each. In one application, one rotor could be used for motoring and the second dedicated to generation; in another, each shaft could be connected to a separate drive axle, allowing independent control of right and left axle.[26]

In the end, the choice of technology depends on the application, size requirements, the needed torque and power, acceptable costs, and any number of other factors. There is no one right answer. But, the options are getting increasingly exciting. While the potential cost and limited availability of rare earth elements is likely to continue to drive innovation in magnet-less options, it is also likely that PM machines will continue to dominate the industry in the near future. Nevertheless, lower cost and robust options like SRMs are too promising to ignore. And, most of all, the capacity of precision digital control has redefined our options and catalyzed a wide array of innovative machines with performance characteristics that were unimaginable just a generation ago. In short, the heart of the electric car of the future is ready to go.

---

[26] Super Motor | Nissan | Technological Development Activities. Available at www.nissan-global.com/EN/TECHNOLOGY/OVERVIEW/super_motor.html

5

Electrified Powertrains

Early automobiles included some with internal combustion engines and some with electric motors. In fact, until the fossil fuel boom that led to low-cost gasoline, it was not clear which would emerge as dominate, if either. If advanced digital control technology had been available at the time, it is possible that a combination of the two, a hybrid, would have led the pack. However, in the absence of the capacity to seamlessly integrate two traction sources into a single drivetrain, and with battery technology in its infancy, the clear winner at the time was the gasoline ICE. Perhaps this is a pity.

Gas versus Electrons?

The internal combustion engine has the remarkable advantage of tremendous flexibility of operation and range, but it also has its limits. It is easy to refuel, and the fuel is incredibly energy dense; so the effective range off a small tank of gas is large, and unlimited in a world that has ready available gasoline virtually on every block. In fact, it is only thanks to the incredible energy density of gasoline that the basic ICE is able to produce adequate power, as the typical **thermal efficiency** of an engine is only about 25%. Toyota recently unveiled a new engine that can achieve 40% thermal efficiency.[1] This is a remarkable achievement, but it still represents less than half the energy input resulting in useful power. And no matter how much we improve on engine efficiency, it is by nature based on combustion and inevitably results in emissions. Some of these emissions cause immediate health and environmental threats, and others threaten the entirety of the planet through climate change. The emissions resulting from the manufacture of the gasoline only adds to the problem.

Conversely, the great advantage of the electric vehicle is that it does not necessarily entail emissions. There are no emissions from the vehicle itself, and depending on how the electricity is generated, potentially no or little total effective emission. Even when electricity is generated from fossil fuels, the **well to wheel** efficiency of an EV is much higher than its gasoline-powered counterpart. However, the range and flexibility of current EVs are limited. Even the best batteries available cannot match the energy density of gasoline, and charging them takes time.

Combining these two systems, and benefiting from the best either has to offer, can be a match made in automotive heaven. The low-end torque and efficiency of electric machines complements the flexibility and power of internal combustion engines. Electrification of the drivetrain can be seen as an enabler of continued improvement in the internal combustion engine, not necessarily a competitor.

---

[1] D. Carney, "Toyota Unveils More New Gasoline ICEs with 40% Thermal Efficiency." *Automotive Engineering* SAE, April, 2018.

However, as technology advances, the option of tossing the ICE completely and going with a battery electric vehicle can be increasingly viable for certain application. The challenge of electric vehicles over the past two decades has been twofold: first, the cost of batteries means a pure electric vehicle that requires higher energy density and greater capacity has been significantly more expensive. Second, because even the best batteries have notably lower energy density than gasoline, electric vehicles exhibit a more limited range. Both of these challenges are real, but both are also diminishing and the pure electric option is beginning to look more practical.

In any case, the advantages of electrification are certain to define the future of the automobile. Multiple manufactures have plans to incorporate more electric drives into their lineup. Volvo has committed to transitioning its entire lineup to electric or hybrid drive by 2019. GM has announced plans for an 'all electric future' with no precise date. Ford has promised 13 new all-electric models by 2023. Mercedes-Benz has announced plans to electrify its entire fleet by 2022. In fact, virtually all major manufacturers have committed to the expansion of their hybrid and electric options over the next decade.[2] Nevertheless, predictions of the death of the ICE can be greatly exaggerated. The energy density of gasoline makes the internal combustion engine continuingly vital; and the improvement in combustion performance and efficiency we talked about in the first two chapters promises to ensure its relevance into the next generation. But, the advantage of electric drive makes this an ideal accompaniment; and we are likely to see future fleets of electric and hybrid vehicles, in a variety of configurations. Technologically, there really is no best option, just a range of innovative possibilities that can help achieve the targeted goals of the vehicle.

Nevertheless, we are clearly approaching a tipping point in EV market penetration, marked by a significant uptick in the rate of EV adoption. Some see this shift coming globally within the next few years.[3] In the US, where fuel prices remain low and eased emissions standards may be on the political horizon, this point may be a bit further off. However, within a decade, the average EV is expected to be cheaper in Western countries than a comparable internal combustion car.[4] In fact, globally, electric vehicles are projected to make up more than half of all light vehicle sales by 2040.[5]

Electrifying the powertrain isn't just about mixing and matching traction drives, it is also about seeking opportunities for synergy and efficiency wherever they lie. This means capturing the wasted energy of braking to use for subsequent acceleration or maybe just to power the air conditioner. It can mean downsizing to a smaller, lighter, and more efficient engine, and allowing it to operate at its peak efficiency. Shutting the ICE down when it's not needed, maybe just at stoplights, maybe while cruising. Replacing inefficient engine-driven pumps with more efficient electric pumps; and perhaps upgrading the entire electric system to facilitate this and other features such as driver assistance, safety, navigation and entertainment. And, of course we want to maintain responsiveness, acceleration, handling, and general drivability while we do all this.

---

[2] A.C. Madrigal, "All the Promises Automakers Have Made about the Future of Cars." *The Atlantic* July 7, 2017.

[3] "Q-Series: UBS Evidence Lab Electric Car Teardown—Disruption Ahead?" UBS Limited May 18, 2017.

[4] N. Soulopoulos, "When Will Electric Vehicles Be Cheaper than Conventional Vehicles?" *Bloomberg New Energy Finance* April 12, 2017.

[5] "Electric Vehicle Outlook 2017: Bloomberg New Energy Finance's Annual Long-Term Forecast." *World's Electric Vehicle Market* July, 2017.

Hybrid Drive

The performance characteristics of the ICE and electric motor complement each other admirably. Electric motors produce their maximum torque at low speeds and are able to maintain that torque through a significant speed range. As we have seen, as the motor speeds up, torque production eventually drops off though power is maintained for quite a while. On the other hand, internal combustion engines have a limited range of operation. At the lower end, torque is limited, increases with speed to a plateau then drops off relatively quickly. Power production increases with speed more compellingly, but drops off quickly once maximum power is achieved. As we have discussed, these characteristics mean engines require a complex set of gearing to enable efficient operation; and innovative but intricate mechanics are required to allow the engine to operate efficiently in a wider range of conditions and vehicle speeds. Melding electric motors, with their low-end performance and wide operating range with the power capacity of a fuel-fed internal combustion engine can offer us the best of both worlds (Image 5.1). In fact, if electric drive and internal combustion had been melded sooner, it is possible that some of the recent innovations in combustion control, as exciting as they are, might never have been necessary.

Most notably, bringing these two power plants together has clear efficiency advantages. Internal combustion engines are sized to produce needed power for acceleration, hill climbing, and overtaking, which means the majority of the time the engine is significantly oversized for the needs of the car. So, while only 30 horsepower or so are needed to keep a car cruising down the road, the typical engine has about five times that amount of power available. By integrating an electric motor that can provide a torque boost when needed, engines can be downsized significantly, as they no longer need to be sized for the maximum torque conditions.

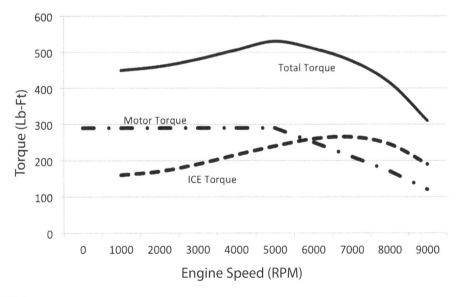

IMAGE 5.1
Combined torque.

Combining the low-speed torque of an electric machine with the higher speed torque of an ICE, can offer the performance advantage of a very fat torque curve.

The effect on the combustion engine is called **load leveling**. Combining the power of an electric motor and combustion engine means we can manage with a smaller combustion engine with little performance loss. And a smaller engine means less weight and less friction loss. More than this, taking the peak load burden off the combustion engine allows it to operate closer to its peak efficiency. At light loads, a given engine may consume less fuel, but that's not the way to think about efficiency, because the engine's production of power can drop even more than its fuel consumption. As discussed in Chapter 3, an engine achieves lower BSFC closer to its peak load and typically on the lower side of its speed range (Image 5.2). So, since a typical car spends most of its time cruising, a smaller engine can operate much closer to optimal load and speed if an electric motor is available to help meet momentary high loads, providing an efficiency improvement that is more than just about a smaller displacement.

There are multiple modes and methods for the integration of the combustion engine and electric motor. The overall architecture of the hybrid system can be understood as a power circuit. The ICE and electric machine can be connected in series, parallel, or various combinations of the two. And, the relative balance of tractive capacity from the two drive sources can vary.

Similarly, the details of the mechanical connection that allows the traction motor to be incorporated into the powertrain can vary greatly. Initially, carmakers wondered whether a new form of transmission was needed to enable hybrid powertrains, or could hybrid drive simply tap conventional transmissions. The early go-to was the CVT, however, this quickly gave way to the development of a **dedicated hybrid transmission** (DHT). DHTs integrate the electric machines fully into the transmission defining a new sort of power

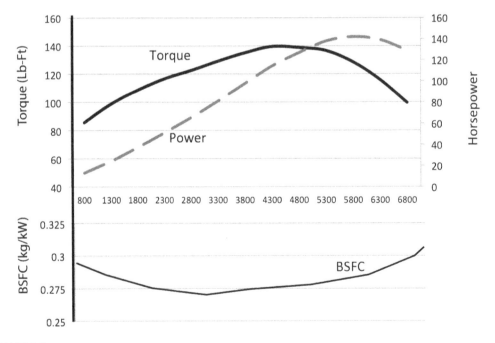

IMAGE 5.2
Power, torque, and efficiency.

An internal combustion engine can operate closer to its optimal efficiency if an electric motor is used to reduce momentary high loads, and allow the engine to remain on the lower side of its speed range.

coupling unit. While a dedicated unit now represents the majority of current full hybrid systems, the high cost of reengineering the entire powertrain is making more modest options attractive as hybrid drives grow more widespread. Many more modest hybrid systems rely on modified conventional transmissions. For example, dual clutch transmissions are becoming a more popular choice, with VW, Porsche, and Hyundai leading the way. The electric drive can help address the lag when starting off that can plague some DCTs. Alternatively, using a modestly modified conventional automatic transmission can allow for a simple replacement of the torque converter with a traction motor to define a low-cost hybrid option, and now probably represents about a third of all hybrid systems. AMTs, on the other hand, are somewhat problematic, as the torque interruption in low gears can be challenging to accommodate.[6] In sum, which system makes the most sense depends on the type of hybrid system being considered.

In fact, 'hybrid' can mean a great many things: At one end might be a powerful electric motor that provides all the propulsion for the vehicle, with a small ICE that can drive a generator and provide extended range. On the other end might be what is essentially a conventional ICE automobile, but with an added capacity to harvest, store and reuse wasted kinetic energy through the use of a small motor-generator. To better understand what this spectrum entails, we need to examine these variations in steps.

Baby Steps

We might start with the most modest electric-drive augmentation of the ICE, so-called **micro hybrid** systems. At their most simple, micro hybrids incorporate two systems: one that allows engine shutdown when it is not needed and another that allows us to capture lost kinetic energy in braking. The former utilizes a more robust starter and higher capacity battery to allow the vehicle to be automatically shut down when not needed in normal operation, say at a stop light, and restarted seamlessly when again needed. Called **start–stop** technology, this can improve fuel efficiency by as much as 10%, depending on driving conditions.

This is not as simple as just linking a starter-kill switch to the brake pedal. Shutting the engine down regularly during normal operation presents some complications. An electrically driven compressor may be needed for cabin cooling, for example, to provide greater efficiency and to enable operation when engine is off. A small additional electric pump is needed to keep transmission fluid flowing during engine stop. And power electronics may need continuous cooling and necessitate an additional electric pump. And there are a whole lot of functions that are essentially provided as peripheral product of the combustion engine's inefficiency that need to be otherwise met. Cabin heat for example, is provided through the engine cooling system. If the engine is stopped, an electric heater may be needed to ensure cabin heat. Vacuum pressure is used for a great many things, so an electric vacuum pump may be needed if the engine is going to be shut down while driving.

A simple system might stop the car when the brake application indicates an imminent stop and restart the engine as the brakes are released. The process needs to be seamless and quiet, to preserve drive quality. With a low cost, good consumer acceptance, and a

---

[6] C. Guile, "The Effect of Vehicle Electrification on Transmissions and the Transmission Market." *CTI Magazine* December, 2016.

good return on investment in fuel economy, the application of basic start–stop systems is growing, and now used in a wide array of vehicles, not all of which are true hybrids.

Defining a true hybrid entails incorporating **regenerative braking** capacity. The idea is so simple, it's brilliant: capture the excess kinetic energy when braking, and use that energy to augment acceleration when the vehicle speeds back up. Braking is unavoidable, of course; but it is also a significant form of energy waste. If you recall the basic equation for the energy of a moving body, **kinetic energy** is defined by half the product of the mass and the square of the velocity. So, with increasing vehicle speed, the energy available for recovery (or the energy wasted) increases exponentially. Each time the brakes are applied, valuable forward movement, paid for with expensive fuel and hazardous emissions, is converted to unwanted heat. In fact, the heat is so excessive and unwanted we design brakes to dissipate it quickly, literally tossing energy to the wind. Typically, regenerative braking allows us to capture about a third of that energy and; as a result, in city driving regenerative braking alone can increase fuel efficiency by about 20% (Image 5.3).[7]

The regenerative braking system is integrated with friction brakes to ensure braking performance while maximizing recovered energy. So, with a two-wheel drive vehicle, regeneration is limited to one drive axle. A **brake-by-wire** system that replaced hydraulic actuation with electromechanical actuators simplifies the mixing of braking elements, called **blended braking**. How much energy is actually recaptured depends on a variety of factors, including motor and battery capacity, the current state of charge (SOC), and the rate of braking demanded. In some vehicles, driver input can often vary the regenerative braking effect. In the Chevy Bolt electric vehicle, for example, a 'regen' paddle allows the driver to apply strong regenerative braking, enough to bring the vehicle to a complete stop in fact; or drop the transmission into Low to allow what Chevy calls 'one pedal' driving with strong regenerative braking whenever the accelerator is released. Of course, in

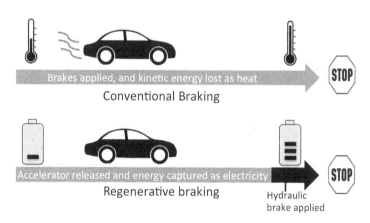

IMAGE 5.3
Regenerative braking.

Regenerative braking recaptures a portion of the car's kinetic energy during deceleration, typically using it to drive an electric machine so the energy can be saved in a battery and used to supplement acceleration later.

---

[7] G. Xu, W. Li, K. Xu and Z. Song, An intelligent regenerative braking strategy for electric vehicles. *Energies* 4, 2011, 1461–1477; and *Cost, Effectiveness, and Deployment of Fuel Economy Technologies for Light-Duty Vehicles.* Committee on the Assessment of Technologies for Improving Fuel Economy of Light-Duty Vehicles, Phase 2; Board on Energy and Environmental Systems; Division on Engineering and Physical Sciences; National Research Council, National Acadamies Press, Washington, DC, 2015.

IMAGE 5.4
48-volt boost recuperation machine.

This compact boost recuperation machine produced by SEG Automotive provides 12 kW generating power and up to 40 lb ft (55 N m) of torque. Small electric machines can allow for simple, low-cost hybridization of a conventional ICE with minor modifications to the electrical system and engine, and significant improvement in fuel economy and engine starting, while enabling the advantages of a 48-volt system.

the future such features could be applied automatically; but the opportunity for driver engagement and variable driving modes is considered desirable, particularly as the driving experience moves away from the conventional.

Defining a modest hybrid system, with start–stop and regenerative capacity, doesn't require a total redesign of the powertrain. This makes it an attractive option for carmakers. A modest version of this might just rely on a hardy starter that can also function as a generator (Image 5.4). General Motors' belt alternator starter (BAS) system offers a useful example. Connecting a robust starter/alternator through a high-tension serpentine belt to the crankshaft offers a clever way to address braking waste without costly modifications to the ICE or drivetrain. The belt allows two-way transmission of mechanical power; so power can to be delivered to the crankshaft and drawn from the crankshaft. A BAS system typically uses a 36–48 volt machine for starting as well as recuperative braking and torque assist. However, the high torque required at cold starts can often exceed the friction limits of the belt causing slip; so, the conventional 12-volt starter is retained. Improved belt connections are being developed to address this; though the BAS system has received limited praise and is unlikely to continue.

Mild Hybrid

A somewhat more aggressive incorporation of electric drive is defined by what is typically called a **mild-hybrid** system. By scaling up the coupled electric machine and improving the mechanical interconnection, an internal combustion engine can be downsized significantly and rely on the attached motor to provide augmented torque. Regenerative capacity is no longer limited to reducing the accessory load and supporting a start–stop capacity, it can now add some tractive force to the drivetrain. Still, in a mild application, the electric motor is limited to a supporting role, with no exclusive electric-drive mode. Nevertheless, this provides load leveling and allows for more complete regenerative braking that can apply recovered power to assist the ICE. Start–stop can kick in at idle and potentially while moving for sailing capacity.

Such functions require a 48-volt system, but we should note that the presence of a 48-volt system does not necessarily mean the use of a hybrid drive. For example, Porsche relies on a 48-volt system solely to power the active roll stabilization system on its new Cayenne (More on this in Chapter 7). Audi offers a more complete example of a 48-volt mild-hybrid system at work. The carmaker taps a 48-volt system for a mix of efficiency and power, designating the 48-volt system as primary and using a 12-volt system for low-load accessories such as the lighting and audio. The 48-volt system uses a belt-driven motor to provide seamless starting when the engine is warm, offers load leveling acceleration, some sailing capacity at cruising, and takes on some of the accessory load. The motor and particularly the power electronics receive active cooling from the engine cooling system. The fuel savings is modest; however, the system is relatively light and offers smooth operation and some efficiency gains, all while maintaining power performance. It is now incorporated into all of Audi's A8 variants (Image 5.5). In the diesel SQ7, the 48-volt system is used to operate a modest supercharger that addresses turbo lag.

Such systems may become more common soon. The growing need for higher voltage capacity to service increasing accessories raises the attractiveness of this simple option.

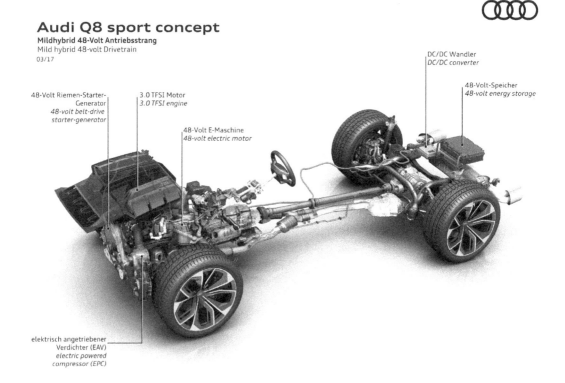

Audi Q8 sport concept
Mildhybrid 48-Volt Antriebsstrang
Mild hybrid 48-volt Drivetrain
03/17

DC/DC Wandler
DC/DC converter

48-Volt-Speicher
48-volt energy storage

48-Volt Riemen-Starter-
Generator
*48-volt belt-drive
starter-generator*

3.0 TFSI Motor
3.0 TFSI engine

48-Volt E-Maschine
48-volt electric motor

elektrisch angetriebener
Verdichter (EAV)
*electric powered
compressor (EPC)*

IMAGE 5.5
Audi mild hybrid.

Audi has announced a plan to incorporate some variant of this 48-volt technology in every car it makes.

Image: Audi

Now an established technology, versions of 48-volt systems have been adopted by numerous manufactures; and multiple 48-volt hybrid systems are available as low-cost add-ons by industry suppliers. So, the impact on fuel economy is modest, but the cost is low. As a growing number of manufacturers commit to the integration of electric drive, some version of a simple 48-volt motor-generator drive with starter capacity may become very common in the future. Forty-eight volts is a bit of a sweet spot, since it significantly enhances opportunities for mild-hybrid features and enhanced electronics, but it is not so high that it requires complex and expensive high-voltage safety features.

The mechanical connection of the electric drive to the ICE in this mild application as well as any other hybrid system can vary greatly. The spectrum of possibilities is sometimes talked about as position variants, labeled P0 through P4, with the simple belt-driven starter generators previously discussed defining a P0 systems (Image 5.6). An electric motor mounted directly on the crankshaft is known as a P1 system. While initially attractive because of its simplicity, the benefit is limited. The principle challenge is that, with the two mechanically linked, either the motor or the ICE has to be dragged along by the other if either is to operate alone. This increases drag when the electric assist is not operating and it makes electric-only operation impossible. It also reduces the efficiency of regenerative braking, as the engine drag cannot be detached and so absorbs deceleration energy through engine braking.

Honda's early **integrated motor assist** (IMA) system used in the early Civic and Insight hybrids offers a typical example. A significantly downsized 1.3-liter (93 HP) engine coupled with a 13 HP PM machine replaced the standard 1.7-liter (115 HP) powertrain in the Civic. The great advantage over a belt-drive system is the lack of the belt and the slip it entails. However, the increased cost and complexity on the existing vehicle are significant, and generally seen not to be worth the return. So, both Mercedes and Honda have dropped their flirtation with P1 systems, and it seems unlikely anyone will pick it up.

A more modest version of this approach was defined by General Motors when they essentially placed a robust alternator in the transmission bell housing. Called an **integrated starter alternator damper** (ISAD), the system was defined by integrating a motor between the flywheel and the torque converter, with the stator attached to the housing and the rotor assembly fixed to the engine shaft. A variation on this system places the ISAD on the side of the transmission. The additional cost or such a system is modest, and there were a few positive effects, such as quiet and smooth engine starting, start–stop performance,

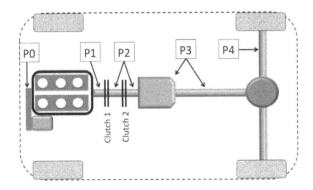

IMAGE 5.6
Mild-hybrid configurations.

The spectrum of possible locations for the integration of an electric machine to define a mild hybrid are sometimes discussed as P0–P4 systems.

driveline vibration dampening, and modest regenerative braking for accessory loads.[8] Used in the GM pickup line, a notable advantage for a working truck application was the capacity to generate 14 kW of electric power connected to an inverter and multiple 120-volt outlets. However, the efficiency improvements were a modest 2 miles per gallon.

More recently, an effective approach is defined by a P2 hybrid configuration that places the electric drive between the engine and the transmission but includes a clutching mechanism on either side of the electric motor. So, the engine can be mechanically uncoupled, allowing for electric-only propulsion with no engine drag. This provides a capacity for all electric very low-speed operation, or creep, increased energy recovery, and energy recovery during coasting, making it a more appealing option than the P1 systems. Typically, the electric motor is the sole source of propulsion only at very low speeds or light cruising. Above this, or when the battery runs low, the ICE engages to provide additional power.

Porsche offers a good example. The Porsche Cayenne hybrid is equipped with an Audi direct-injection, supercharged V6 rated at 333 HP, and a 95 HP electric machine placed between the engine and an eight-speed transmission. As a result, despite its formidable size and weight, the car reaches 60 mph in just over 5 s. Reaching 60 mph takes an extra 2.5 s if you rely only on the combustion engine and nearly an extra 4 s when only the electric motor is used. Selecting the 'E Power' mode allows the car to run off electric only with modest acceleration and a range of 14 miles.

The most significant advantage of the P2 system is its low cost. The arrangement is relatively simple compared to more extensive hybrid architectures we'll discuss next, requiring only one electric machine and two clutches. So, it can be used with a conventional transmission and engine (Images 5.7 and 5.8). The demands of the control system increase,

IMAGE 5.7
Four-wheel drive efficiency.

Magna's 48-volt transfer case simplifies the incorporation of mild-hybrid capacity into a four-wheel drive vehicle. This system actually allows a four-wheel drive system to offer better fuel economy than its two-wheel drive counterpart.

Image: Magna

---

[8] I.A. Viorel, L. Szabó, L. LöWenstein and C. Şteţ, Integrated starter-generators for automotive applications. *Acta Electrotehnica* 45 (3), 2004, 255–260; and C. Cho, W. Wylam and R. Johnston, The integrated starter alternator damper: The first step toward hybrid electric vehicles. SAE Technical Paper 2000-01-1571, 2000.

IMAGE 5.8
Eight-speed hybrid automatic transmission.

Adding an electric drive within a conventional transmission housing can offer an array of alternatives. A modest motor can be placed in the bell housing for a mild hybrid. Or a larger motor can be used to offer the all-electric drive capacity of a full hybrid. For example, the engineering firm ZF provides a modular transmission design that can include a 15 kW motor for mild-hybrid performance with very little mechanical modification of the drivetrain, or the unit above which integrates a 90 kW motor and can provide up to 30 miles of all-electric travel.

Source: ZF Friedrichshafen AG

but this cost increase is modest compared to full hybrid systems. Applications vary, with some utilizing a torque converter and others relying on the electric motor for torque rendition, and some using a separate starter and others relying on the integrated motor for engine starting. Generally, this system is popular for manufacturers who are not looking to redefine a model, but to add a few MPG and a hybrid logo on the side. Production cars such as the Hyundai Sonata Hybrid, the Infiniti Direct Response Hybrid system used in the M35, Bosch's system in Volkswagen Touareg and Porsche Cayenne all use variations on this theme.

Full Hybrid

The step beyond these modest P2 systems would define a so-called **full hybrid**, which is characterized by an increased capacity for all-electric drive, an associated higher operational voltage and greater battery capacity. Higher voltage means that not only is an electric-only mode possible, defining a key characteristic of a full hybrid, but also energy recovery potential is increased. Generally, a larger battery capacity, in particular a high-power density, allows regenerative braking to be reapplied to the drivetrain more efficiently.

Once again, the basic architecture of hybrid systems can be generally described as a power circuit. A parallel system allows the electric drive, the combustion engine or both to drive the wheels, something that can be possible in the P2 configuration we've discussed as well as more complex systems we'll discuss next. Linking the combustion engine and electric machine in sequence defines a series architecture in which only the electric motor drives the wheels, and the ICE then either drives a generator to charge the batteries or directly powers the electric motor (Image 5.9). The great advantage of this, of course, is that with only the electric motor(s) providing direct propulsion, we can forgo much of the complex mechanical coupling and complicated control logic of parallel systems.

So, the series architecture can offer the advantage of simplicity. With no direct mechanical connection between the engine and wheels, the ICE can be operated much more closely to optimal efficiency. And, like an electric vehicle, a series hybrid can place drive motors on multiple axles or multiple wheels, allowing for the potential for enhanced traction control and all-wheel drive. There are many variations on this basic theme, but the common denominator is that the engine never directly drives the wheels in a series hybrid.

Although the series configuration might be considered simply a variation on an electric vehicle, it has several advantages over the basic EV: Because gasoline has an energy density that is still much higher than the best lithium-based batteries now in production, a series hybrid can offer significantly greater range. In the same vein, the resulting ability to use a smaller battery than a pure EV, means appreciably lower cost. Lastly, general consumer sensibilities related to **range anxiety**, refueling and basic operation could make a

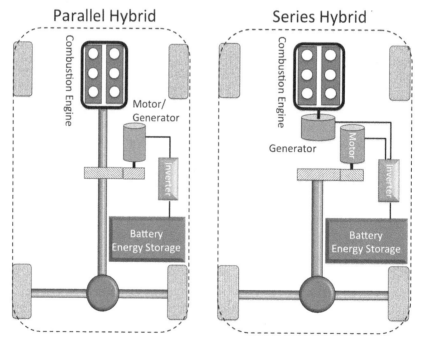

IMAGE 5.9
Series and parallel hybrid architecture.

series hybrid more familiar and comfortable to the average customer than an EV.[9] Though, as we will see, these advantages may be diminishing.

The principle challenge of the series architecture is inefficiency. In a series system, energy is converted from one form to another in multiple steps, and with each added step, some energy is lost. Gasoline is converted to rotational energy by the combustion engine; this is then converted to electricity by the generator. This energy is in turn stored as chemical energy in the battery. And the battery delivers the energy that remains back to the electric motor through an inverter, with yet more energy loss.

A parallel system that can offer direct drive from the combustion engine can therefore offer higher potential efficiency. P3 positioning with an architecture that links an electric machine and engine together through a mechanical coupler provides a common option. This design can enable either the combustion engine or the electric motor or both to provide the source of propulsion. Because the electric machine is connected to the driveline, it can't easily be used for start–stop. So a second, smaller, machine can be mounted on the engine side, defining a combined P2 and P3 positioning. The resulting improvements in fuel economy over a similarly sized ICE generally orbit 20%.

Combined P2/P3 dual motor configurations can allow us to tap into the best of both worlds and defined a so-called **series-parallel hybrid** (Image 5.10). Using both a mechanical

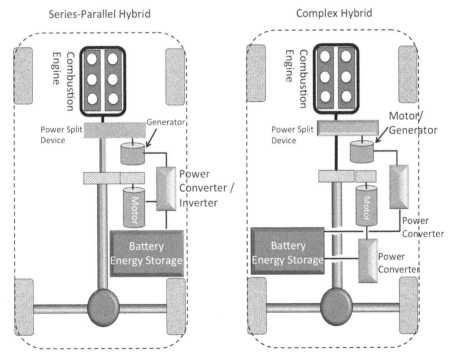

IMAGE 5.10
Series-parallel and complex hybrid architecture.

---

[9] *Hybrid and Electric Vehicles: The Electric Drive Delivers.* International Energy Agency, Implementing Agreement for Co-operation on Hybrid and Electric Vehicle Technologies and Programmes, 2015; and T. Altenburg, *From Combustion Engines to Electric Vehicles: A Study of Technological Path Creation and Disruption.* Deutsches Institut für Entwicklungspolitik, Bonn, Germany, 2014.

and electric coupler, we can enable an electric-drive mode with the combustion engine driving the electric machine to recharge the batteries. This is a classic series circuit. Or we can operate as a parallel system, with the combustion engine directly contributing to the drive of the wheels along with the electric motor. Managing this requires a mechanical connection between the electric machines, combustion engine and drivetrain that can allow for variable integration of torque, something called a **power-split** device.

A close cousin to this architecture is the so-called **complex hybrid**, which enables a route from the combustion engine to the batteries. The addition of a power converter between the motor and the battery effectively allows the battery to serve as the electric coupling. A key difference is that the power flow of the electric motor in complex hybrid goes in both directions, while the series-parallel hybrid provides only a unidirectional power flow through the generator.

The technological heart of this system is the power-split transmission and the associated digital controller. As mentioned in Chapter 3, when the notion of a hybrid powertrain was younger, the go-to transmission was a continuously variable unit. With relatively few parts, the compact design of a CVT made it a good fit with the compact car typical of early hybrids. And the generally low torque of the small combustion engine meant push-belt slip was limited. Most importantly, the ICE could operate as close as possible to its optimal BSFC, maximizing the primary intent of the hybrid, fuel economy.

However, this use of conventional CVT in hybrids has largely given way to an innovative deployment of DHTs using planetary gearsets. With enhanced digital control to link the combustion engine and motors, a basic gearset can provide a range of power configurations. In fact, series-parallel hybrids are often called power-split hybrids because the capacity to allow combustion power to go to both the generator and the driveline is the defining element. The planetary gearset in a conventional transmission defines a single input and output path in any given gear. However, in a power-split device, two inputs can be defined, allowing one drive source to provide traction drive while the other provides low-speed rotation to modify the operative output, effectively defining continuous gear ratios. So, the unit works as a sort of continuously variable transmission, without the burden of belts and pulleys. This sort of varying of gear ratios though a planetary gearset by varying the input rotation from a secondary motor defines an **electronically controlled CVT** or eCVT. From the driver's perspective, the ride feels a lot like any other CVT.

Perhaps the most well recognized example is Toyota's **Synergy Drive** (TSD). The TSD system uses two electric machines, a relatively small unit used for starting and regenerative braking, called motor-generator 1 (MG1) and a 80 HP (60 kW) traction motor, called MG2 (Image 5.11). The combustion engine is mechanically linked to the drive axle and two motors through a simple planetary gearset. MG2 is connected to the ring gear of the gearset, which drives the final drivetrain. The combustion engine drives the carrier. And the smaller starter-generator motor, MG1, connects to the sun gear. The beauty of this configuration is that any of the three propulsion devices can drive the system or be driven by it; and by combining multiple motor drives, an array of drive ratios can be achieved.

The operation of the power-split device is straightforward, but not exactly simple (Image 5.12). At initial low-speed start, the vehicle is driven only by MG2, and the carrier is fixed since the engine is not running. With increased speed, the combustion engine is started by MG1 driving the sun gear. Since the planet gears are made to walk around the rotating ring gear, a high drive ratio is defined, allowing seamless engine integration. With the engine started, MG1's role is generation, as the carrier now spins it while combining with MG2 to provide propulsion. With modest acceleration, engine torque both propels the car and drives the sun gear, which causes MG1 to rotate and feed energy to MG2 at varying

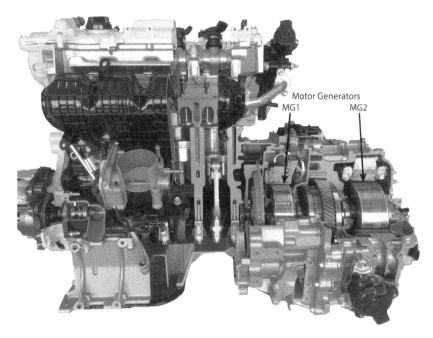

IMAGE 5.11
Toyota synergy drive.

Toyota's synergy drive utilizes two motor generators, MG1 used for starting and regenerative braking and MG2 as the primary traction motor. The two are linked to the ICE through a power-split device.

Image: Toyota Motor Corporation

levels. At cruise, the load on the engine decreases, and MG1 may provide more power to MG2, rotating the ring gear and reducing the load on the combustion engine. At high load, if overtaking is needed for example, MG2 is supplied with increased battery power, and combines with the combustion engine to provide high acceleration. At deceleration, the control system may shut off the combustion engine in certain conditions, holding the carrier, and enabling MG2 to absorb forward energy to charge the batteries. Throughout operation, with variable rotation supplied by MG1, the effective drive ratio can be changed continuously.

Ford's power-split device is not very different. At low load and low speed, or when the batteries are at a low SOC, the motor drives the wheels and the engine drives a generator, defining a series power circuit, or what Ford called the positive split mode. At higher loads and speeds, a negative-split mode is defined, with the combustion engine providing the main propulsion power, and the electric motor defining the drive ratio through a planetary gearset and either contributing traction or charging the batteries as conditions require. At modest speeds and high SOC, the electric motor alone can drive the wheels in what is called the electric mode.

This use of an electric drive to define a variable transmission through a planetary gearset can be applied in multiple ways. For example, a cooperative effort by GM and BMW has developed a full hybrid drive system that can be incorporated into just about any larger rear-wheel-drive vehicle (Image 5.13). The system combines two 60 kW electric machines with three planetary gearsets and four multi-disk clutches into what GM calls an **electrically variable transmission** (EVT). The whole unit fits into a standard bell housing.

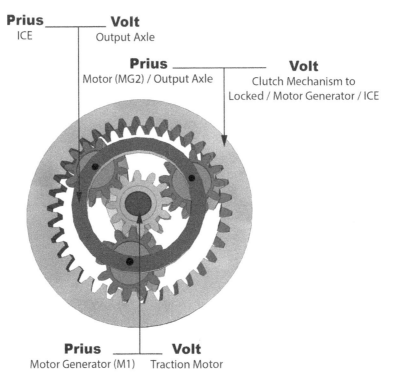

Prius _____ **Volt**
ICE   Output Axle

Prius _____ **Volt**
Motor (MG2) / Output Axle   Clutch Mechanism to
                                 Locked / Motor Generator / ICE

Prius _____ **Volt**
Motor Generator (M1)   Traction Motor

IMAGE 5.12
Power-split device.

While the Toyota Prius and Chevrolet Volt both use a simple planetary gearset at the heart of the power-split device, they rely on very different overall configurations.

IMAGE 5.13
BMW hybrid transmission.

The BMW 7-series hybrid incorporates a 15 kW electric motor between the torque converter and the engine in the housing for an eight-speed automatic transmission.

Image: BMW

The system provides two modes of operation: Under light load, the unit operates as a variable transmission; but when more torque is required, the gearing remains fixed, operating with set gear ratios. Now available in a variety of GM trucks and SUVs, the unit can reduce fuel consumption by about a quarter.[10]

Varying configurations of this basic approach can define differing combinations in electric and combustion drive, blurring the line between series and parallel drivetrains in new and interesting ways. The Chevy Volt's multimode DHT (MM-DHT) drivetrain includes two planetary gearsets. The engine is connected to the ring of the first gearset through a one-way clutch. A small, 48 kW machine is connected to the sun gear in that same gearset. A larger 87 kW machine is connected to the sun gear of a second gearset. The two gearsets are linked, with the sun gear of the first gearset clutched to the carrier of the second. The final drive is connected to the two carriers.

Connecting the smaller motor to the ring gear of the second gearset allows for an electronically adjusted variable character to the gearset. By varying the drive speed of the ring gear, the gearset operates as a variable transmission for the primary electric machine. In normal operation, the electric motors provide sole propulsion for the first 25–50 miles. When the battery pack is depleted, the ICE can drive the smaller machine and provide charging to the 18.4 kWh battery pack, defining a series architecture. However, at high speeds and under certain conditions, the ICE can function in parallel with the electric drives, with the engine mechanically linked to the planetary gearset to provide direct supplementary power. This can allow vehicle speeds over 70 mph when the efficiency of the larger machine drops due to the high required motor speed. Billed by GM as an extended-range electric vehicle, the Volt might be more correctly understood as a series hybrid, most of the time.

Clearly, the control demands of these systems are high. Processing speeds need to be fast and capable of responding to data as it develops, necessitating a complex **real time operating system** (RTOS) that has raised the bar for digital control logic. In addition to the obvious parameters that inform drive control, such as throttle position, engine speed, and vehicle speed, the SOC, voltage, current flow and temperature of the batteries need to be monitored to manage battery cooling and adjust drive control parameters. A controller area network (CAN) integrates the multiple control modules and componentry, and enables the processing of several thousand data points a second. The system must integrate the charge management of the battery, operation of the combustion engine, control of the electric machines as motors and generators as well as transmission control inputs, braking control system, electric-assist power steering, cabin heating and cooling, and a great many other functions as driving conditions continuously change.

The level of complexity is orders of magnitude greater than the basic vehicle of just a generation or two ago. Consider something as simple as braking. When the accelerator pedal is released, the controller shifts the traction motor into generator operation. When the brake is applied, the engine may be turned off, depending on the battery SOC; or at high speed, it may be left on depending on the constraints of the gearset. The controller determines the needed braking force depending on brake pedal pressure, vehicle speed, and input from the skid control electronic control unit; and the generative action of an electric machine will be varied to provide calculated resistive load. If this is inadequate to meet braking demand, hydraulic brakes will be applied. Alternatively, if engine braking is

---

[10] G.S. Vasilash, "Pacifica Hybrid Explained". *Automotive Design and Production*, December 19, 2016. Available at www.adandp.media/blog/post/pacifica-hybrid-explained; and C. Guile, "The Effect of Vehicle Electrification on Transmissions and the Transmission Market." *CTI Magazine*, December, 2016.

called for by selecting low gear, or 'B' drive in a Prius for example, one machine can charge the battery pack while the other rotates the engine to supply continuous braking effect without hydraulic brake application. This description is highly simplified, but it makes the point clearly: even an apparently simple action by the driver, or a modest change in driving conditions, requires a great deal of sophisticated calculations by the electronic control system.

Adding a Plug

Gains in fuel economy from a hybrid drive are impressive, but even more impressive gains can be achieved by increasing the potential for all-electric propulsion. A change in the control logic and an expanded battery capacity can add the ability to accept an external battery charge in a full hybrid. This defines a **plug-in hybrid** (PHEV) and offers further gains in efficiency and emissions reduction. But it's not just the size of the batteries that needs to be changed. As we will discuss in the next chapter, a hybrid vehicle battery pack typically prioritizes power density over energy density. This is because the ability to absorb a quick charge with regenerative braking and quick discharge for subsequent acceleration are key to load leveling in a hybrid, and that requires power density. However, with plug-in capacity, the all-electric range of the vehicle becomes important and a more energy-dense, rather than power-dense, battery option is needed. So, every major manufacture of a plug-in hybrid has replaced the power-dense nickel metal hydride batteries normally used in hybrids with energy-dense lithium-based batteries to provide increased range.

The result is not the sort of **all-electric range** (AER) you would expect from a respectable pure electric vehicle; but it's getting better. Typical battery capacities of PHEVs orbit a bit over 10 kWh, allowing an AER of anywhere from 12 to 50 miles. By comparison, the early Prius plug-in offered half the capacity of the current battery pack and an AER of 6 miles. The deployment of this technology may have been more strongly motivated by the need for compliance with California's PEV requirements rather than genuine emissions reduction or driver demand. In fact, it seems a large fraction of plug-in drivers did not actually charge their vehicles regularly.[11] However, more recent plug-in hybrids have significantly increased that range, making all-electric commuting viable for many drivers and defining a worthwhile fusion of all-electric efficiency with combustion-powered range and flexibility (Table 5.1).

Power

Of course, driving isn't always about improved fuel economy, sometimes it's about power; and hybrids can deliver that too. At times called **performance hybrids**, a growing number of performance cars have incorporated some element of electric drive with the principle aim of enhanced acceleration and high-speed handling. The low-end torque that can be

---

[11] Transportation Research Board and National Research Council, *Overcoming Barriers to Deployment of Plug-in Electric Vehicles*. The National Academies Press, Washington, DC, 2015.

TABLE 5.1

Comparing Hybrids

	All Elect Mileage (MPGe/ Le/100 km)	Battery Capacity (kWh)	AER (mile/km)	Gas Mileage City/ Hwy Combined (MPG/L/100 km)	Charge Time (h@240V)	MSRP ($)
Prius Prime Plus	133	8.8	25	54/4.36	2.1	27,100
Chevrolet Volt	106	18.4	53	42	4.5	33,220
Chrysler Pacifica	84	16	33	32	2	41,995
Mini Cooper SE Countryman	65	7.6	12	27	2	36,800
Sonata	99	9.8	27	39	2.6	34,600
Cadilac CT6	62	18.4	31	25	4.5	75,095
Kia Optima	103	9.8	29	40	2.7	35,210
BMW 330e	71	7.6	14	30	2	45,600
Cayenne	47	10.8	14	22	3	60,600

provided by an electric machine can help address turbo lag and enhance acceleration; and the ability to deliver this power though precisely controlled traction motors at each axle or even wheel can offer greatly improved handling (Image 5.14).

Of course, it doesn't have to be entirely about power. These gains can be incorporated with some regard for preserving fuel efficiency, as BMW has done. Not exactly a supercar, the BMW i8 offers a notable example of the performance hybrid that maintains respectable fuel efficiency. A 1.5-liter turbocharged 3-cylinder engine over the rear axle produces 228 HP. An AC synchronous electric machine produces 141 HP at the front axle. With a combined 369 HP, the i8 reaches 60 mph in just about 4 s; and defines a top speed of 155 mph.

IMAGE 5.14
Fast hybrid.

As the fastest hybrid in production, the Porsche 918 Spyder offers a striking example. A 4.6-liter V8 provides 608 HP. This is coupled with two AC permanent magnet synchronous machines, one on each axle, for a total of 887 HP, providing a 0–60 time of 2.5 s and a top speed of 214 mph. And, just in case it matters, it's still capable of producing 24 mpg and 18 miles of all-electric range.

Image: Mario Roberto Durán Ortiz / CC BY-SA 4.0

But it still manages a respectable 36 mpg.[12] The MINI Cooper SE Countryman All4 plug-in hybrid defines a similar approach. An 87 HP electric motor powering the rear axle can provide drive in town when the batteries are changed, and the 134 HP turbocharged 1.5-liter inline 3-cylinder engine driving the front wheels kicks in for greater acceleration or when the battery charge runs low.

A key for these cars is a particular form of parallel architecture with no mechanical combining of torque within the drivetrain. The BMW leaves the rear axle to the combustion engine and the front to the traction motor. The Mini is similar, but turns this around. The Porsche 918 has one electric machine augment the combustion drive at the rear axle, the other powers the front axle. In all cases, the two traction forces are integrated digitally and meet mechanically only at the pavement, earning the name **through-the-road** (TTR) hybrid, also called an electric axle or P4 system. The potential for precisely controlled traction delivery to all four wheels and the advantage of relative simplicity is great. It makes this architecture a simple and attractive addition to a performance platform (Image 5.15).

A TTR configuration is not simply for sports cars. For example, the Volvo XC60 plug-in hybrid 'twin-engine' crossover is a case in point. A 313 HP 2.0-liter turbo supercharged 4-cylinder engine drives the front axle while an additional 87 HP is delivered to the rear

IMAGE 5.15
Electric drive simplified.

The engineering firm ZF has placed an electric machine and transmission into a common housing with integrated water cooling, defining a mechanism with reduced complexity and associated friction.

Source: ZF Friedrichshafen AG

---

[12] J. Meiners, "2019 BMW i8 Roadster: i-Opener." *Car and Driver*, November, 2017.

axle by an electric motor. The result is the fuel economy and emissions of a 4 cylinder, the power of a 6 cylinder, and the advantage of all-wheel drive.

The technologies are not mutually exclusive, of course. A power-split configuration on one axle can be combined with a TTR electric traction drive on the other, as done in the Spyder. The Lexus RX 450 hybrid offers another impressive example. In order to achieve what Lexus calls, dynamic torque control all-wheel drive, the normally front-wheel-drive RX450 is fitted with an additional electric machine at the rear axle. Two motor generators are used at the front axle in a power-split architecture, with the additional traction at the rear coordinated electronically. A voltage boost to the normal battery pack helps ensure extra power is available when the front motor-generator (MG2) and rear motor-generator (MG1) combine with the combustion engine to deliver up to 308 total horsepower.

The TTR architecture is not perfect, of course. Since the combustion engine remains linked to the drivetrain, it is not possible to gain the full advantage of load leveling and operation at highest BSFC. With the drivetrain on both axles, design and chassis configuration can get tricky. And, while regenerative braking remains viable, battery charging can't happen at a stop.

However, there are several advantages to a TTR architecture that make it a likely future choice for certain production cars. The first is probably the basic simplicity of the configuration. Within a broader effort on the part of manufactures to consolidate and simplify control and drivetrain systems, the increasing complexity of power-split architecture rest uneasy. The idea is to lower costs and enhance reliability by seeking mechanical simplicity and digital consolidation when possible; and a TTR system makes this feasible in a hybrid. When power is a priority, this architecture becomes even more favorable. Delivering high torque from two sources through a complex system can be unwieldy. Delivering it more directly to the pavement is not only somewhat more straightforward, but the ability to do this at all four wheels with precise traction control makes it phenomenal. In short, if your only priority were performance, you would want this system.

Electric Vehicle

As impressive as these hybrid systems may be, combining a combustion engine and one or multiple electric motors in a drivetrain can't help but add complexity and weight. So, there are significant advantages to dumping the combustion engine in favor of a pure electric drive. The environmental benefits are clear: emission can be a lot lower; and a broad-based shift to electric cars could be coupled with the expansion of renewable energy to effectively produce a solar and wind powered national transportation system. Electric cars can be much less expensive to operate. And with far fewer moving parts and far less componentry, EVs promise a generally more reliable, simpler, and more easily maintained vehicle. For example, the Chevy Bolt's powertrain includes 24 moving parts, or about six times fewer parts than an analogous combustion-based system (Image 5.16).[13] Still, don't expect the death of the combustion engine anytime soon. Technological challenges, established industry practices, and consumer expectations are going to keep the combustion engine relevant for some time.

---

[13] "Q-Series: UBS Evidence Lab Electric Car Teardown—Disruption Ahead?" UBS Limited May 18, 2017.

IMAGE 5.16
Chevy bolt motor and transmission.

A simple single-speed gearbox with only four gear wheels is integrated into the motor housing of the Chevy Bolt. An electric machine's ability to provide maximum torque from one to several thousand rpm makes multiple gear ratios unnecessary.

Like HEVs, there are multiple possible electric vehicle architectures. In fact, one of the great potential advantages of electric drive is the flexibility of possible configurations. The simplest option replaces the ICE with a single electric motor. While this is the most common EV configuration to date, it may be just a bit too influenced by past design limitations and expectations defined by the bounds of combustion engines. After all, it is also possible to utilize a separate motor for each drive axle, much like a TTR hybrid, but without the combustion engine. Similarly, one could imagine a two- or four-wheel drive vehicle with a separate motor inboard of each driven wheel, much like inboard brakes are mounted. Alternatively, if they can be made light and robust enough, traction motors could be integrated into each driven wheel hub. The latter options could allow for precise traction delivery based on the relative slip of each wheel, and thus provide significantly enhanced handling possibilities. Moreover, with the need for a central drivetrain gone, the resulting flexibility in chassis and body design opens a range of possibilities.

With all these prospects in mind, the most common configuration is simply a single electric machine connected to a simple drivetrain through single-ratio gearing. The availability of maximum torque at initial spin-up means that in most cases there is no need for a clutch, torque converter or complicated transmission. As a result, the BMW i3, Mitsubishi i-MIEV, Nissan Leaf, and Tesla Model S and Model 3 use a single-speed drivetrain. With the ability of electric machines to produce all their torque at 1 rpm but also spin happily at upwards of 20 k rpm, the potential for a much fatter torque curve simplifies the drivetrain. Gearing can be tuned to the targeted performance of the car, offering desired acceleration or high-speed capability without a great compromise in overall performance over the machine's range.

With that said, some increased motor efficiency and therefore range can be achieved with multiple gear ratios that narrow the motor's operational speed range. Even a simple two-speed transmission can offer notable improvement.[14] The problem is the resulting increase in transmission losses and added weight can make this a fool's economy. Nevertheless, a simple two-speed transmission or CVT can make sense in certain applications, particularly when power is a consideration. High-end acceleration can deteriorate as the motor moves into its constant power region and is unable to produce needed torque. A simple two-speed gearbox can allow the motor to maintain high torque capacity at high vehicle speeds. So, cars like the announced Tesla Roadster are likely to tap a two-speed transmission to achieve the blistering 0–60 time of less than 2 s. However, none of this negates the initial point: electric cars are much simpler; after all, absolutely no one is suggesting an eight-speed transmission on an EV.

Similarly, the general popularity of the simple single-motor EV does not mean it's the only EV configuration out there. In fact, several manufactures are pushing the boundaries of innovation to take advantage of the full scope of possibilities with electric drive (Image 5.17). The Tesla Models S 75D, 100D and P100D, with the D meaning dual motors, are nice examples. All of the 'D' configurations incorporate all-wheel drive through dual induction motors that provide independent traction at each axle. The performance model P100D offers 184 lb ft (249 N m) at the front and another 479 lb ft (649 N·m) at the rear. Coupled with a 350 V 100 kWh battery pack, the result is 0–60 speeds of 2.5 s when demanded or a range of more than 300 miles off a single charge. Impressive even by supercar standards;

IMAGE 5.17
Electric axle.

GKN's eTwinster prototype electric axle system with two-speed transmission for electric drives. Note the motor to the right, two-speed e-Transmission on the upper center, and torque vectoring twin-clutch to the left. The increased efficiency of a simple, lightweight and efficient transmission such as this could significantly improve EV range.

Image: GKN ePowertrain

---

[14] J. Ruan, P.D. Walker, J. Wu, N. Zhang and B. Zhang, Development of continuously variable transmission and multi-speed dual-clutch transmission for pure electric vehicle. *Advances in Mechanical Engineering* 10(2), 2018, 1–15.

and not bad at all, for a 5 seat sedan. Jaguar has announced plans to follow suit with its own electric sedan. In Jaguar's case, two 200 HP, 258 lb ft motors are expected to provide a 0–60 time of about 4 s.

The Croatian carmaker Rimac Automobili has taken this a step further with its highly innovative Concept One supercar (Image 5.18). The vehicle relies on a motor, gearbox and independent inverter for each wheel. The traction delivery system adjusts hundreds of times a second to provide precise and highly responsive power delivery at each corner. The front wheels are driven by twin oil-cooled PM motors, offering a combined peak 670 HP (500 kW), each connected to single-speed gearboxes. The rear wheels are driven by a similar two-motor unit providing a peak 805 HP (600 kW) and connected to the wheels through a unique two-speed double-clutch gearbox. The package provides 1,224 HP for a 0–60 time of 2.5 s and a top speed of 355 km/h (Table 5.2). Are you starting to see why electric cars might be sort of cool?

An often-discussed but never fully implemented alternative approach to the EV architecture could place the motors not just *at* each wheel, but *in* each wheel. Called a wheel hub or in-wheel motor, the approach could greatly simplify electric vehicle design by placing all the key moving parts within the wheel itself. The resulting flexibility in chassis and body design are obvious. But so are some of the challenges. First among them is weight. Producing an electric motor and gearing mechanism that is small enough to fit inside a wheel hub is tough; making it light enough to function as **unsprung weight** without destroying the vehicle's handling and ride is even tougher. (We'll get back to unsprung weight in Chapter 7.)

If this nut can be cracked, some think it will define an entirely new horizon of possibilities for EVs.[15] The potential weight savings could mean much longer ranges. The responsiveness could be phenomenal. And designers could reimagine vehicles with no constraints on the body or chassis design that are now required to allow for transmissions, drive shafts, axles, or other burdensome hardware that have defined the shape and function of cars since the beginning. Round cars, cubes, side-crabbing cars, anything is possible if all the

IMAGE 5.18
The Concept One.

The Rimac Concept One is not only stunning, but also innovative. The carmaker places a motor, gearbox and independent inverter at each wheel for precise, high-performance handling.

---

[15] Y. Sun, M. Li and C. Liao, Analysis of wheel hub motor drive application in electric vehicles. *MATEC Web of Conferences* 100 (01004), 2017, 1–6; M. Li, F. Gu and Y. Zhang, Ride comfort optimization of in-wheel-motor electric vehicles with in-wheel vibration absorbers. *Energies* 10, 2017, 1647; and J. Thorton, "Circular Precision." *Electric & Hybrid Vehicle Technology International* July, 2013.

TABLE 5.2

Electric Car Performance

	Power (HP/kW)	Torque (lb ft/N m)	0–60 (s)	Top Speed (mph/kph)	Range (mile/km)
Tesla Model 3	221	203 lb ft	5.1	141	220
Chevrolet Bolt	200/150	266/350	6.5	93	238
Mercedes B-Class EV	177	251 lb ft	7.9	100	87
BMW i3	170	184 lb ft	7.2	93	81
Performance Cars					
Mercedes-Benz SLS AMG Coupe	740	738 lb ft	3.9	155	155
Tesla Model S P100D	762	792	2.5	155	315
Porsche Mission E Concept	600/440		3.5	124	>300
Ridiculous					
NextEV Nio EP9	1341	1480	2.7	194	265
Rimac C Two	1914/1408	2300 N m	1.86	258	400

key moving parts are in the wheels and the rest is just a conveyance cabin. But again, that first step, defining a wheel-hub drive system that is light enough and powerful enough to make this possible is a doozy.

While not yet ready for prime time, there are some early possibilities in the field. The German high-performance manufacturer Brabus has assembled a four-wheel drive electric concept car from an E-Class Mercedes and four electric wheel-hubs from Protean Electric. With 430 total horsepower and over 2,000 lb ft of torque, the car's a powerhouse to be sure; though a 0–60 time of nearly 7 s may indicate the limited gearing possibilities available in a constrained hub configuration. Similarly, Toyota's been playing with the notion of a wheel-hub drive for a while. Their i-ROAD concept car is a three-foot wide 'personal mobility vehicle' that looks a bit like an enclosed motorcycle with two tilting wheels up front and one in the back and weighing in at only 660 pounds. However Toyota's ME.WE concept car looks like a futuristic fun-machine for the beach-bound, seating five in a boxy polypropylene body. None of these ideas are road ready.

However, this doesn't mean real units aren't ready for deployment. The Japanese engineering firm NSK, known principally for high-performance bearings, has successfully demonstrated a wheel-hub motor. The key challenge to producing a viable wheel-hub drive unit for a full sized car is having the torque needed for initial acceleration while maintain the high-speed capability in a very small unit. NSK approached this by providing two motors and two integrated planetary gearsets in its prototype. The unit offers 2 speeds: a low speed with each motor rotating in opposite directions and a high speed with both motors operating in the same direction. And the whole thing fits in a 16-inch wheel (Image 5.19).

Electric Car Viability

While the adoption of these technologies may be a few years off, the general growth of electric vehicles is clearly imminent. The key factor that has limited the appeal of electric cars is about battery capacity, particularly as it relates to four key variables: range, charge

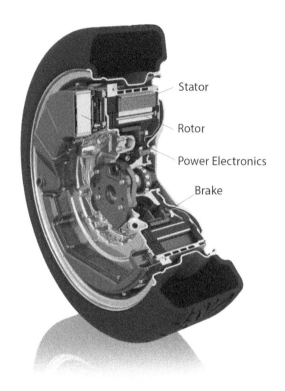

Stator

Rotor

Power Electronics

Brake

IMAGE 5.19
Wheel-hub motor.

Several wheel-hub motors are under development for EVs. Schaeffler's E-Wheel Drive electric wheel hub is an innovative example. This highly integrated drive system incorporates the motor, brakes, and power electronics into a compact package.

Image: Schaeffler AG

time, performance, and cost. However, as we will see, with the major battery improvements discussed in the next chapter on the horizon, we can expect range, recharge speed, and cost to all continue to improve. And, as we've seen, lighter weights and improved motor capacity mean the performance of electric cars borders on the unbelievable. So, even for those that have absolutely no interest in saving the planet, the performance of electric cars should make them tempting.

Similarly, the range of EVs is clearly on the rise. This is not necessarily the result of improving battery technology, though increasing energy density is a clear factor. It's also the result of manufacturers' efforts to produce the range consumers demand by fitting more batteries in vehicles, lightening vehicle systems, and prioritizing efficiency in control systems.

For an increasing number of EV owners, range anxiety is not nearly as concerning as it once was. A full charge in a Tesla Model S takes just over an hour at a Tesla supercharger (Image 5.20). But this is largely because of the reduced charging rate needed to manage cell balancing as the battery nears a complete charge. A half charge can add about 130 miles to range and takes about 20 min. So, for example, starting with a fully charged battery from home, you could take a 400 mile trip with one 20 min stop for a partial charge. Similarly, a Chevy Bolt gives you about 90 miles of added range on 30 min of fast

IMAGE 5.20
EV charging.

The Tesla Model S has an EPA range of over 300 miles off a single charge. Tesla has also taken the unusual step of building dedicated charging stations across the US. Using a proprietary connector that made the stations only useful to Tesla drivers, charging was once available for free to all Tesla owners, and now entails a fee for some.

charging. And the trend is moving toward faster, more convenient charging. A prototype charger jointly developed by BMW and Porsche, for example, could potentially provide a 450 kW jolt of charge, more than three times Tesla's supercharger, offering about 60 miles of range in less than 3 min.[16]

Until recently, EV ranges above 100 miles were limited to the Tesla and the luxury car market. The high cost of batteries made the integration of extensive battery capacity cost-prohibitive in a moderately priced vehicle. So, the early Nissan Leaf or electric Ford Focus had more modest ranges around 80 miles. However, recent declines in battery costs has changed this. The Leaf's range has climbed to 150 miles. More notably, the Chevy Bolt offers a range of 238 miles and the Tesla Model 3 provides 220 miles of driving off a full charge (Image 5.21). Both sell for about $35,000.

So, the performance of electric vehicles is impressive, and the range is increasing, leaving us with the remaining potential relative disadvantage of pure electrics: charging time. A gasoline-powered car can be gassed up in a matter of minutes at any number of ubiquitous stations across the globe. Electric cars are different. As we will see in the next chapter, bringing a battery system to a full charge can be tricky, and takes time. And the number of charging stations, while growing, is much smaller than the number of gas stations. However, this common observation can be a bit misleading, as virtually any home or business can be a vehicle charging station, making the number of potential charging stations for electric cars far, far more common and convenient than the number of gas stations. After all, you can't gas up your car in your home garage (or at least you really shouldn't), or in the parking lot at work, or at a shopping center. In fact, it's clear that if all the potential charging spots for electric cars were fully realized, keeping your electric car charged would be far more convenient and cheaper that having to stop at a gas station constantly to tediously fill an old-fashioned gas tank.

---

[16] C. Reiter, "BMW, Porsche Boast Three-Minute Charging Jolt for EVs." *Automotive News* December 13, 2018.

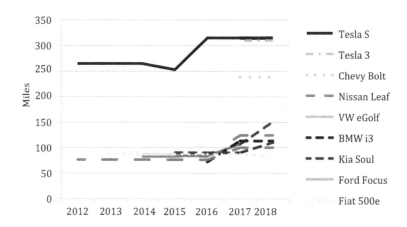

Note: In 2014, the EPA changed the way it estimates EV range by no longer using a full change to define the normal range of the vehicle, as they noted that many drivers do not ensure a 100% SOC for normal driving. The result was a small drop in some EV ranges, even though the actual range at 100% SOC increased.

IMAGE 5.21
Increasing EV range.

Charging can occur at differing levels and through differing modes. The simplest charging can occur through a humble 120-volt wall outlet, called Level 1 charging. A specially designed plug connects with a dedicated socket on the car (called an SAE J1772) and can allow a rate of up to 12 amps, resulting in about 4–8 miles of range per hour of charge. While this might work in an emergency, it is clearly not adequate as a daily charging method. A Level 2 charge utilizes a 240-volt, split-phase circuit, and typically provides about 30 amps, though could provide up to 80 depending on the vehicle capacity (and the limits of the circuitry in the home or business). The result is generally 20–30 miles of added range for each hour of charging. Direct DC charging, on the other hand, can utilize high-voltage DC to provide up to 200 amps and usually 208 or 480 volts, to add about 70 or more miles of range in 20 min. As mentioned, Tesla uses their own supercharging system, with unique proprietary plug, to provide higher rates of charge at its superchargers. Other manufacturers are still working on standardizing their charging systems and plugs. Leading U.S. and German automakers have developed a DC Combined Charging System using the SAE J1772 plug, while Japanese automakers have defined an alternative fast-charging system with a so-called CHAdeMO plug.

So, aside from the small annoyance of differing plugs, what's the problem? The challenge is with the cumbersome and unusual nature of plug-in charging. Drivers can forget to plug in. In public spaces, the charger is vulnerable to vandalism and theft. And in general, charging requires cumbersome wires and a change in behavior that doesn't always happen. So, automakers are looking at a simpler option. Enter contactless charging called **wireless power transfer** (WPT).

The common form of WPT now used is electromagnetic induction. The basic system functions much like a transformer or even an induction motor, and is a regular feature of many consumer electronics. An electromagnetic coil located in a stationary pad generates a magnetic field that induces a current in a coil on the vehicle. The trick is to achieve good alignment of the coils and closeness. The further the coils are away from each other, the lower the efficiency. Ideally, something in the neighborhood of 4–6 inches would allow a good transfer of power. This is not hard for your cell phone that can be laid directly on a charging pad, but it's a bit tricky for a car.

Innovators are looking at an improvement on this system that can allow for multiple vehicle orientations, a larger air gap, and higher rates of energy transfer. In other words, it could allow a car to be charged without the owner having to think about it. The key is magnetic resonance, or more precisely, defining magnetically coupled resonators. Initially developed by a team of researchers at MIT, the idea is to match inductance and capacitance of the sending and receiving coil to achieve identical magnetic resonance, and enhance the efficiency of power transfer.

Effectively the receiving coil in the vehicle is designed to respond in a synchronous phase, or resonate, at the frequency of the primary coil, creating a strong coupling. As a result, the sending field does not need to produce far reaching electromagnetic radiation, but can simply fill the area around it with a non-radiative magnetic field limited to what is called the near field region. Because the resonant nature of the interaction provides a tuned coupling between the two coils, power transfer is still viable and the power not picked up by the receiver stays close to the sending unit minimizing interaction with other objects in the area for reduced leakage. The resulting efficiency is well beyond a typical transformer, with coil-to-coil efficiencies of over 90%. The efficiency of whole charging system is slightly lower, but still in the mid-80%. WiTricity, the company established by the MIT researchers who first created this technology, claims 11 kW or better transfer rates and more than 90% overall efficiency. With tech like this now viable, several car companies are moving toward the incorporation of wireless charging as an optional feature; Audi, BMW and Mercedes have announced wireless charging options that use WiTricity technology (Image 5.22).

IMAGE 5.22
Wireless charging.

BMW is the first carmaker to offer a factory-fitted inductive charging system. With 3.2 kW of power, about 3.5 h is needed to charge the 530e iPerformance, with a claimed efficiency of about 85%. The driver is guided into optimal position by a 360° parking camera.

Image: BMW

So far, this is impressive and pretty straightforward. However, this so-called static system, that is to say charging when the car is standing still, isn't the end; some envision a *dynamic* system that could allow for charging while the car is traveling. The basic idea is to imbed coils in the roadway itself, charging the vehicle as it moves. This could enable virtually unlimited range for electric vehicles. Delivering the charge directly to the motor, rather than through the battery, could significantly improve overall transfer efficiency. It'll be a while before this is ready for prime time.

Using Energy Effectively

Batteries are not always an ideal way to store energy. Particularly when rapid, high-power charges and discharge are needed and the required energy density is low. In racing conditions or even in heavy traffic, rapid stopping and starting is normal, and the power-restricted battery may not be an ideal short-term storage mode. Systems that can manage this, and allow the quick capture of high-power burst and the controlled reapplication of that power to the drivetrain were first exploited in Formula 1 racing. Dubbed **kinetic energy recovery systems** (KERS), the key to these systems is a small, light storage device that, while perhaps ineffective at longer term energy storage, can absorb power effectively in short burst and deliver it on demand. Super capacitors and flywheels fit the bill.

A capacitor is a pretty straightforward device. Imagine two conductive plates with a thin insulator between them, like a dielectric sandwich. A charge at one plate will induce an opposing charge in the other plate, in accordance with Coulomb's Law discussed in the previous chapter. This provides an ability to absorb and store a charge in the plates. The larger the mated surface area, the larger the charge capacity, or **capacitance**; so large thin paired plates are typically rolled to provide a compact electronic component. But the real story is the dramatic increase in capacitance made possible with nanotechnology. Instead of using solid metal conductors, the conductive plates are coated with activated nanocarbon material that defines a sponge like surface with a dramatic increase in active surface area. An electrolyte with dissolved charged particles is used in the middle to maximize the interface between opposing charges. Or, extremely fine tube-shaped scaffolds of carbon molecules can be used. About ten-thousandth the size of a human hair, these nanotubes and nanofibers are finding remarkable applications in power electronics.[17] With these and other nanomaterials, so-called ultra-or **supercapacitor** can provide an active surface area per weight that is orders of magnitude greater than past materials, producing capacitance that can be a hundred times greater than their predecessors as well as the ability to charge and discharge very quickly. Small, light, powerful, and increasingly affordable supercapacitors are a very promising mode of short-term energy storage that can accommodate the intense power surges associated with automotive operation. For example, they are ideally suited to absorb regenerative braking energy and redeliver it for subsequent acceleration.

Mazda offered the first production application of this idea a few years ago. The manufacturer's i-ELOOP is fairly straightforward and modest (Image 5.23). A variable

---

[17] T. Chen and L. Dai, Carbon nanomaterials for high-performance supercapacitors. *Materials Today* 16 (7–8), 2013, 272–280.

IMAGE 5.23
Mazda i-ELOOP.

Mazda reports that their capacitor energy storage system can provide an up to a 10% improvement in fuel economy in traffic conditions. Mounted under the hood, at about 14 inches (350 mm) long and less than 5 inches (120 mm) in diameter, this is an impressive potential savings for a small space and weight cost.

Image: Mazda

Source: Mazda introduces supercapacitor-type regenerative braking. *Automotive Engineering* **SAE February, 2013.**

voltage alternator is driven by the drivetrain when braking, and this charges a capacitor. That energy can then flow through a DC–DC converter to 12-volt components, taking some load off the combustion engine by powering cabin cooling, lights, audio, or other devices.

A capacitor could also be attractive for simple start–stop systems. Twelve volt batteries are not ideally suited to producing the high power needed for city driving. Repeated starting and frequent high-power burst without fully recharging can shorten their life considerably. A supercapacitor can supplement the battery, drawing a charge from regenerative braking and applying that at subsequent acceleration, while stabilizing the impact on the battery. The **voltage stabilization system** (VSS) used in Cadillac's ATS and CTS sedans does just this. With two ultra-capacitors providing a quick burst of power, the restart is smooth and the vehicle electric system held stable. With an ability to deliver high power in a fraction of a second, the capacitors can also serve as an additional power source to stabilize the electrical system voltage as power demand varies.

Besides capacitors, an additional form of short-term energy storage could capture the vehicle's energy not in electrical form but as kinetic energy in a flywheel. This is not a new idea, the use of flywheels in transportation has been around at least since the 1950s in various innovative applications, most notably the Swiss Gyrobus that used overhead connections at each bus stop to spin up a flywheel for power to the next stop, eliminating the need for continuous overhead wires. Various other applications of flywheels followed through the 1970s and 1980s.[18]

The theoretical efficiency advantage of a flywheel is notable. With each conversion of energy from one form to another, there is an unavoidable loss in energy, since no system offers perfect efficiency.[19] A typical battery-based system converts kinetic energy to electrical energy, then to chemical energy for storage, only to convert back to kinetic energy a few moments later. The conversion of AC to DC and back to AC only adds to the losses. Instead, in braking conditions, kinetic energy could be transferred to a flywheel through a double toroidal CVT. It can then be stored as kinetic energy and reapplied to the driveline in subsequent acceleration. Systems might function as series or parallel configurations and at high or relatively low speeds. Although there will be unavoidable loss through the transmission, avoiding a long string of energy conversions can still offer a promising overall efficiency.

With this in mind, Volvo has experimented with a mechanically driven flywheel attached to the rear axle of its front-wheel-drive S60 prototype. The flywheel is a 13-pound carbon fiber disk about 20 cm across. During braking, the flywheel spins up through reduction gearing to about 60,000 rpm. The connection is made with the Torotrack CVT discussed in Chapter 3. Volvo claims that the system adds 80 HP, while providing improved fuel economy.[20]

A variation on this idea developed by Porsche a decade ago for race applications uses electrically driven flywheels. An electrical machine powered the flywheel, spinning it to several thousand rpm. The energy could then be applied on demand to the drivetrain. With the car's kinetic energy converted to electrical energy and back to kinetic energy for storage, only to reverse the process a few moments later, this system did not promise much of an efficiency boost. But, it could provide another kind of boost: In the racing variant of Porsche's 918 Spyder it added 162 horsepower for up to 8 s. When brakes were applied, the front electric motors spun a flywheel in the form of a permanent magnet rotor surrounded by a stator up to about 40,000 rpm. With a push of a button the driver applied this power back to the electric drive for a power boost. As for production cars, few drivers need to slam on their brakes at 200 mph and then rapidly accelerate 3 s later; so a flywheel may not offer the added value needed to justify the mechanical complexity and weight. But, come on, who wouldn't want that button on their steering wheel?

Peugeot Citroën, known as Groupe PSA, played with an especially inventive mode of energy storage a few years ago. The French carmaker linked a hydraulic pump and motor to an internal combustion engine. During braking, the hydraulic pump recovered the energy and stored it as compressed air in a hydraulic accumulator at the back of the chassis. During acceleration, this compressed air pushed hydraulic fluid into a low-pressure tank, driving the hydraulic motor along the way. A hydraulic accumulator isn't as efficient as lithium batteries, and the energy density is much lower; but the system can absorb a charge more quickly and operate more efficiently for short power turn-around, say in urban traffic. And, as PSA points out, all the parts are readily available and easily recyclable, defining a low-impact hybrid option that they claim can nearly halve urban fuel

---

[18] A. Dhand and K. Pullen, Review of flywheel based internal combustion engine hybrid vehicles. *International Journal of Automotive Technology* 14 (5), 2013, 797–804.

[19] A. Dhand and K. Pullen, Review of flywheel based internal combustion engine hybrid vehicles. *International Journal of Automotive Technology* 14 (5), 2013, 797–804.

[20] Volvo car and Flybrid vehicle testing showing flywheel KERS can deliver fuel savings up to 25%, with significant performance boost. *Green Car Congress*, March 26, 2014. Available at www.greencarcongress.com/2014/03/20140326-kers.html

consumption. However, perhaps due to the high upfront cost of engineering such a system, a production version does not seem to be on the horizon.

Short-term, high-power storage is not the only space for innovation within a hybrid architecture. The basic ability of any electric hybrid to store electric energy and use it to offset combustion energy opens up a new realm of possibilities, for example, exhaust heat recovery. The ICE generates a lot of waste heat. As mentioned, depending on how you measure it, an ICE is only about 20% efficient, with most of the waste escaping as heat. The HEV's potential for electric storage and the capacity to deliver that as useful energy either to the drivetrain or accessories opens up the possibility of using that heat to generate electricity and so recapturing some of the loss.

One option is applying a **thermoelectric generator** that can harvest waste heat without a major impact on performance. This device is essentially a chain of semiconductors linked in series in a way that they can also be exposed to heat in parallel (Image 5.24). The semiconductors alternate between materials that are manufactured to have extra electrons, and so have a negative charge and are thus called n-type, and material with an absence of electrons and therefore a relative positive charge, or so-called p-type semiconductor. When exposed to a heat difference in parallel, so each opposing material experiences an opposite heat gradient, the mobile electrons in the n-type material and gaps caused by the lack of electrons, or 'holes', in the p-type material, collectively called charge carriers, move from the high energy hot side to the relatively lower energy cooler side. You might recall from basic physics that energy flows from high to low temperature, a process described by the second law of thermodynamics. The result is an electric current driven by the excess heat of the engine. The ideal material is electrically highly conductive but with very low thermal conductivity, so that the temperature gradient does not quickly degrade and the electron flow persists.

Until recently materials that could be made into semiconductors and had the requisite low thermal conductivity were scarce and expensive; notably bismuth telluride (Bi_2Te_3), lead telluride (PbTe), and silicon germanium (SiGe) were available, with only the latter really viable for high-temperature exhaust gases. Additionally, the typical percentage efficiency of energy recovery was in the low single digits.[21] However, researchers are

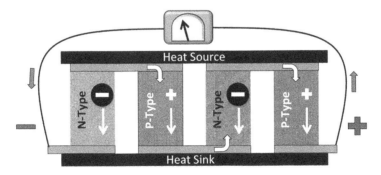

IMAGE 5.24
Thermal generator.

Semiconductors linked in series can be used to generate electric current from waste heat.

---

[21] B. Orr, A. Akbarzadeh, M. Mochizuki and R. Singh, A review of car waste heat recovery systems utilizing thermoelectric generators and heat pipes. *Applied Thermal Engineering* 101, 2016, 490–495; and S. Rajoo, A. Romagnoli, R. Martinez-Botas, A. Pesiridis, C. Copeland and A.M.I. Bin Mamat, Automotive exhaust power and waste heat recovery technologies. In Apostolos Pesiridis (ed) *Automotive Exhaust Emissions and Energy Recovery.* Nova Science Publishers, New York, 2014, 265–282.

trying to change that. The use of a phonon glass electron crystal (PGEC), which combines glass-like thermal conductivity with crystal-like electronic properties, is a possibility.[22] Notably, nano-structured materials that incorporate quantum dots (micro particles of semiconducting materials) or nanowires into more common semiconductor materials can dramatically lower the cost and improve the effectiveness of possible materials. These nanowires can define a latticework at the nanoscale that scatters units of heat energy, or phonons, and so decrease thermal conductivity and increase the power factor while maintaining low cost.[23] Still, making a thermoelectric generator is not easy. The semi-conductor material must be exposed to a high-temperature gradient, causing thermal expansion and contraction and associated mechanical stress. So, all the materials need to be carefully engineered to have very close thermal expansion characteristics. Even then, engineering the heat flow, thermal gradient, and conductivity of the system requires careful balancing.

Of course, this isn't the only way to recapture exhaust energy. Exhaust gases have not only heat energy, but also kinetic energy; you'll recall that's what drives a turbocharger. This power can be harvested not only to provide a turbo boost; it can be used to improve efficiency by recapturing lost energy. It's pretty easy to imagine a small turbine in the exhaust stream of a combustion engine linked to a high-speed electric generator. As the exhaust gases spin the turbine, the energy is recaptured to charge a battery. The idea is called turbo compounding and it's not new; it's been used on aircraft jet engines since the 1940s. The problem is that in an internal combustion engine this would restrict exhaust flow and cost as much or more energy as it saves. The solution is enhanced control that allows for multiple exhaust pathways coupled with a turbocharger, and energy harvesting only during periods when power demand is low or declining. This makes particular sense for large diesel engines with plenty of airflow, prompting Volvo to offer a turbo compounding option for its D13 engine. But such a system could make sense in a variety of applications.[24]

In fact, Porsche has developed a remarkable application of this idea. Rather than a normal wastegate on their turbocharged 2.0-liter powered 919 hybrid race car, they placed a turbo compound unit that allows the conversion of surplus exhaust energy into electricity. Along with a KERS on the front axle to make maximum use of regenerative braking, this system can harvest surplus energy on deceleration, with about 60% coming from the front axle and 40% from the exhaust. The energy is stored in a battery and can be directed to an electric motor for an on-demand momentary boost.[25]

Not all of these ideas are ready for primetime. Some may prove viable for production cars, others may not. But they all reflect the tremendous innovation and possibilities made available by electrification of the drivetrain. As hybrid and electric technologies redefine the automobile, all sorts of new things become possible.

---

[22] Z.G. Chen, G. Han, L. Yang, L. Cheng and J. Zou, Nanostructured thermoelectric materials: Current research and future challenge. *Progress in Natural Science: Materials International* 22(6), 2012, 535–549.

[23] Z.G. Chen, G. Han, L. Yang, L. Cheng and J. Zou, Nanostructured thermoelectric materials: Current research and future challenge. *Progress in Natural Science: Materials International* 22(6), 2012, 535–549.

[24] R. Chiriac, A. Chiru and O. Condrea, New turbo compound systems in automotive industry for internal combustion engine to recover energy. *IOP Conference Series: Materials Science and Engineering*, Volume 252, Conference 1, 2017.

[25] "How the Technology of the 919 Hybrid Works." July 21, 2016. Available at newsroom.porsche.com/en/motorsports/porsche-motorsport-fiawec-919-hybrid-technology-lmp1-race-car-12724.html

6

The Electric Fuel Tank

Recent innovations in electric machines and hybrid drivetrains are undeniably impressive. But, if electrification is ever going to challenge the reign of the combustion drivetrain, we're going to have to have more than just excellent motors. We're going to have to provide energy storage capacity that delivers EV and HEV performance and range that gives their gas-powered counterparts a run for their money.

In fact, recent innovations in battery energy storage technology have revolutionized the automotive industry, even if that revolution has been a bit slower than some would like. Battery capacity has improved at a rate of about 5%–8% a year over the past two decades (Image 6.1).[1] This is impressive, though it may not be as grand as the steady announcements of battery 'breakthroughs' would have you expect. Still, dramatic improvements in battery technology have revolutionized a whole array of industries. For example, the

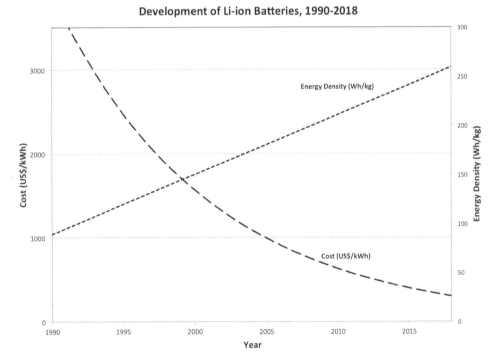

IMAGE 6.1
Trends in lithium-ion battery development.

As the cost of lithium-ion batteries has dropped, their capacity has improved markedly, though the rate of improvement has slowed.

---

[1] Quadrennial Technology Review 2015, Chapter 8: *Advancing Clean Transportation and Vehicle Systems and Technologies: Technology Assessments.* US Department of Energy, Washington, DC, 2015.

ability to store nearly 20 Wh of energy in a tiny, secure, and stable packet that can fit easily in your pocket and power a complex processor for a day is remarkable and a true breakthrough that has made cellular phones possible. Or, consider cordless tools, from drills to circular saws to leaf blowers. Just a generation ago, you would have been lucky to get half an hour of modest use out of a cordless hand tool; now, you can use a tool with far more torque all day without running out of juice. In fact, the amount of energy a battery can store as a proportion of its size and weight has roughly tripled in the past two decades, while the price has dropped dramatically.[2] The result has been a transformation in multiple industries, not least of which is cars.

But all of this comes with challenges. Batteries are sensitive creatures; they degrade over time and with use; they can be damaged by overcharging, excessive discharge, and a host of other factors, some easily controllable others not. And their performance is defined by multiple complex components, each defining an aspect in a series of partial reactions. Changing any one aspect has implications for all the others, so the process is tricky, and progress is incremental. Still, development continues at a notable pace on a wide array of battery chemistries and configurations. Similarly, control technology and monitoring capacity are improving rapidly, as are cooling and management systems.

In short, it's an exciting time for batteries, and so for electric and hybrid automobiles; the future seems to be unfolding before us at an extraordinary pace. However, we aren't there yet. Battery technology still falls short of what is needed to compete with the internal combustion engine. Charge time will need to get shorter and range will need to get longer, before the petroleum-powered car really has something to worry about. Still, make no mistake, it's coming.

What's a Battery?

People tend to imagine a battery as a sort of container, like a gas tank; you can put electricity in and draw electricity out. While this is not an entirely incorrect analogy, it is not quite right. Batteries don't really store electric charge, they generate it. Unlike a capacitor, a battery generates electricity by chemically converting its substance to electric current; and when an electric charge is applied, the chemical reaction can be reversed, allowing the battery to store electricity as chemical energy.

So, the fundamental operation of a battery is a chemical reaction, or more precisely, a paired coupling of two half reactions. We call this a **redox** reaction, because one half of the reaction is **reduction**, defined as a reaction that causes a gain of electrons, and the other half is **oxidation**, a reaction that causes a loss of electrons. The two halves of the reaction take place separately, but as a paired set. You might already see where this is going. If we have the loss of electrons in one place and the gain of electrons in another, then we can connect these two together with a conductor and have the electrons travel from one half of the reaction to the other. Electron charge traveling through a conductor is electricity, and that is our goal.

In its simplest form, a redox reaction looks like Image 6.2.

---

[2] P. Slowik, N. Pavlenko and N. Lutsey, Assessment of next-generation electric vehicle technologies. White Paper, International Council on Clean Transportation, Washington, DC, 2016.

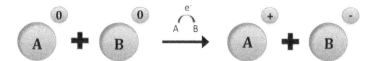

IMAGE 6.2
Redox reaction.

No text, image integrated into main text

The two starting components, for simplicity, let's call them A and B, react causing a change of charge. Neither of them is charged initially, so the upper notation is a zero. But, when we allow them to react together, component A loses an electron. Since electrons are negatively charged, that means A ends up with a positive charge. Component B, on the other hand, gains an electron, making it more negatively charged. So far, even though electron charge is moving, it would be really hard to harvest that as electricity. But, if we pull this reaction apart, physically separating A and B, we can channel the electron movement from one to the other. You might think of it like Image 6.3.

When an electron is pulled away it leaves behind an atom or molecule with a relative positive charge. Charged particles like this are called ions. And since this one is positively charged, it's called a **cation**. (A negatively charged particle is called an **anion**.) So, our reaction, or two half reactions, causes a flow of electrons in one direction and cations in the other direction.

Clearly, to understand the mechanics of how this actually takes shape in the form of a battery, we need to define the parts a bit more. A battery is what is called an **electrochemical cell**. In this cell, there are two electrodes; these are the two components for the reaction below that we called A and B. These electrodes are typically plates of metals or metal composites that are prone to react in the way we want. One of them is prone to oxidation. Think of iron, for example, which oxidizes, or rusts, pretty easily in the right environment. While we don't use solid iron electrodes in batteries, we do use metals like zinc, aluminum, and magnesium, which all oxidize even more easily than iron. The other electrode is made of a material prone to reduction, say gold or copper, which doesn't oxidize easily at all. (So if your 'gold' ring is rusting on our finger, you may have been cheated.) We place these electrodes in a solution, a liquid or maybe a gel, called an electrolyte that can conduct the flow of ions from one electrode to the other.

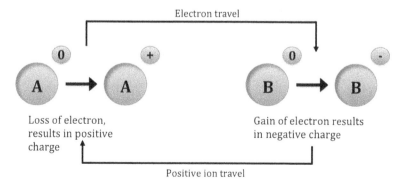

IMAGE 6.3
Reaction separated.

When we put all this together we have a readily oxidized electrode, an electrode prone to reduction, and an electrolytic solution that allows for the movement of ions. When we provide a conductive path for electrons (by connecting the two posts of our battery to a circuit), electrons start to flow. That, in a nutshell, is a very simple battery.

Electrode materials are defined by their ease of oxidation. As you can see from the table below, lithium oxidizes very easily (Image 6.4). Gold, platinum, or mercury does not. Looking at this table, you'll recognize metals that are commonly used in batteries you've heard of, like lithium, nickel, cadmium, lead, copper, zinc. All batteries utilize two different materials, each with a different ease of oxidation. And we tend to name the batteries by the materials we use.

Let's consider a more precise example of a battery. If we used zinc as our oxidizing electrode, for example, we could submerge it in a copper sulfate solution as the electrolyte. Why this particular solution makes sense will become clear in a minute. The zinc would start to erode or oxidize in the electrolyte. The basic result of this oxidizing reaction is that electrons are left behind, so this can be thought of as zinc dissolving and releasing electrons. We can write it like this:

$$Zn \rightarrow Zn^{2+} + 2e^-$$

It's worth noting that this reaction is slightly **exothermic**, meaning it generates heat, and so the electrolyte will get a bit warm. If we add a conductive path for these electrons to another electrode prone to reduction, copper, for example, the copper in the electrolyte will combine with the electrons we send, and we could get a reaction that looks like this:

$$Cu^{2+} + 2e^- \rightarrow Cu$$

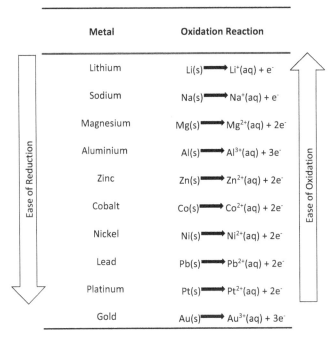

IMAGE 6.4
Ease of oxidation or reduction.

The copper in the copper sulfate solution breaks away and bonds with the electrons from the zinc reaction. The copper then deposits on the copper electrode. So, while the zinc electrode diminishes, as zinc dissolves in the electrolyte, the copper electrode grows with the deposits of copper particles from the electrolyte. If we disconnect the conductive path the reaction stops. That's good. It means when we stop using the battery, it won't lose energy.

We need the resulting charged particles, or ions, to connect to allow the reaction to continue. So, when the copper is deposited on the copper cathode, it leaves behind sulfate, SO_4, which now has a couple of extra electrons, so we write it as SO_4^{2-}. Because this electrode is drawing cations, we call this electrode the **cathode**. And, when the zinc anode oxidizes in the electrolyte it attracts negatively charged anions, so we call this electrode the **anode**. If these two charged particles could not travel through the electrolyte, the reaction couldn't continue very long. The positively charged zinc ions resulting from the first reaction would build up around the zinc electrode and the development of positively charged ions from the electrolyte around the copper electrode will soon lead to a growing negative charge around that electrode. Both diminish the cell's ability to keep producing a reaction. That's why what is happening should be understood as two half reaction, the connection is needed to allow each half of the total reaction to continue. We need an **electrolyte** to allow ions to travel between electrodes. Electrolytes need to be an acid, a base, or a salt, since any of these can conduct charged particles. This sometimes is reflected in the name of the battery as well, as a 'lead-acid' battery that has an acidic electrolyte, or an 'alkaline' battery that, you guessed it, has an alkaline electrolyte.

While the electrolyte needs to be ion-conducting, it is also important that the electrolyte be electrically insulating. If not, the electrons would flow through the electrolyte rather than through our circuit, shorting the circuit. A **separator** is used to ensure this. The separator is typically a porous polymer film or mat that is saturated with electrolytic fluid, so ions can pass through but it stops electrons. If this weren't so, electrons could pass through the electrolyte, allowing the redox reaction to take place without producing usable electric current. This is called **self-discharge**, and every battery has varying degrees of this.

So, let's summarize: We put a zinc anode and copper cathode in a sulfate solution. The oxidation reaction at the anode causes a release of positively charged zinc ions into the solution. These cations flow to the negative charge resulting from the reaction at the cathode. Similarly, the anions at the cathode can now travel to the relative positive charge at the anode. We enable a complete redox reaction when a conductor is provided to allow electrons to travel from the anode to the cathode to complete the reaction, and this flow of electrons is our targeted outcome. We could write the combined reaction like this:

$$Zn(Solid) + CuSO_4(liquid) \rightarrow ZnSO_4(liquid) + Cu(Solid) deposited$$

But, again, it might be better thought of as two half reactions. The reaction is reversible. If we can move the electrons in the opposite direction, by applying an electric charge, we can undo the chemical reaction at each electrode. This effectively allows us to store electric energy as chemical energy in a battery, and then harvest that energy by allowing the redox reaction later.

The basics of this simple battery are really the core elements of any battery. A battery has an electrode made of a material that tends toward oxidation, or the loss of electrons, and another electrode made of a material that tends toward reduction, or gaining electrons. They are immersed in an electrolyte that can conduct ions, and this sets up the possibility for reduction at one electrode and oxidation at the other. When we connect

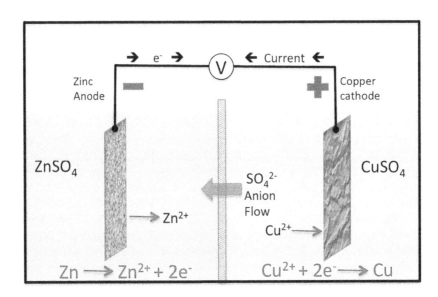

IMAGE 6.5
Electrochemical cell.

An oxidation reaction at the zinc anode releases a positive zinc ion while a reduction reaction at the copper cathode releases an electron. That electron travels through a conductor as it is drawn to the anode. Of course, the flow of these electrons through the conductor is exactly what we're after, electric current.

the two electrodes with a conductor, we allow the reaction to take place by allowing the electrons generated by oxidation to move toward the reduction reaction. Because electrons are (somewhat arbitrarily) said to be negatively charged, we say that the electric current (which is positive) is traveling the opposite direction of the electrons. So, the zinc electrode, in our case, is the negative side and the copper provides the positive side (Image 6.5).

But this isn't really the end of the story. The copper sulfate electrolyte needs to be aqueous, meaning it's mostly water, and that water doesn't just sit idly by. When a voltage difference is created by the battery, it defines a voltage difference across the electrolyte as well and this can undo water molecules. The hydrogen and the oxygen in water are held together with an electromagnetic attraction, but the charge difference at the cathode and anode can create an electromagnetic pull that exceeds this force. The positively charged hydrogen ions will be drawn to the cathode and the negatively charged oxygen ions will be drawn to the anode, and the water molecules can be separated into their hydrogen and oxygen components, a process called **electrolysis**.

The result of this electrolysis is a release of gas from the battery electrolyte. When the hydrogen cation reaches the copper cathode, it will combine with the available electrons and become hydrogen. Similarly, at the anode, oxygen is produced. The possible buildup of hydrogen is particularly problematic, not only because it interferes with the battery's operation but also because hydrogen is potentially explosive. So, we have to have some safe means of releasing or containing the hydrogen gas that is produced. This is why it used to be necessary to add water to old lead-acid batteries. The batteries would vent hydrogen, and over time this would diminish the electrolyte in the battery, leaving a more concentrated solution. So, car owners would add a bit of distilled water from time to time. Newer lead-acid batteries are sealed and vented and no longer need added water, but the reaction still takes place.

The rest of battery technology is really all based on this simple theme. Batteries are simply collections of these electrochemical cells. Adding more cells can increase battery capacity. Variations on anodes, cathodes, and electrolytes can offer different advantages and challenges. And as battery capacity grows, so do the generation of heat and possibly hydrogen.

Battery Performance

All electrochemical cells have a voltage, and when connected together in series their sum defines the voltage of the battery. However, the open circuit voltage alone is not definitive; as a load is placed on the battery and current is drawn out, the voltage will drop. In fact, a key characteristic of a battery is how much voltage drops when producing a current, and this is a function of the battery's **internal resistance**. The lower this resistance, the better the battery is able to maintain its voltage. Clearly, a battery that can only maintain a high voltage when it's not connected to a load isn't much good as a power source.

A typical 12-volt automotive battery, for examples, is made up of six cells. Each of these cells is connected together in series, so the positive electrode of one is connected to the negative electrode of another. So, while each cell typically has a nominal voltage of 2.2 volts, the voltage of the battery overall is the result of all these cells connected together (6 × 2.2) for a total of 13.2 volts. We call them 12-volt batteries because the battery's voltage will drop closer to 12 volts when a load is placed on it.

Typically batteries are rated by their capacity to store and supply energy, expressed as how many amperes of current they can provide for how many hours, called **amphour capacity** and often represented with the letter C. If it can produce 20 amps for 2 h, we might say it has a 40 amphour capacity or $C = 40$. But in the real world, this gets a bit fiddly. Batteries generally have an easier time producing a lower current for a longer period of time than a high current for a short period of time. So, a battery with 20 amphour capacity may be able to produce 2 amps of current for 10 h, but it would probably not be capable of producing 20 amps of current for a full hour, even though both are equal to 20 amphours. So, we need to think about how much energy a battery can store and deliver, or its **energy capacity**; and how quickly it can provide that energy, or its **power capacity**.

Because the space for batteries is limited in an automobile, and the weight of the vehicle will define the efficiency, performance, and range of a hybrid or electric vehicle, we're even more interested in the amount of energy a battery can store as a function of its weight, called **specific energy**, or, to a lesser degree as a function of its size or volume, called the **energy density**. Weight is critical, as the battery pack is frequently the heaviest single component of an EV. However, we'd also like to know the speed that a battery can provide that energy, that is to say the power of the battery. And once again, we talk about that as **specific power**, indicating the maximum watts of power a battery can provide as a unit of its weight, or sometimes **power density** as a unit of volume (these terms are often used improperly or taken as synonymous; they are not). While we generally never operate a battery at its maximum power capacity, this measure allows us to compare differing batteries for their ability to provide burst of current when needed.

The distinction between power and energy is critical here, as the two meet very different needs in automotive applications. Start–stop technology, for example, requires an ability to deliver and absorb energy quickly, and so benefits from a higher specific power. Specific energy, on the other hand, would be more desirable in an electric vehicle, where the ability to store energy per unit weight will fundamentally define the range of the car. Very often, the two are inversely related. The better a battery is at storing energy typically the worse it is at providing instantaneous power for immediate burst of current. As we will see, this trade-off is fundamental to the path of battery innovation.

Another key characteristic of our battery is its ability to hold onto a charge. A lot of factors help define the **self-discharge rate**, including temperature, the level of charge, and type of battery. And the importance of this variable depends on the battery's use. For example, for a micro-hybrid that aims at using energy from regenerative braking to augment acceleration usually only a few moments after receiving the charge, a higher self-discharge rate can be tolerated. But, for an EV that may be parked for a day, a week, or a month between uses, a high self-discharge rate could be a big problem.

Similarly, in selecting the proper battery for a vehicle application we need to know the battery's ability to return the energy provided to it, sometimes called **round trip efficiency** (since the electrons make a round trip from anode to cathode and back). No battery does this perfectly. And this factor will vary depending on the battery's **state of health**. To make it trickier, a battery's ability to accept a charge, or **charge efficiency**, varies throughout the discharge cycle and drops off dramatically in the last 20% or so of a battery's charge. Putting all this together we sometimes talk about a battery's **state of function** reflecting the battery's ability to meet the demands placed on it.

Of course, all of this changes as the battery ages. From their very first use, batteries degrade. Active materials dissolve into the electrolyte. The electrolyte oxidizes. Expansion and contraction of active materials during charge and discharge can accelerate this. And in general, both mechanical and chemical changes degrade battery performance. Both time and usage contribute to this, so we often talk about **calendar life** and **cycle life**, indicating the number of years or charge cycles respectively a battery can manage before its capacity is seriously diminished. Typically degradation of no more than 20% in 10 years is a benchmark. Since excessive **depth of discharge** (DoD) can significantly impact this life, batteries are sometimes oversized to avoid over discharge and excessive heating and enabling a longer life.

Battery Management

Cells can be connected together in various forms and configurations to define batteries and modules. They can be prismatic cells in a rectangular hard case. They can be stacked with a gel and placed in a pouch. Or we can make an electrode, electrolyte and separator sandwich and roll it up into a so-called jellyroll to form a cylindrical cell. At the moment, Tesla is the only car company that uses these, assembling thousands of cylindrical cells, defined by a rolled anode, separator, cathode, and electrolyte all rolled up just a bit larger than a AA battery, to define its battery packs (Image 6.6). Whatever shape is used, batteries are combined together into a module that provides physical protection, and allows for the incorporation of a **battery management system** (BMS) and **thermal management system** (TMS) to service the **battery pack** (Image 6.7). This is not as easy as it may appear, as the pack needs to provide protection from road debris, stability in variable driving conditions, allowances for the expansion and contraction of batteries due to temperature and state of charge (SoC), venting of hazardous gas, and safety shutdown provisions.

IMAGE 6.6
Tesla battery modules.

The lithium-ion cells used in the Tesla Model-S module are each just a bit larger than an AA battery. A total of 7,104 of them are aggregated into 16 modules of 444 cells to define the total pack. The entire pack weighs 1,200 pounds.

Image: Jason Hughes

IMAGE 6.7
Bolt battery pack.

The Chevy Bolt uses LG Chem prismatic pouch cells and is made up of ten modules, eight with 30 cells and two with 24 cells. An active thermal management system uses liquid cooling flowing through a base plate under the cells and connected fins extending between the cells. All of this is neatly organized into a battery pack designed to provide an optimized fit and secure structure within the vehicle.

High-performance batteries require high-performance control and management. As we'll soon see, lithium-ion batteries, in particular, are vulnerable to overheating, overcharging, excessive discharge and overcurrent during discharge, all of which can lead to battery deterioration or failure. And adverse operating conditions, such as over or under temperature and unbalanced charge levels, can significantly reduce range. The BMS integrates a range of battery performance parameters and tracks, assesses and controls battery pack components with four key tasks: first, evaluating the charge, voltage, temperature, and health of each cell and the battery module overall; second, identifying and diagnosing possible battery faults; third, managing current flow among the cells to ensure balance; and lastly monitoring battery temperatures and control of the cooling system.

To accomplish this, battery performance data must be precisely measured and aggregated at multiple levels, from individual cell supervision circuit (CSC) that monitor cell voltage, temperature and other parameters, to wider battery module balancing, assessment, and control, to the overall battery pack management and communication with external embedded systems. So, the BMS must be able to identify hot or cold spots and charge imbalances in individual cells, control battery operation overall, such as managing the charge and discharge current, and enables the production of needed information such as SoC or remaining range. In addition, the BMS must monitor for potentially hazardous conditions and activate disconnect relays or shutdown when needed.

This is not as easy as you might imagine. For example, something as deceptively simple as a battery's SoC can in fact be quite complicated. This is, of course, a critical parameter as it defines the expected remaining range of an electric vehicle and therefore impacts basic operation and safety. You might expect that a battery's voltage would be a good way to measure the level of charge, but this is not always so. The voltage of a lead-acid battery changes in proportion to the level of charge, so a reasonable assessment of SOC can be made based on voltage in this case. However, lithium-ion batteries have a very flat voltage curve on discharge (Image 6.8). So, looking at the voltage alone would give us a very poor sense of the SoC. In fact, a great advantage of lithium batteries is that they maintain their voltage and performance over a much greater range of their discharge. However, this makes SOC more difficult to measure. Instead close and precise monitoring of battery parameters is used to carefully track the history of charging, discharging, and battery operation in a process

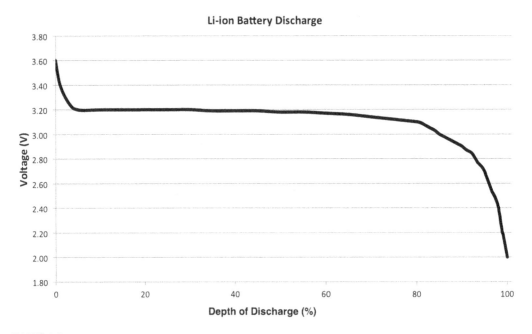

IMAGE 6.8
Lithium battery discharge.

The voltage upon discharge of a lithium-ion battery is relatively flat, making it difficult to estimate state of charge.

called **coulomb counting**. This method must be informed by an ongoing assessment of the battery's state of health, based on current, voltage and internal resistance data, adjusted for the battery's self-discharge, the effects of temperature, and other parameters. Fuzzy logic and quadratic estimation are used to try to compensate for variations in the data and derive the best possible model of battery performance, no simple task.

Cell Balancing

The fact that each cell in a battery can be slightly different due to manufacturing variations, heating and cooling, and the rate and nature of degradation, makes their management and charging very complicated. Every cell will perform slightly differently, with different internal resistance and variations in self-discharge rate. So, if a battery is discharged, say, by 40%, the SOC of some cells may be at 60%, while others may be a bit lower at 55%. During the subsequent recharging, when the first cell has achieved a full SOC, the diminished cell may only be at 95%. With another discharge, the same variation will occur, and with each recharge–discharge cycle, the difference between the two cells will grow greater (Image 6.9). As the cycle continues, the diminished cell will go flat, and effectively act only as a form of resistance in the battery's series circuit, significantly compromising its performance and likely leading to battery failure.

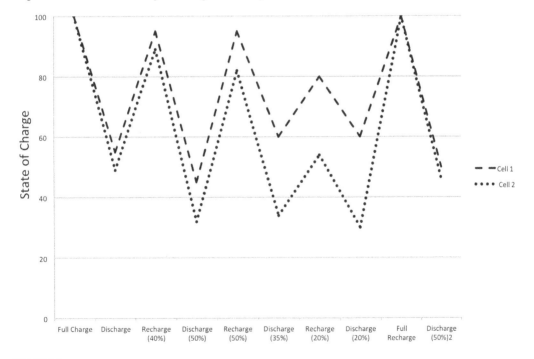

IMAGE 6.9
Potential cell imbalances with repeated partial charging.

As a battery discharges, each cell will exhibit a slightly different rate of discharge, resulting in an imbalance. With repeated discharges and partial recharges, this imbalance will accumulate, as this conceptual graph indicates. This can damage the battery. A full charge can bring all the cells back into balance.

So, proper recharging of a battery requires that we ensure that each cell receives a full charge in order to achieve **charge equalization**. This means that we need to continue to apply a charging current until all the cells are fully charged, which means some cells must continue to receive a current well after they have reached full charge capacity. But cells don't deal well with being overcharged at high current. Typically, a cell at 100% SoC will only be able to manage a modest current, say C/10, that is to say one-tenth its amphour capacity. This is why bringing a battery to 90% SOC can take much less time than bringing it to 100%. It is also the reason why it is vital that battery modules receive a full and complete charge frequently.

Hybrid drivetrains can make this more difficult. In order to maintain the capacity to absorb energy from regenerative braking, a hybrid vehicle typically does not operate the battery module at full charge. So, the BMS must periodically charge the batteries to a complete charge to allow for equalization. Regenerative braking can be tricky too. The kinetic energy of a vehicle moving at highway speeds is formidable. Strong braking might require a battery to absorb a 100 kW of power in a very short time. Mechanical brakes are used to aid in braking and ensure a safe level of battery operation. However, the more the mechanical brakes work, the more energy is wasted; so significant power density is needed to support regenerative braking.

Cooling Systems

Temperature is one of the major factors shaping battery performance. Heat influences the chemical reaction, and that reaction is what defines power, energy, chargeability, and round trip efficiency. In fact, a temperature rise as little as 20°F can more than double the speed of the reactions.[3] At best, excess heat can cause aging, but it's very likely to cause performance deterioration and can lead to catastrophic failure if not addressed. As a matter of fact, an active cooling system can extend battery life beyond an uncooled system by as much as a factor of six.[4]

Proper thermal management would be easier if cars only needed to operate in one climate, but they don't. A car must be able to accommodate a wide range of environmental conditions, from Alaska to South Texas, Norway to Greece. This is a challenge in hot climates in particular, as a BMS must ensure against chronic elevated temperatures. A battery system that might offer 10 years of service with 20% degradation in Maine may exhibit twice that power loss if it lived in Arizona for example. If battery temperature can't be managed, battery size needs to be increased to ensure performance. And cooling systems are often less expensive than a battery upgrade. As a result, all contemporary electric and full hybrid cars maintain some form of cooling system, either a simple air-cooled system, a more robust liquid-cooled system, or more recently developed refrigerant based systems (Image 6.10). Because batteries can be degraded by repeated heating while charging, a standby cooling system that can provide cooling while the vehicle is plugged in has advantages as well and is absolutely necessary for high speed three-phase charging.

---

[3] C. Huber and R. Kuhn, Thermal management of batteries for electric vehicles. In Bruno Scrosati (ed) *Advances in Battery Technologies for Electric Vehicles.* Woodhead Publishing, Cambridge, 2015, 327–358.

[4] T. Yuksel, S. Litster, V. Viswanathan and J.J. Michalek, Plug-in hybrid electric vehicle LiFePO4 battery life implications of thermal management, driving conditions, and regional climate. *Journal of Power Sources* 338 (January), 2017, 49–64.

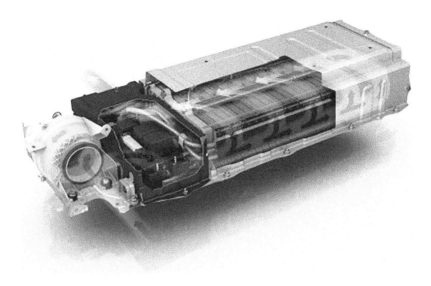

IMAGE 6.10
Synergy drive battery cooling.

The Toyota Prius supplies conditioned air from the cabin to manage battery cooling.

Image: Toyota Motor Corporation

Excess heat isn't the only challenge; low temperatures can damage batteries as well. So, heating is required to enable high-speed charging and aggressive discharge. Some management systems provide fluid heating during cold operation, such as those used by Chevy or Audi (Image 6.11). Some offer both heating and cooling through a refrigeration system, such as Tesla. Some offer resistive heating while plugged in for battery charging, such as the BMW i3, Chevy Bolt, or Nissan Leaf.

Air-cooling has the advantage of being simple and light. A fan and some well-defined channels can shape airflow across the cells. Unconditioned outside air can be used; or the system might use conditioned air, either by redirecting air from the cabin or maintaining a separate evaporator unit. Using cabin air can make sense, as people and batteries tend to like the same temperature, but it requires some precautions to make sure vented gasses from the batteries do not enter the cabin. The advantage of any of these is simplicity. Servicing of the battery pack is less complicated, maintenance is minimal, and energy demand is very low. However, this system has some clear limitations. The amount of heat energy a given unit of air can absorb, called **specific heat capacity**, is low. So, a large volume of air is required to ensure adequate cooling, and this means larger ducts and fans. Such a system, particularly if it is connected with the cabin's conditioned air, is likely to be noisy. In general, an air-based system offers limited cooling capacity, an inadequate ability to cope with wide-ranging ambient temperatures, bulky ducts, and high noise. Nevertheless, vehicles with more modest demands on a smaller battery pack may be well served by an ambient-air cooled system. Nissan utilizes such a system with its Leaf, though the battery capacity is modest and performance can deteriorate in hot climates. The Kia, on the other hand, chose to oversize the battery pack on their Soul EV to reduce battery heat generation. Kia utilizes cabin air for their cooling system; as mentioned, passengers and batteries tend to prefer the same temperatures.

Using liquid for cooling offers some inherent advantages. The specific heat capacity of liquid coolant is 3–5 times greater than that of air and offers better heat conduction. Fluid

Audi Q7 e-tron 3.0 TDI quattro

Hochvolt Batterie
High-voltage battery
03/15

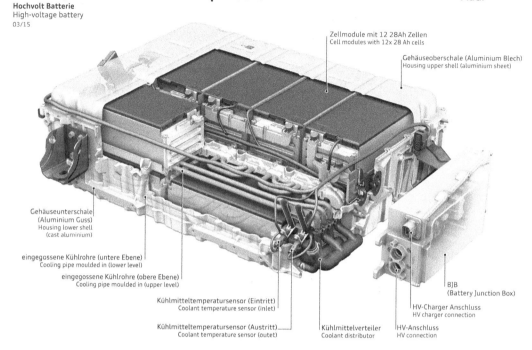

Zellmodule mit 12 28Ah Zellen
Cell modules with 12x 28 Ah cells

Gehäuseoberschale (Aluminium Blech)
Housing upper shell (aluminium sheet)

Gehäuseunterschale
(Aluminium Guss)
Housing lower shell
(cast aluminium)

eingegossene Kühlrohre (untere Ebene)
Cooling pipe moulded in (lower level)

eingegossene Kühlrohre (obere Ebene)
Cooling pipe moulded in (upper level)

Kühlmitteltemperatursensor (Eintritt)
Coolant temperature sensor (inlet)

Kühlmitteltemperatursensor (Austritt)
Coolant temperature sensor (outet)

Kühlmittelverteiler
Coolant distributor

BJB
(Battery Junction Box)

HV-Charger Anschluss
HV charger connection

HV-Anschluss
HV connection

IMAGE 6.11
Audi battery module.

Audi utilizes a liquid cooling system with integrated cooling plates to regulate battery module temperature. The power electronics and charger are integrated into the cooling loop to provide cooling when needed.

Image: Audi

cooling can use ambient cooling by simply running the coolant through a radiator. Or it can use an active system that relies on an air conditioning heat exchanger. These systems are not mutually exclusive. The Chevy Volt plug-in hybrid, for example, can route battery coolant to a passive radiator, cabin air, a chiller, or even a heat exchanger with the engine coolant if heating is needed. Other systems may use an electric heater to provide heat when batteries are cold. Of course, the added complexity and componentry makes fluid cooling heavier than a simple air system, but the added control and cooling capacity is thought to be worth it for most manufactures, and absolutely indispensible for larger, high-performing battery packs such as Tesla's.

Of course, the effectiveness of a cooling system also depends on good thermal conduction with the cells. One approach could be to submerge the battery module as a whole in a dielectric fluid. While this approach is not utilized by any major manufacturer, it is regularly used with other electronic cooling systems, and sometimes discussed as an alternative for future EVs. Though a glycol-based system would not allow submerging of the battery components, the coolant could be channeled through cooling passages that offer good thermal contact with cells. For example, Tesla uses a serpentine ribbon-shaped cooling tube that snakes through the cell pack, offering simplicity and few points for leakage. However, these coolant lines take space, making the battery pack larger. So, BMW and

GM use cooling plates under the battery, with thin fins extending between the cells for improved thermal conduction while allowing cells to be packed tighter.

Even greater control and capacity can be achieved by integrating the cooling system directly into a refrigeration circuit. BMW's direct expansion, or DX, cooling system offers an impressive example. Refrigerant tubes inside the cooling plates allow for marked improvement on the heat transfer characteristics of a glycol system. With no need for a separate coolant loop and heat exchangers, such systems can also be more compact. This may also allow for more precise and responsive control, and a more even distribution of cooling across the pack. The system is more complex, of course; and a secondary source for battery heating must be added, but BMW clearly thinks the added capacity and control is worth it. Alternatively, with proper design and refrigerant selection, the system could be reversed to operate as a heat pump and provide heating and cooling as needed.

Any battery TMS must be able to account for multiple variables. The load on the battery matters greatly, as the principle predictor of heat generation is current. Ambient conditions matter, of course, most notably outside temperature. The state of health and basic characteristics of the battery matter as well, since differing chemistries have different heat patterns and these can change over the life of the battery. One of the less obvious factors that can impact battery thermal performance is the size, or more precisely the mass, of the battery pack. A heavier pack has greater **thermal inertia** and thus temperature fluctuations are dampened. A lighter pack, with less thermal mass, is therefore at greater risk of overheating. This is a bit of a dilemma since a main concern in battery design is to make the pack as light as possible, yet this can lead to greater temperature fluctuations.

One possible solution that allows for moderation of temperature fluctuations without excessive added weight would be the incorporation of **phase changing materials**, or PCMs. When materials change phases as a result of cooling or warming, that is to say when a substance freezes or melts, it absorbs a high amount of heat energy without changing temperature, this is called **latent heat**. So, if we connected battery packs to a material that 'freezes' at the targeted operating temperature of the battery, the absorption of thermal energy would help dampen any thermal fluctuation and help keep the pack within a targeted temperature range (Image 6.12). Typically, a petrochemical fluid is used, such as a specifically engineered paraffin wax. Including graphene in the PCM can more than double heat conduction while maintaining heat storage ability.[5] To be clear, PCMs do not dissipate heat, so they can't replace cooling systems; but they do protect batteries by buffering fluctuations in temperatures and reducing the load on the cooling system.

Battery Chemistry

The dominant battery in the automotive field has been the lead-acid battery, so-called because both its electrodes are made of lead, with a spongy lead used for the anode and a lead-oxide used for the cathode. The electrolyte is sulfuric acid, completing the name.

---

[5] P. Goli, S. Legedza, A. Dhar, R. Salgado, J. Renteria and A.A. Balandin, *Graphene-Enhanced Hybrid Phase Change Materials for Thermal Management of Li-Ion Batteries*. Nano-Device Laboratory, UC, Riverside, CA, 2013.

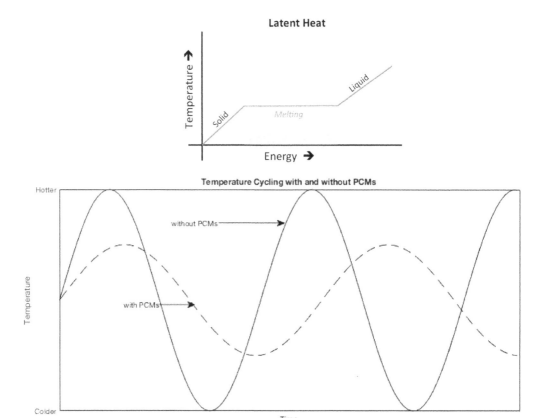

IMAGE 6.12
Phase changing materials for thermal management.

As the graph on the left indicates, as a material changes phase from solid to liquid, or melts, it absorbs significant heat energy without increasing temperature, allowing increased absorption of energy at this targeted temperature. On the right, the effect of the PCM is evident, dampening the temperature fluctuations near the PCM melting point.

The overall electrochemical redox reaction combines the lead and lead oxide through the sulfuric acid, creating lead sulfate at both the anode and the cathode. It looks like this:

$$\underset{\text{Spongy Lead Negative Plate}}{\text{Pb}} + \underset{\text{Lead Oxide Positive Plate}}{\text{PbO}_2} + \underset{\text{Sulfuric Acid Electrolyte}}{2\text{H}_2\text{SO}_4} \leftrightarrow \underset{\text{Lead Sulfate}}{2\text{PbSO}_4} + \underset{\text{Water}}{2\text{H}_2\text{O}}$$

Discharge goes from left to right in the equation above; and when the battery is being charged, the reaction reverses, moving from right to left.

Because the battery's internal resistance is remarkably low it's able to produce significant current without suffering a major drop in voltage. The resulting ability to provide a momentary boost of current is made to order for starting an internal combustion engine. This, and low cost, have allowed lead-acid batteries to remain the dominant choice for SLI (starter, lights, ignition) applications.

But, the lead-acid battery has several weaknesses that make it a problematic choice for advanced vehicle power applications. The lead in these batteries decomposes over time, resulting in a self-discharge of the battery. This degradation speeds up with the heat caused by higher discharge rates. And because this discharge does not occur evenly across all the battery's cells, the resulting imbalance in the cells makes full recharging necessary. This reaction also breaks down the electrolyte, releasing hydrogen and oxygen gas. As previously mentioned, in the past you'd add water to a battery periodically to make up for this loss. Modern sealed batteries trap this hydrogen gas in the case and don't require added water; but the reaction still occurs, so there is a resulting degradation. The recent development of absorbed glass mat (AGM) or gel lead-acid batteries, that suspend the electrolyte in a physical matrix to provide a more robust and stable battery, offer lower internal resistance so better cold weather performance and deeper discharge capabilities. Nevertheless, the battery's overall vulnerabilities, combined with an unfavorable specific energy (they're heavy), makes conventional lead-acid a poor choice for full hybrid or electric cars.

In addition, lead-acid batteries are vulnerable to **sulfation**, defined by small sulfate crystals forming on the anode. These are usually not a problem for an SLI battery, as the sulfate buildup is reversed with full recharging. However, if a car is driven frequently on short trips, with a heavy accessories load, and full recharge is not achieved, these crystals can form into a more durable deposit that covers the electrode and significantly reduces cycle life.[6] This limits the basic lead-acid battery's utility for start–stop applications.

However, recent work has demonstrated that adding carbon to the anode could address partial-charge sulfation. The basic problem is that the anode is not sufficiently cleared. Adding carbon enables the battery to act as a sort of capacitor, absorbing electrostatic charge without chemically converting it. This Advanced Lead Carbon (ALC) battery may offer a very promising alternative to the 12-volt lead-acid SLI battery now in use. For micro-hybrid and start–stop systems, this battery is low cost and can quickly absorb and supply current bursts making it the leading option for this application. It can operate at various states of charge consistently without sulfation and it has a wide temperature tolerance and so does not need the expensive thermal management that more sensitive lithium-based batteries need. This battery could also be made to produce 48 volts for the high voltage systems that are likely to be adopted in the future. Some challenges still exist, like addressing water loss. But the increasing interest in dual battery systems where a 48-volt Li-ion system functions in conjunction for a lead-based battery supporting micro-hybrid functions, as well as the common use of 12-volt lead-acid battery for supplemental accessories and redundancy in electric vehicles, means that the use of lead-acid batteries is nowhere near over. And the combination of fast charging capacity and ability to operate at partial SOC is likely to make the ALC battery a useful option for years to come.

Nickel-Based Batteries

Nickel-based batteries once dominated the battery industry. The nickel-cadmium, or NiCd, battery was the go-to power source for countless portable devices just a couple of decades ago. However, limited **cycle life**, the environmental concerns of disposing of heavy metal cadmium and something called the **memory effect**, made this a poor choice for any automotive application. The memory effect describes the battery's tendency to lose

---

[6] J. Büngelera, E. Cattaneoa, B. Riegela and D.U. Sauerb, Advantages in energy efficiency of flooded lead-acid batteries when using partial state of charge operation. *Journal of Power Sources* 375 (January), 2018, 53–58.

its maximum energy capacity if it is repeatedly only partially discharged before recharging. The battery seems to 'remember' past charge patterns and limit its capacity to this charge level. Happily, nickel-metal hydride (NiMH) batteries have displaced this early front-runner, offering a much higher specific energy, lower environmental harm, and overall greater robustness and utility, though a form of the memory effect is still present but much diminished.

The key to the NiMH battery is an anode composed of rare earth elements bonded to a metal. When combined, these elements form a stable electrode that can easily absorb and release hydrogen compounds. Rare earth elements share a very strong affinity to hydrogen. So, when we combine rare earth elements such as lanthanum, cerium, praseodymium, and neodymium with common metals like nickel, cobalt, manganese or aluminum they form an intermetallic alloy that can store hydrogen at a density that is even higher than pure liquid hydrogen. More importantly, they absorb and release the hydrogen easily because the hydrogen atom occupies the space between atoms, or interstice, rather than bonding firmly with the atom. The movement of this hydrogen is the key agent of the battery's operation. Placed in an electrolyte of potassium hydroxide (KOH), at the anode, hydroxide ions (OH⁻) react with the hydrogen in the metal hydride (MH) to release electrons (Image 6.13). It looks like this:

$$MH + OH^- \rightarrow M + H_2O + e^-$$

The positive electrode is nickel hydroxide, NiO(OH), which has a history of use in batteries dating back to Thomas Edison. The electrons allow a reaction with the water in the electrolyte to generate the same hydroxide ion that was absorbed in the anode reaction:

$$NiO(OH) + H_2O + e^- \rightarrow Ni(OH)_2 + OH^-$$

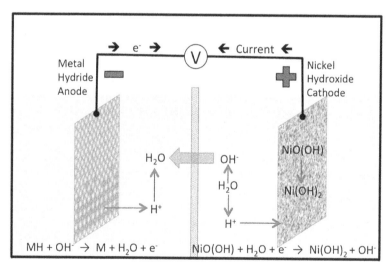

IMAGE 6.13
Nickel-metal hydride electrochemical cell.

A NiMH reaction is not as complicated as may seem. The anode is defined by hydrogen held in a metal alloy. This hydrogen reacts in the electrolyte to form water. This releases an electron that is drawn to the cathode where it forms nickel hydroxide.

If we put the two halves together, the overall reaction when discharging the battery looks like this:

$$NiO(OH) + MH \rightarrow Ni(OH)_2 + M$$

If that seems complicated, it's not. It comes down to this: The hydrogen held by the metal alloy of the anode is released into the electrolyte where it reacts with hydroxide ions to form water $(H^+ + OH^- \rightarrow H_2O + e^-)$. This also releases an electron that travels to the cathode through our circuit (remember, that's the point of a battery). At the cathode, this electron combines with hydrogen ions in the electrolyte to form nickel hydroxide. Not surprisingly, the reaction is exothermic, so it generates some heat; and, of course, the reaction reverses when the battery is charging.

If you are battery-savvy, you might wonder why any of this is relevant. Nickel-based batteries have been largely superseded by lithium batteries in recent years. Once popular in portable devices, NiMH batteries are now pretty much out of date, and the electronic devices out there that still use NiMH are few and far between. However, cars use batteries very differently than a phone or laptop. Portable devices need batteries that can store significant energy and provide a large portion of that full charge predictably and completely. But hybrid cars do not rely on a full charge and discharge cycle from their batteries, rather they require an ongoing rapid fluctuation of partial charges and recharges. Every few minutes regenerative braking offers a partial recharge, and hard acceleration requires a partial discharge, defining a cycle of current pulses. The ability of nickel-based batteries to provide high current discharge and accept a high current surge made them well suited to hybrid vehicles, and this kept them relevant in the automotive world longer than other applications. Fully electric vehicles require the highest specific energy possible to provide maximum range with lowest weight, making NiMH a poor choice, but NiMH has been more than able to meet the needs of hybrid electric vehicles at significantly lower cost and improved durability, leading car makers like Toyota to continue to rely on them past their use in many other fields (Image 6.14). So, they continue to be used in full hybrids and at times as supplemental batteries to allow start–stop feature in micro-hybrid applications.

The decreasing cost of lithium batteries probably means the expansion of NiMH batteries is unlikely. As we'll see, lithium-based batteries are improving their power density

IMAGE 6.14
Toyota nickel-metal hydride and Li-ion batteries.

The Toyota lineup includes both nickel-metal hydride (left) and lithium-ion batteries (right) as integrated elements of Toyota's New Global Architecture (TNGA), a platform the carmaker is expanding throughout its product lineup. As shown here, the two options are very similar from the outside. While the lithium-ion offers higher energy density, NiMH batteries continue to offer stable performance, good durability, and lower cost.

Image: Toyota Motor Corporation

and charge capacity, making NiMH seem increasingly heavy, low in energy capacity, and generally out of date.[7]

Lithium

Although NiMH still maintains a strong presence in the hybrid vehicle market, lithium-based batteries have become the dominant choice for advanced EVs. Lithium has the admirable combination of being the lightest of all metal and also the metal with the greatest amount of energy in its outer electrons, called **electrochemical potential**. For EVs or PHEVs, the resulting high specific energy is key (Image 6.15). Low specific energy is acceptable in hybrid vehicles, as long as the battery has the power capacity to support the internal combustion engine. However, when the only propulsion comes from the electric drive, a battery's capacity to store energy and produce a reasonable range is vital.

The basic operation of the battery is a lot like any other battery, but also a bit different. When charged, the lithium ions (Li^+, like the hydrogen in NiMH) leaves the positive electrode drawn to the negative electrode through a lithium salt electrolyte. Because lithium salts are dissolved in the electrolyte, the ions need not actually travel the full length between electrodes to complete the circuit, like electrons in the flow of current. The movement of ions defines an ionic charge through the electrolyte, so as one ion is released at the

IMAGE 6.15
BMW 330e battery.

Despite the continued presence of NiMH batteries in hybrid systems, lithium-ion batteries increasingly make sense in high-performance hybrid applications. The BMW 330e relies on a 7.6 kWh, 293-volt lithium-ion system to provide a bit over 20 miles of all-electric range.

Image: BMW

---

[7] C. Iclodean, B. Varga, N. Burnete, D. Cimerdean and B. Jurchiş, Comparison of different battery types for electric vehicles. *IOP Conference Series: Materials Science and Engineering* 252 (1), 2017.

anode another is absorbed at the cathode. When discharged, the Li ions are drawn back to the positive electrode. These are sometimes called 'rocking chair' batteries, since the back and forth movement of lithium ions defines their operation.

Beyond this, lithium-based batteries can vary greatly. In fact, variations of electrode composition and architecture have increased significantly in recent years as manufacturers seek improved performance and safety. In particular, manufacturers have tried to balance three general characteristics that are often in tension: safety, power, and energy. A battery that can readily absorb and provide a powerful current enables faster charging and improved acceleration. The problem is that very often a battery that is able to move electron charge quickly also tends to present challenges with stability, requiring enhanced care and management to ensure battery safety. And a battery with high specific power such as this can often tend toward low specific energy and declined cycle life, since the rapid chemical reactions that can enable fast movement of ions can also deteriorate the integrity of the electrodes. So, as in so many other technologies we've discussed, designing a battery is often a game of trade-offs.

Early lithium metal alloy batteries, for example, offered very high specific power but also posed a serious challenge. With aggressive charging and discharging, small bits of lithium could break loose from the electrode and reassemble during the reaction, defining whisker-like extensions called **dendrites** (Image 6.16). These branches could extend enough to penetrate the separator, contact the cathode, and cause a short circuit within the battery. This could quickly lead to an unrestrained reaction that produces a lot of heat. The heat would in turn accelerate the reaction rate, and cause more heat. The resulting positive feedback loop, called **thermal runaway**, can melt a battery or engulf it in flame. In case it's not clear, that's bad, and it's one of the principle reasons we've moved away from lithium metal batteries for automotive use. However, the problem is far from solved, all lithium-based batteries have at least some potential for a runaway reaction, caused by excessive current, due to either overcharging, an extreme discharge rate, or a short circuit.

Instead of a solid lithium electrode, a graphite anode is now used that allows for the movement of lithium ions but addresses the growth of dangerous dendrites. This anode defines a layered lattice structure of graphite that can hold lithium ions, a process called **intercalation**. Because they are small and light, a lot of lithium ions can be stored in the intercalation, giving the battery its remarkable energy density, and its name: **lithium ion**. Graphite is ideal for the job. Made of carbon, but structured into microscopic hexagonal crystals ordered into a regular pattern, graphite provides conductive non-metal layered sheets that readily store lithium between the layers as LiC_6. So, lithium ions are able to readily move to and from the interlayer of the intercalated anode (Image 6.17). Although intercalation involves somewhat of a drop in specific energy compared to lithium metal, the added safety is considered well worth it.

With good safety, power, high energy density, and comparatively long cycle life, lithium-ion batteries have become tremendously popular. Graphite is readily available at low cost,

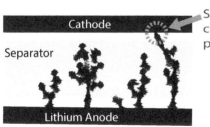

IMAGE 6.16
Lithium dendrites.

Small particles of lithium break loose and stack into loose, tree-like whiskers called dendrites. These can penetrate the separator; and if they reach the cathode, can produce a short circuit that can produce an unrestrained reaction and thermal runaway.

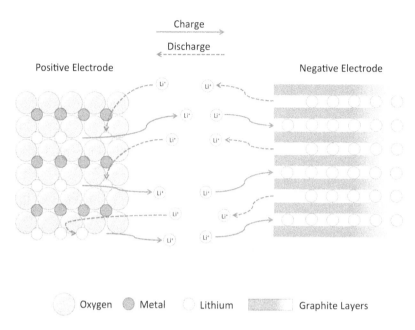

IMAGE 6.17
Lithium-ion intercalation.

A lithium-ion battery does not define the same sort of redox reaction of other batteries. The lithium ions are tied to an electron within the anode structure and when released are intercalated at the cathode, allowing ion charge to shuffle back and forth between anode and cathode.

after all it's the same stuff that's in your pencil, and its electrochemical performance is good. But there are still trade-offs. For example, when the flow of ions to the anode is too high, say during regenerative braking, ions are unable to intercalate with the graphite adequately and may instead collect on the anode as metallic lithium, sometimes called plating. This can degrade the battery's life and potentially cause a short circuit, so it is vital that the battery be managed carefully. In fact, an ongoing theme for Li-ion batteries is the need for precise management.

This isn't the only degradation lithium-ion batteries suffer. In fact, the main mechanism of degradation in lithium batteries in most cases is the irreversible side reactions with the electrolyte solvent and salt at the graphite surface of the anode that forms a **solid-electrolyte interphase** (SEI) comprised of a wide variety of compounds.[8] Some SEI is desirable and necessary to help protect the electrode from dissolving in the electrolyte, and the SEI is initially easily penetrated by ions and so does not impact performance. But as the layer thickens, it reduces access to the anode and defines a major source of performance degradation. While this can occur even when the battery is not in use, the deterioration during cycling is more significant because SEI forms in the cracks caused by expansion and contraction when cycling. And SEI can increase with temperature and aggressive charging and discharge, once again underscoring the need for precise monitoring and control.

Current battery research is focused on addressing these challenges and improving battery energy and power density. There are basically only two ways to do this. You can

---

[8] S.J. An, J. Li, C. Daniel, D. Mohanty, S. Nagpure and D.L. Wood III, The state of understanding of the lithium-ion-battery graphite solid electrolyte interphase (SEI) and its relationship to formation cycling. *Carbon* 105 (August), 2016, 52–76.

increase the cell voltage or increase the current capacity. Increasing cell voltage requires changing an electrode. The threat of plating tends to limit our options at the anode. However, as we will see, multiple variations of cathode materials, architectures and surface modifications are demonstrating promise, with some current technologies possibly leading toward a 5-volt cell potential in the future. On the other hand, increasing current capacity may require we turn our focus back to the anode, where plating with lithium metal, alloying with silicon, and other innovations have led to greater power capacity. Once again, all of this is complicated by the multiple components and reactants at play. Changing one variable invariably affect the others, possibly leading to undesirable consequences. Electrolyte deterioration, dendrite growth, and excess heating, all have to be carefully considered and balanced with the desire for performance and need for safety.

As mentioned, the most significant variation in lithium battery construction is in the composition of the cathode. A common cathode material, particularly for mobile phones and laptop computers, is Lithium Cobalt Oxide ($LiCoO_2$ or LCO). The material is defined by a layered structure, with the lithium ions held between the layers of cobalt oxide. This configuration can hold a fair amount of ions, so offers a high specific energy. However, these batteries can also be susceptible to plating as lithium ions can have trouble accessing the layered structure. So, a high charge rate is problematic and can lead to a significantly diminished service life; and the high cost of cobalt doesn't help. For these reasons, these batteries are not generally utilized in the automotive industry.

An alternative cathode material is lithium manganese oxide (LMO). Rather than a layered structure, LMO defines a three-dimensional manganese cubic crystal, or spinel, that greatly improves ion access to the electrode by allowing multi-sided access, rather than the two-dimensional access enabled by the layered LCO architecture (Image 6.18). So, the ions are able to flow much more quickly from one electrode to the other, increasing current capacity and lowering internal resistance, and thus making plating far less of a problem. Because LMO batteries can absorb a fast charge and provide a fast discharge, this battery is useful in hybrid vehicles; though the energy capacity is significantly lower than other options due to the limited space for ions within the spinel structure, so they are not an ideal option for PHEV or EVs.

The search for higher performing batteries that can offer a targeted balance of specific energy and durability has defined varying cocktails of added metals, in particular lightweight 'transition' metals, which offer better specific energy. The addition of cobalt and nickel, for example, defines a lithium nickel manganese cobalt oxide battery ($LiNiMnCoO_2$ or NMC). NMC forms a so-called layered-layered structure, with composite layers that enable some of both worlds, high current for acceleration, and energy density for longer

 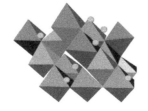 

Layered LiCoO₂　　　　　Spinel LiMn₂O₄　　　　　Olivine LiFePO₄

IMAGE 6.18
Structures of electrode materials.

range.[9] They are more expensive than LMO because they require nickel and cobalt, which is pricy; and they are a little more challenging to manage for safety, as high rate batteries often are. Jaguar opted for 432 NMC cells in a liquid-cooled pack to power its I-Pace. Rimac uses a 90 kWh NMC system in its Concept One vehicle, allowing it to deliver a full mega-Watt of energy in acceleration, and absorb 400 kW of braking energy.[10] Adding aluminum to create a lithium nickel cobalt aluminum oxide battery, or NCA, can offer improved specific energy and longer cycle life. The Tesla 100 D is powered by a 350-volt 100 kWh NCA pack, for example.

All transition metal oxides have the potential to be unstable, particularly at high temperatures and NCA presents particular challenges.[11] Still, both NMC and NCA have now become widely used options for automotive applications. The cost of NCAs and need for careful management and safety precautions exclude their use for portable devices, but their long lifespan makes them attractive for use in cars. And since they require less active material due to a higher specific energy and cell voltage, the additional cost is largely offset (Table 6.1).[12]

Manufacturers are tuning an ever growing and sophisticated range of proprietary alloys in the cathode to achieve the right combination of power and energy performance for their application. Nickel, manganese, and cobalt are combined in differing proportions to balance thermal stability, and therefore safety and cycle life, against increased power performance. Nickel is often a primary component because it offers high capacity, but used alone it demonstrates low thermal stability, and can lead to degradation of the layered structure as the rate increases.[13] Cobalt offers good power capacity, but it's expensive, so manufacturers avoid its extensive use. Manganese is more stable, and offers good cycle life, but because it's not electrochemically active it means lower capacity. Once again, there are compromises and trade-offs, with no one best answer.

TABLE 6.1

Battery Chemistry Comparison

Cathode Composition	Common Acronym	Cell Potential (V, vs Li/Li$^+$)	Specific Energy (Wh/kg)	Specific Capacity (mAh/g)	Structure
$LiCoO_2$	LCO	3.7–3.9	518–546	140–165	Layered
$LiNi_{0.8}Co_{0.15}Al_{0.05}O_2$	NCA	3.7–3.8	650–760	175–200	Layered
$LiNi_{1/3}Mn_{1/3}Co_{1/3}O_2$	NMC	3.6–3.8	590–650	155–160	Layered
$LiMn_2O_4$	LMO	3.9–4.1	400–492	100–115	Spinel
$LiFePO_4$	LFP	3.3–3.45	500–587	150–170	Olivine

[9]   K.C. Kama, A. Mehtab, J.T. Heronc and M.M. Doeffa, Electrochemical and physical properties of ti-substituted layered nickel manganese cobalt oxide (NMC) cathode. *Materials Journal Electrochemical Society* 159 (8), 2012, A1383–A1392.

[10] Battery System. Avaialble at www.rimac-automobili.com/en/supercars/concept_one/

[11] C. Arbizzani; F. De Giorgio and M. Mastragostino, Battery parameters for hybrid electric vehicles. In W. Tillmetz, J. Garche and B. Scrosati (eds) *Advances in Battery Technologies for Electric Vehicles*. Woodhead Publishing, Cambridge, 2015, 55–72.

[12] "Batteries for Electric Cars: Challenges, Opportunities, and the Outlook to 2020". The Boston Consulting Group, 2010.

[13] H. Kim, S.-M. Oh, B. Scrosati and Y.-K. Sun, High-performance electrode materials for lithium-ion batteries. In W. Tillmetz, J. Garche and B. Scrosati (eds) *Advances in Battery Technologies for Electric Vehicles*. Woodhead Publishing, Cambridge, 2015, 191–233; and P. Hou, J. Yin, M. Ding, J. Huang and X. Xu, Surface/interfacial structure and chemistry of high-energy nickel-rich layered oxide cathodes: Advances and perspectives. *Small* 13(45), 2017, 1–29.

The ongoing challenge of thermal and structural instability has led to a more significant shift in cathode composition with the development of lithium iron phosphate batteries ($LiFePO_4$ or LFP). Very stable, due to a dense tetrahedral crystal (or olivine) structure, LFPs offer excellent safety and have defined a leading presence in electric car innovation and research.[14] In addition to offering greater stability, the replacement of problematic metals with iron means an improved environmental profile and lower cost. While the battery offers good power density, the principle challenge is a lower energy density tied to the lower cell voltage of 3.2 volts. However, coating the cathode with a single-molecule-thick lattice network of carbon atoms, called graphene, enhances the surface area and conductivity and can help address this weakness.[15] In the future, a possibly better answer may come from work on a lithium manganese phosphate ($LiMnPO_4$) battery that shares a similar olivine structure, and so similar stability, but with a cell voltage of 4 V (Table 6.2).[16]

As mentioned previously, variations in Li-ion configurations are principally about the cathode; but they are not limited to the cathode. The graphite anode can also be varied to improve performance. Adding certain high-energy-density metals can improve the ion holding capacity of the graphite lattice. Silicone offers a key example. Consider this, when silicon combines with lithium it forms Li_4Si, but when graphite combines with lithium it forms LiC_6. This means a single atom of silicon can bind four lithium ions but it takes six graphite atoms to hold a single lithium ion. More critically, the discharge capacity of Li_4Si is nearly six times that of LiC_6. So, you would expect that adding some silicon to the graphite would improve performance, and it does, but silicon's use is limited by the fact that it can also greatly increase the propensity of the material to expand and contract when charging or discharging. This problem is apparent with other metals as well. So, while a small amount of silicon, constituting only a few percent of the anode material, can be added, adding more makes the mechanical management of the battery difficult. Once again, we face a trade-off. Adding silicone to the anode of an NMC can provide extra power, but at the cost of a shorter cycle life due to the stress of significant expansion and contraction (Image 6.19).

TABLE 6.2

Battery Performance Comparison

	Cost	Safety	Cycle Stability	Lifespan	Challenges	Advantages
NCA	High	Poor	Good	Moderate	Combustibility and cost	High specific energy
NMC	High	Moderate	Good	Good	Cost and scarcity of elements	Higher voltage
LMO	Low	Moderate	Low	Low	Short life	High voltage and low cost
LTO	High	Excellent	Good	Excellent	High cost	High voltage
LFP	Moderate	Excellent	Excellent	High	Low voltage and energy density	Low cost and abundance of material

[14] Boston Consulting. And T. Yuksel, S. Litster, V. Viswanathan and J.J. Michalel, Plug-in hybrid electric vehicle LiFePO4 battery life implications of thermal management, driving conditions, and regional climate. *Journal of Power Sources* 338 (January), 2017, 49–64.

[15] K.C. Kam and M.M. Doeff, Electrode materials for lithium ion batteries. *Material Matters* 7(4), 2012, 182–187

[16] C. Liu, Z.G. Neale and G. Cao, Understanding electrochemical potentials of cathode materials in rechargeable batteries. *Materials Today* 19 (2), 2016, 109–123.

IMAGE 6.19
BMW i3.

BMW designed their i3 electric hatchback to be upgradable, with a battery pack that can easily be removed and replaced. The first available upgrade in 2017 offered a bump from a 22k Wh LFP pack to 33.4 kWh pack of the same dimensions. The removed batteries can then be used as home backup power units. Such practices make solid environmental sense, and with decreasing battery cost, are likely to make good economic sense for customers in the future.

Image: BMW

Future Possibilities

If the scope of change in battery technology seems overwhelming, that's because it is. If you blink, you may miss a breakthrough, or at least an announcement of a 'breakthrough'. Which technological advances will actually redefine the industry is still a question. For example, a potentially promising development was announced by Toshiba. They're calling it a next generation of their Super Charged Ion Battery (SCiB). With an anode of titanium-niobium-oxide, Toshiba claims that less than 10 min of charge is enough to provide nearly 200 miles of range in a small vehicle.[17] Meanwhile, Samsung is developing a battery using 'graphene balls' applied to both anode and cathode. The result is a highly stable battery that it is claimed can be fully recharged in 12 min.[18] Predicting which technology will define a breakthrough and which will fall flat is not easy, but some recent trends seem ripe for adoption.

---

[17] D.E. Zoia, "For EV Batteries, Future is Now." *Wards Auto* November, 2017.
[18] Samsung Develops Battery Material with 5× Faster Charging Speed. November 28, 2017. Available at news. samsung.com/global/samsung-develops-battery-material-with-5x-faster-charging-speed

One is the continued development of nanostructures that enhance lithium-based performance. For example, recent innovations allow us not only to define the elements of the cathode, but the physical construction of the electrode material at the particle level, for example defining one composition at the center of a particle and another on the surface, called core-shell structured materials. This could allow a nickel-rich center for the particle to enhance power performance, and a reduced presence of nickel at the surface, allowing greater stability and cycle life.[19] Alternatively, a core-shell microfiber separator can maintain a non-reactive polymer outer surface with an inner-core made of a flame retardant that will be released when the polymer melts with overheating. By keeping the flame retardant sequestered until needed, it doesn't interfere with battery performance the way previous efforts have, but it will be there when needed.

Similarly, encouraging developments in nano-engineering of the electrode promise to enhance lithium-based performance. An important example in anode composition is defined by the use of lithium alloys such as $Li_4Ti_5O_{12}$, called LTO. The lithium titanate is placed as nanocrystals on the surface of the anode. This microscopic spinel crystalline material provides a far larger surface area than carbon, about 30 times more, for the same weight.[20] This allows ions to join and leave the anode quickly, making much faster recharging possible with no threat of plating.

Research on lithium-sulfur batteries may be even more promising . . . maybe. Sulfur offers better specific energy, much lower cost, and exceptional durability; but it is also electrically insulating, expands greatly, and can dissolve in electrolytes. So, there are some challenges to be solved. Combining sulfur with fine graphite can address conductivity, but addressing mechanical expansion is not as simple. Sulfur expands to four times its original size when it absorbs lithium, causing sulfur film or particles on electrodes to break apart quickly. However, a solution may once again come from nanotechnology.[21] Cathodes of precisely manufactured nanofibers containing sulfur, carbon, and a polymer binder can offer extremely high electrode surface area, good conduction of charge, and a short pathway for ions. In addition, when properly spaced and designed, they can manage expansion and contraction without the mechanical damage that occurs to the film and particle coatings. Think of a shag carpet with strands that can grow thick with lithium, but still release it readily at discharge.

Sulfur may be useful at the cathode as well. A lithium sulfide cathode has the theoretical potential to offer nearly three times higher specific energy than existing lithium batteries. However, the sulfur is vulnerable to dissolution in the electrolyte, and demonstrates poor conductivity. Research on a cathode of graphene coated with lithium sulfide nanoparticles may signal a solution, and has already produced a laboratory battery with a specific energy considerably higher than any lithium battery previously discussed.[22]

[19] H. Kim, S.M. Oh, B. Scrosati and Y.K. Sun, High-performance electrode materials for lithium-ion batteries for electric vehicles. In B. Scrosati (ed) *Advances in Battery Technologies for Electric Vehicles*. Woodhead Publishing, Cambridge, 2015, 191–241.

[20] J. Coelho, A. Pokle, S.H. Park, N. McEvoy, N.C. Berner, G.S. Duesberg and V. Nicolosi, "Lithium Titanate/Carbon Nanotubes Composites Processed by Ultrasound Irradiation as Anodes for Lithium Ion Batteries." Scientific Reports 7 (Article 7614), 2017.

[21] C.K. Chan, H. Peng, G. Liu, K. McIlwrath, X.F. Zhang, R.A. Huggins and Y. Cui, High-performance lithium battery anodes using silicon nanowires. *Nature Nanotechnology* 3, 2008, 31–35; and C.K. Chan, R. Ruffo, S.S. Hong and Y. Cui, Surface chemistry and morphology of the solid electrolyte interphase on silicon nanowire lithium-ion battery anodes. *Journal of Power Sources* 189(2), 2009, 1132–1140.

[22] H. Wu, G. Zheng, N. Liu, T.J. Carney, Y. Yang and Y. Cui, Engineering empty space between si nanoparticles for lithium-ion battery anodes. *Nano Letters* 12 (2), 2012, 904–909.

Lithium-based batteries are likely to be the future of EV batteries for the reasonable future, but other options are also being looked at closely. The chief reason is the scarcity and the resulting high cost of lithium. Only available from a small region of South America, and controlled by a few mining companies, the price of lithium is high and likely to go higher. The search for a possible competitor to lithium-based technology may well bring us back to nickel. In particular, a nickel-zinc battery. Zinc is readily available on several continents. It's cheap. It's robust, and it doesn't require the high-cost manufacturing practices of lithium. It also doesn't present the same potential volatility as lithium, making the battery more stable and requiring less expensive battery management and cooling. The key to its relevance for automotive applications is the recent development of a three-dimensional sponge-like zinc electrode. So, when the outside of the zinc gets coated, the inside is still available for reaction. However, the continuing challenge of electrode stability during charge and discharge means there is still work to be done. EnZinc, the principle firm working on an advanced zinc anode, is aiming for a low-cost battery that can match the performance of lithium.[23] But the future of nickel-zinc batteries seems less than certain.

On the other hand, the development of a solid-state battery seems imminent and very promising. Replacing the liquid electrolyte with a solid ceramic-like material could provide a battery that cannot produce dendrites. This means a battery that is far more stable, less dangerous, and does not require expensive battery management. The anode could then be made of higher performing lithium metal, allowing for a higher current capacity while maintaining safety at higher operating temperatures. While polymer and gel electrolytes are common, and can offer more compact and lighter batteries in flexible packets instead of a hard case, true solid-state batteries would far exceed these advantages. Ceramic or glass electrolytes would not be subject to the degradation at high voltages demonstrated by liquid electrolytes, allowing for a boost in cell voltage and performance. The larger temperature range could also simplify cooling needs. Moreover, without requiring individual casings for each cell, solid-state cells can provide a compact, lighter, more robust battery that could be three or more times more powerful than existing rivals at far less cost.[24] Toyota, Honda, and BMW have all announced plans to incorporate solid-state batteries in their vehicles within the next decade.[25] So, while Fisker's recent announcement of a 500+ mile range solid-state battery with a charging time of a few minutes may have been taken as crazy just a few years ago, it now seems like a real possibility.[26]

Battery innovation may also go beyond the battery itself. For example, using a high-performance supercapacitor in parallel with the battery can help protect the battery from undue stress by absorbing current pulses. By storing electrostatic energy rather than converting that energy to a chemical form, and with surface area far greater than any battery electrode, capacitors can more readily manage high power pulses. So, for example, a supercapacitor could absorb much of the fast charge of regenerative braking and quickly provide it upon acceleration, providing better performance with less stress on the battery; or it can absorb some of the stress of start–stop systems with an effective load leveling that

[23] D.E. Zoia, "For EV batteries, future is now." *Wards Auto* November, 2017; and J.F. Parker, C.N. Chervin, I.R. Pala, M. Machler, M.F. Burz, J.W. Long and D.R. Rolison, Rechargeable nickel–3D zinc batteries: An energy-dense, safer alternative to lithium-ion. *Science* 356 (April), 2017, 415–418.

[24] "California's Advanced Clean Cars Midterm Review Appendix C: Zero Emission Vehicle and Plug-in Hybrid Electric Vehicle Technology Assessment". California Environmental Protection Agency Air Resource Board January 18, 2017.

[25] D Stringer and K. Buckland, "Inside the race for next-generation EV battery supremacy." Automotive News January 8 2019; and www.libtec.or.jp

[26] "Fisker files patents on solid-state battery technology; anticipating automotive-ready from 2023" November 13, 2017. Available at www.greencarcongress.com/2017/11/20171113-fisker.html

protects the battery.[27] So, when used with start–stop applications, a capacitor can extend the battery's life and allow for a smaller, lighter battery.[28] However, while supercapacitors have much longer life expectancies than batteries, they also have much lower energy density. So, the price will have to come down below the value of the battery savings before we see widespread usage.[29]

An even more significant variation from the conventional battery is the promise of a metal-air battery. Since metals like zinc, for example, can be oxidized with oxygen directly, it is possible to make a battery that uses air as a cathode. Using ambient air as one of the electrodes removes a major component of the battery; and since the cathode is often the heaviest part in a battery, this promises a reduction of more than a third of the battery weight. Of course, the battery would require an air management system to supply the oxygen, but this is far lighter and less expensive than a second electrode. The potential is remarkable, as zinc-air batteries can produce much greater energy density than existing lithium batteries.[30]

The technology is provocative, but not all that different from other batteries. A mass of metal particles are saturated in an electrolyte, defining a sort of anode-electrolyte paste. Oxygen is supplied at a conductive cathode and reacts to form hydroxyl ions that travel through the paste. The resulting reaction defines a zinc hydroxide that releases electrons to travel to the cathode conductor. So, in the end, we still have an anode providing electron flow to a cathode. The reaction cannot be easily reversed, though this is a focus of ongoing research. One option would be to replace the zinc physically after discharge, using this relatively plentiful metal as a sort of fuel.

This same idea can be used to produce a lithium-air battery that has the low weight of the zinc-air battery but can be more easily recharged.[31] While the research is probably more than a decade away from commercial use, major automotive manufacturers like GM are actively working on a lithium-air battery that could offer performance that is orders of magnitude greater than existing lithium batteries.[32] But this is nowhere near ready . . . yet.

Current air batteries tend to be short lived, with limited cycle lives. But they represent one of the many possible innovations that could fundamentally redefine automotive energy storage. So, the possibility is a reliable metal-air battery that can challenge the dominance of lithium is real, if not quite ready…yet.

---

[27] C.G. Hochgraf, J.K. Basco, T.P. Bohn and I. Bloom, Effect of ultracapacitor-modified PHEV protocol on performance degradation in lithium-ion cells. *Journal of Power Sources* 246 (January), 2014, 965–969; and L. Kouchachvili, W. Yaïci and E. Entchev, Hybrid battery/supercapacitor energy storage system for the electric vehicles. *Journal of Power Sources* 374, 2018, 237–248.

[28] M.A.M. Mahmudi and A.A. Gazwi, Battery/Supercapacitor combinations for supplying vehicle electrical and electronic loads. *International Journal of Electronics and Electrical Engineering* 2(2), 2014, 153–162.

[29] A. Burke and H. Zhao, "Applications of Supercapacitors in Electric and Hybrid Vehicles." Institute of Transportation Studies University of California-Davis, Research Report—UCD-ITS-RR-15-09, April 2015.

[30] J.S. Lee, S. Tai Kim, R. Cao, N.S. Choi, M. Liu, K.T. Lee and J. Cho, Metal–air batteries with high energy density: Li–air versus zn–air. *Advanced Energy Materials* 1(1), 2011, 34–50; and Y. Li and J. Lu, Metal–Air batteries: Will they be the future electrochemical energy storage device of choice? *ACS Energy Letters* 2 (6), 2017, 1370–1377.

[31] T. Vegge, J. Maria Garcia-Lastra and D.J. Siegel, Review Article Lithium–oxygen batteries: At a crossroads? *Current Opinion in Electrochemistry* 6, 2017, 100–107.

[32] J.S. Lee, S. Tai Kim, R. Cao, N.S. Choi, M. Liu, K.T. Lee and J. Cho, Metal–Air batteries with high energy density: Li–air versus zn–air. *Advanced Energy Materials* 1 (1), 2011, 34–50.

7

Automotive Architecture

The technological innovations we've examined so far are impressive. Advanced electric machines, high-performing batteries, and sophisticated hybrid powertrains, all heralding a revolutionary future for the automobile. But, we can't forget the basics. The car of the future will still need a chassis, wheels, some way of connecting the two, and an enclosure for passengers. In a hundred years, if there is still such a thing as cars, they will need a chassis, wheels, some way of connecting the two, and an enclosure for passengers. You might think that this basic structure is the last place to expect the application of innovative technology and advanced design. You might imagine that the chassis is simply the metal skeleton that holds the drivetrain, axles, and body together, what room is there for innovation? But you'd be wrong. In fact, ingenuity in design, materials, manufacturing, and control technology have fundamentally redefined the automotive chassis, and made cars stronger and safer than ever before, while enabling impressive efficiency improvements and remarkable performance.

Like so many fields we've examined, chassis design has long been determined by a game of trade-offs. A car's frame needs to be strong and rigid, but adding strength often means adding weight and bulk, a problem for fuel economy and performance. And efforts to reduce weight often threaten to deteriorate the ability of a vehicle to protect the occupants in a crash. Of course, this is an intolerable trade-off. Finding the right balance has become increasingly challenging in recent years due to the steady increase in consumer safety and performance expectations and rising regulatory requirements. Many of the fantastic features we have discussed add weight; and the space for such trade-offs is even tighter with electric and hybrid vehicles, where heavy battery packs and controllers already make it tough to achieve an acceptable range. In short, as customers demand more and more infotainment, safety and performance features, carmakers have had to demonstrate tremendous creativity with materials and design to avoid a very, very heavy car.

As you might have come to expect by now, recent technological innovations defy these long-standing trade-offs. New designs ensure crashworthiness without bulk. New materials allow strength and rigidity without undue weight increases. New control technology enables a ride that is forgiving when you want it to be, but stiff and responsive when you need it. In short, this is not your grandfather's chassis.

General Chassis Design

When you strip away the drivetrain, seats, steering wheel, and all the other pesky parts that allow you to drive a car, what you're left with is the fundamental structure of the automobile. This basic assembly, called the Body in White (BiW), typically accounts for 20% of the vehicle's mass. Classically, it defines a strong, stiff structure made of stamped steel components welded together. But, as we will see, this is changing.

A defining element of the chassis, of course, is strength. But, strength isn't just one thing. It includes an ability to withstand multiple sorts of forces, or loads, in multiple directions. A key example is the **bending** due to the weight of the car, and the up and down force, or **dynamic loading**, caused by road bumps. A car also undergoes twisting around its centerline, defining a **torsional load**. In addition, a car can experience a sort of sheer load when the lower body is pulled to one side and the upper body is pulled to another, called a **lateral load**. The most common cause is a turn, when the wheels are held to the ground and the body pulls outward. A similar sort of force is experienced in acceleration and deceleration along the longitudinal axis. A less significant force is the sort of side-to-side sheer caused when one side of the car is pushed forward or backward while the other side isn't, for example, if one wheel hits a curb and the other doesn't. The resulting stress wants to tweak the chassis; you can imagine the rectangular wheelbase sort of jerked into a momentary parallelogram. This gets the strange name horizontal **lozenging**. Of course, all of these forces are merged in various combinations, making chassis engineering a tricky business.

However, the key goal isn't really strength. After all, it would be pretty easy to make a robust frame that can withstand whatever loads you can throw at it. The real task is to meet strength requirements while ensuring the chassis maintains engineered stiffness and is as light as possible. Stiffness is critical and tricky. The chassis needs to resist deformation enough that it doesn't deteriorate the performance of the vehicle or unduly stress joints and components. It's often said that there are really five springs on any car, one at each wheel and the spring action of the frame. The more that chassis 'spring' can be controlled, the better the other four springs can ensure even tire contact with the road, a key to **road holding**, and provide occupant isolation from rough roads, a key to **ride quality**.

A second principle concern is weight, a key determinant of vehicle performance and efficiency. Reducing weight at the expense of needed strength is not a good thing of course; but if strength and stiffness are maintained, a reduction in weight is always desirable. First, it is typically the most cost-effective way to improve fuel economy. As a rule of thumb, a 10% reduction in a car's weight, can improve fuel economy by 7% or so.[1] And second, by changing the power-to-weight ratio, putting the chassis on a diet can improve multiple aspects of performance significantly. Cornering, acceleration, and braking can all benefit from the process of **lightweighting**, sometimes also called mass decompounding.

It's worth recalling that not all weight reduction is equal. As an example, reducing weight higher on the chassis can lower the center of gravity. This can have a notable benefit for cornering. Reducing what's called unsprung weight, such as the weight of wheels or brakes, can significantly improve ride comfort. And, you'll remember, reducing weigh on rotating components can have a great impact on fuel efficiency.

Moreover, lightweighting can define a virtuous cycle through secondary weight savings. For example, a lighter chassis can be propelled by a smaller and lighter engine and stopped with smaller brakes. This weight savings can then improve fuel economy, and allow a smaller fuel tank, marking further weight savings. If a decrease in chassis weight can allow a drop from a V8 to a V6 and a decrease of 5 gallons in the fuel tank capacity, while maintaining the same or better range and performance, then a relatively small initial weight drop can lead to a major reduction in the final curb weight. Of course, the opposite is also true. For example, increased safety regulations or customer expectation for luxury and performance have at times led to a vicious cycle, with chassis weight increasing accordingly.

---

[1] E. Ghassemieh, Materials in automotive application, state of the art and prospects. InTech. Open Access Publisher, 2011.

So, the required dynamics of a good chassis are complex and elusive and need to balance overall vehicle dynamics. Torsional stiffness is a key factor. But we also need a balanced weight distribution, rigidity, and a responsive geometry to ensure desirable steering behavior, a fast and desirable response on both axles, road isolation, and general characteristics that define the subtle attribute of ride quality.

Frames

For the first 50 years or so of the automobile, the basic structure of the car was defined by a two-dimensional steel frame with a body placed on top, called a **body-on-frame** design. Typically called a **ladder frame**, the steel members could be open C, hat, or boxed channels depending on design and strength requirements. A slight variation on this theme, called a **perimeter frame**, widened the frame near the middle to allow a lower seat position and center of gravity. By adding crossed supporting members in the center, torsional loads could be better handled in what was called a **cruciform frame**. More dramatic variations included GM's X-Frame which supported millions of the carmaker's full-sized cars in the 1950s and 1960s, and was defined by a center cross without parallel side members. Somewhat similarly, the Lotus Elan was one of a few 1960s cars that used a major center box tube to define an "I frame", or torque tube **backbone chassis.** The large rectangular tube running down the center of the car made for a simple and cost-effective structure that allowed a low center of gravity and exhibited good torsional stiffness, though it left a lot to be desired in passenger protection. All of these examples, and more, defined variations of the fundamental body-on-frame approach that dominated automotive structures into the 1960s. And all shared the basic idea that a structural frame under the body would manage the principle loads and provide stiffness, while the body above would largely be coming along for the ride.

This basic body-on-frame design is still around. For trucks that need to accommodate heavy off-road use (and where designers are a little less concerned with high-speed handling, weight and ride comfort) a body-on-frame design is still a viable and cost-effective option. And the advantages of separating the structural frame from the body are hard to ignore. The most notable advantage is production flexibility; it allows a variety of body shapes and box backs to be fitted onto a common frame, suspension and drivetrain configuration. For larger trucks in particular, one frame can be coupled with multiple body options, reducing manufacturing cost, simplifying componentry and reducing parts count. We'll see at the end of this chapter that this idea is still enticing to carmakers as they increasingly adopt modular design concepts.

Nevertheless, overall, the body-on-frame approach has been superseded. The vast majority of today's cars are manufactured as a unitized body, or **unibody**. The idea is to integrate the frame and body together, so the entire assembly serves as an integral load-carrying structure. The result is a more three-dimensional structure that offers improved structural geometry and can therefore forgo heavy beams and beefy cross-members. Eliminating the need for a heavy lower frame significantly reduces vehicle weight. And because the entire body can absorb the impact of a crash, it also offers improves safety. The overall structure also lends itself well to mass production, building on a stamped floor pan and welding on various panels to define the body. Suspension and drivetrain components connect with discrete structural modules, called **subframes** (Image 7.1). These connect at integrated

IMAGE 7.1
Mazda's very light and stiff MX-5 chassis.

Although the Miata utilizes a unibody construction, it also maintains a central substructure to help ensure rigidity, a bit reminiscent of an 'I frame' from the 1960s.

Image: Mazda

hardpoints to spread the load of an axle or engine over a wide area of the body sheet structure, providing strength and NVH isolation. So, unibody construction is cost-effective, able to incorporate crash protection well, capable of accommodating a large and flexible cabin design, and lends itself well to mass production; and as a result, it has been nearly universal among major production cars for quite some time.

Unibody designs are sometimes referred to as a **monocoque** structure, derived from the French term for a single outer shell. Monocoque construction relies on the outer skin as the key structural member that distributes and carries tensile and compression loads throughout the shell with no internal structure. Think of an eggshell. As you apply compressive or torsional force, the shell's symmetrical structure sort of reinforces itself at any given point, providing impressive structural integrity for its weight. Clearly a unibody has some of these characteristics, but the boxed sections, subframes, and reinforcing structures in a unibody system define more of a semi-monocoque structure than a true monocoque. Still, some vehicles get closer to a true monocoque than others, relying more heavily on the outer shell for rigidity and less on major substructure components. The advantage is weight saving and stiffness. Yet the design must be careful to provide adequate strength and crashworthiness, as a true monocoque structure relies on its form integrity.

The Lamborghini Aventador offers an impressive example of this sort of monocoque construction. To save weight, the Aventador maximizes the use of lightweight carbon fiber composites, which we will discuss later. This extremely light and stiff outer skin is composed of precise components that make up the tub, roof and other body elements; but once cured, all are fused together to define a single structural element. Like an eggshell, the result offers exceptional rigidity and low weight. This structure is connected to aluminum subframes on either side to support the drivetrain and suspension. A well-designed monocoque structure such as this is light, rigid, and offers good crash protection, but the engineering must be precise and construction is time-consuming and expensive, which is just fine for a half-million dollar car.

An alternative approach to chassis construction that is typically limited to higher end performance cars is a **space frame** construction. Longitudinal and diagonal tubular members define an interconnected three-dimensional network of triangles that form an overall skeletal frame. So, a space frame provides strength and high stiffness in all directions. A purist might point out that a true space frame, rather than just a tubular frame, has load points reinforced tetrahedrally, providing optimal three-dimensional strength and rigidity; but this can vary. Body panels are added to the frame with little structural function, allowing lighter materials to be used. Properly designed, this system can produce the highest stiffness to weight ratio for any automotive structure. An additional advantage of this sort of structure is the relative ease of assembly in low quantities; making it ideal for specialty cars, and a go-to choice for Ferrari, Lamborghini and some Jaguar models in the past.

A notable example of this idea in mass production is provided by Audi. In the mid-1990s, Audi introduced its Audi Space Frame (ASF) design, and refined it for mass production over the following decade. A collection of extruded members and cast connectors are assembled a bit like a child's model, with aluminum panels added for stiffness (Image 7.2).

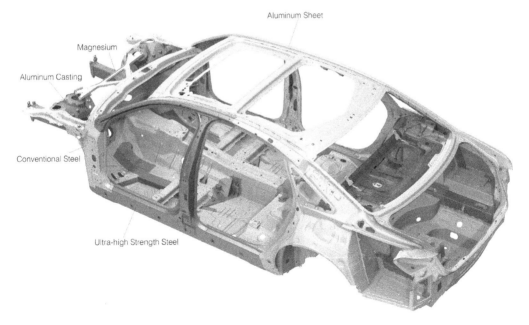

IMAGE 7.2
Multimaterial Audi Space Frame.

The Audi Space Frame (ASF) design utilized in the A8 and multiple other vehicles offers a rigid and lightweight alternative to traditional unibody construction. By replacing the conventional stamping and welding with extrusions, sheets, adhesives and rivets, Audi defined a system with the rigidity of a space frame while maintaining much of the assembly capacity of a conventional unibody. It also exemplifies the use of multimaterial design.

Image: Audi

The system works somewhat modularly, allowing it to be used in multiple vehicles from the A8 sedan to the TT roadster, facilitating the integration of multiple materials and enabling design flexibility. Aluminum sections, sheets, and castings are combined with steel, magnesium and carbon fiber to produce a rigid and lightweight structure.

Crashworthiness

A key element in the design of any chassis is preparation for when things go horribly wrong. A well-designed car has to be ready for the possibility that its stoppable force will hit a less-movable object. The results don't need much pondering. Structures collapse, bonds fail, and materials yield. The key to promoting the safety of the occupants in an impact is not to avoid these material failures, but to design for them, and to ensure that failure occurs in the way we want. Our aim is not the lack of damage, but a well-designed series of mechanical failures that can absorb the energy of a collision and so sacrifice the expendable to protect the priceless. This is called **crashworthiness**, as distinct from the other key safety aspect of an automobile, the capacity for **crash avoidance**, based in handling, braking, and advanced features we'll look at in Chapter 9.

This approach was not always the case. Relatively little thought was give to safety in the early years of the automobile. Society transitioned from horse and buggy, with speeds under 10 mph and little fear of a catastrophic collision, to cars with speeds well above those required to crush bones and sheer flesh, with little consideration for the implications. However, a significant shift began in the late 1960s when, sparked by the public concern catalyzed by books such as Ralf Nader's infamous *Unsafe at Any Speed*, carmakers began to pay at least modest attention to safety. But cars remained effectively metal boxes on wheels through the next decade, and the demand for fuel economy that led to smaller and lighter cars in the '70s didn't help. However, by the 1980s the results of crash testing increasingly drew the attention of regulators, insurance underwriters, and consumers. Within a decade, the vast majority of consumers believe safety to be a basic required characteristic of an automobile.[2] Of course, now safety is consistently ranked as a principle concern when purchasing a car and defines a leading marketing characteristic.

You might imagine that the key to avoiding injury in a crash is to avoid getting crushed, but that's not quite right, or at least it's not the whole story. It's true that the human body doesn't respond well to crushing; but in most accidents, that's not the only threat. The principle hazard is what are called **deceleration injuries**, the effect of going from 70 mph to 0 in under a second. This is the cause of whiplash, the classic automotive injury, but it can also cause other serious muscular trauma, skeletal damage, traumatic brain injury, and even organ failure. Your outer body stops, but the contents maintain inertia that propels them into very unnatural movements. So, the key to crashworthiness is twofold: a strong structure around the occupants that can protect against crushing, and a more compliant structure beyond that to absorb the crash energy and reduce the rate of deceleration (Image 7.3).

---

[2] V. Kaul, S. Singh, K. Rajagopalan and M. Coury, Consumer attitudes and perceptions about safety and their preferences and willingness to pay for safety. SAE International Paper 2010-01-2336. Published October 19, 2010.

IMAGE 7.3
Crash testing.

Vehicles are tested by the National Highway Traffic Safety Administration (NHTSA) as well as the Insurance Institute for Highway Safety (IIHS) in a range of crash geometries. These include the conventional frontal crash, with the full width of the vehicle impacting a barrier, as well as various offsets where only a portion of the car impacts a barrier (as above), and side and rear impact plus rollover testing.

Image: Volvo

A key to crashworthiness therefore is managing what is called the **crash pulse**, the rate and pattern of deceleration during a crash. We typically measure this deceleration in Gs, or the force of acceleration or deceleration in any particular direction expressed as proportions of the typical gravitational force on Earth. So, if a tight turn pushes you sideways at twice the magnitude the Earth is pulling you downward, you're experiencing two Gs of lateral acceleration. Depending on the age and health of the individual and the nature of the impact, the human body can tolerate somewhere in the neighborhood of 20–40 Gs in something like an automotive accident.[3] So, it makes sense that, among many crashworthiness requirements, current US federal regulation requires that a car be able to hit a wall at 35 mph (15.5 m/s) and produce no more than 20 Gs of deceleration on the occupants. The general idea in designing the chassis for a crash is to distribute the deceleration as evenly as possible over the time of the impact (typically about a quarter of a second) to avoid excessive force resulting from a peaked deceleration rate. In other words, we want to spread the deceleration over those 250 ms as fully as we can to avoid injury (Image 7.4).

Energy-absorbing front and rear ends that define **crumple zones** have been the preferred solution for managing the crash pulse. Often, in the public mind, the key to crashworthiness is a strong structure that can sustain a collision with minimal damage. This is wrong. This is why pickup trucks and SUVs, despite their bulk, can be some of the most dangerous vehicles on the road for both occupants and those in other vehicles. Large and midsized cars, with the ample crumple zones they provide, offer a much better survival

---

[3] D.F. Shanahan, Human tolerance and crash survivability. Paper presented at the RTO HFM Lecture Series on *"Pathological Aspects and Associated Biodynamics in Aircraft Accident Investigation"*, Madrid, Spain, October 28–29, 2004; Königsbrück, Germany, 2–3 November 2004, and published as NATO document RTO-EN-HFM-113.

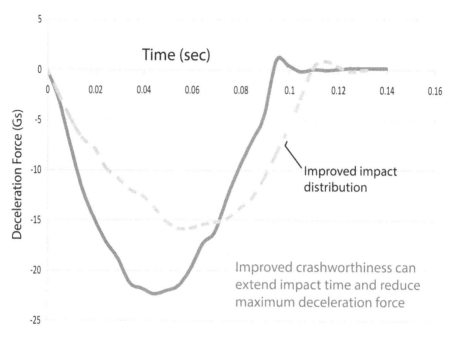

IMAGE 7.4
Dangerous deceleration.

Surviving a crash isn't about how fast you're going, it's how fast you stop. Crashworthiness can be improved by extending the crash pulse with improved structures and materials and thus reducing the maximum deceleration force.

rate for all involved.[4] The progressive crushing of these structures can be engineered to absorbed kinetic energy at a designed rate, resulting in severe and permanent deformation of the body components, as an alternative to the severe deformation of the occupants.

The basic automotive structure is therefore defined in two principle sections. First, a rigid and strong cage to maintain the integrity of the cabin, called a **safety cage**. Fabricated and reinforced with very strong materials, this cage works in combination with the seatbelts and airbags to preserve a **survival space** in the cabin. Second, this is connected to an outer structure that is designed to absorb crash energy and distribute any impact force broadly to the cage to avoid a point of structural failure or intrusion. The outer structure is generally defined in three zones: a softer, compliant zone that can absorb some light initial impact energy and reduce injury to pedestrians in a low speed collision, a primary inner zone that will sacrificially absorb lots of energy in an impact, and a secondary inner zone that will distribute this load to the rest of the car in a way that avoids failure of the safety cage.

It may have already dawned on you that a key problem so far is that not all collisions happen at the front of the car. Designers and safety experts classify crashes as frontal, side impact, rear impact and roll over. Frontal tends to be the easiest to address, as that's where more space is generally available for progressive crushing to provide gradual deceleration. This is one of the reasons electronic stability control systems are more concerned with addressing overseer than understeer. Understeer may cause you to go off the road, but any resulting impact is likely to be frontal, usually the safest sort of crash. This also means that

---

[4] M. Ross and T. Wenzel, An analysis of traffic deaths by vehicle type and model. American Council for an Energy-Efficient Economy, Report Number T021, 2002.

if you're gong to crash, erratic maneuvering that might change a frontal impact into a side impact is not your friend. Try to avoid crashing, of course; but if it's unavoidable, a frontal impact will be the most likely to let you walk away.

Designing for crashworthiness isn't easy. Ideally, in the front and rear, we want material that is deformable but strong. In the rear, an added concern is the protection of the fuel tank. But cars aren't shaped for crashworthiness, they are shaped for purpose, style, and perhaps aerodynamics; and crashworthiness must then be engineered into that shape. Rollover protection means the cage must be three dimensional, defining a strong roof structure. With the higher center of gravity of currently popular pickups, SUVs and cross-overs, this is a key concern. The real challenge is side impact protection for the obvious reason that there is simply less material and space on the sides. Solving all of this requires that we look not only at structures, but also at the materials that define those structures.

Materials

As you might guess, there are several concerns that need to be balanced when selecting automotive materials. Strength, of course; but not just high strength, consistent, predi-cable strength, combined with a known pliability that allows us to design for crash energy absorption with confidence. Good formability and machinability also matter. Weight is always a primary concern. Corrosion resistance is important, and has at times plagued the industry. And compatibility of materials that allows them to be bonded together to provide a strong, durable connection is vital.

Steel, a high-grade alloy of iron and carbon, has been the principle material in automo-tive construction from the late 1920s. The impressive strength of steel is achieved by add-ing carbon to iron at high temperature, allowing the much smaller carbon atoms to fill the spaces, called interstitials, between the large iron atoms. The carbon bonds compress and deform in the small space, and the result is a material that is much harder and stronger than just iron. The manufacturability and strength of steel has allowed it to dominate auto-motive materials for nearly a century. Sure, there have been a few deviations from time to time, the eccentric wood-framed British Morgan comes to mind; but, overall, steel frames with stamped steel body panels has long been the core of car making. Even now, a little over half of the weight of a typical US car is steel.[5] Up until the 1970s, plain carbon steel and cast iron were dominant in the industry. However, new alloying, processing tech-niques, and material innovations have changed that, defining a range of new steel varieties and allowing for more precisely engineered chassis components.

There are several reasons for the continued dominance of steel in car making. The most important is the strength. Or, more particularly, steel's ability to resist deformation when put under mechanical stress without being permanently distorted, what engineers call the **modulus of elasticity**. Called Young's modulus, this is defined by a ratio or tensile force (stress) to amount of elongation (strain). The higher the number, the more force a material can withstand with minimal deformation. Steel's Young's modulus orbits around 30 million pounds per square inch (or 200 GPa). That's about three times the value for aluminum, eight times the value for tin, and even well above the value for titanium. So, it's not only

[5] P.K. Mallick, Advanced materials for automotive applications: an overview. In J. Rowe (ed) *Advanced Materials in Automotive Engineering*. Woodhead Publishing, Cambridge, 2012, 5–27.

strong and rigid, but maintains a great capacity for energy absorption. Even with these notable characteristics, steel is relatively inexpensive. In fact, it is the least expensive major metal used in car production. It's ease of recycling helps, as steel can be reprocessed very easily and repeatedly, allowing for lower material cost as well as an improved environmental footprint. In fact, it is one of the world's most recycled materials, with a recycling rate in the automotive industry just under 90%.[6]

One of the great advantages of steel is versatility. Like any metal, steel varies its characteristics considerably based on its grain structure. At the atomic level, a metal is a matrix of millions of tiny crystalline grains of differing shapes, orientations, and sizes depending on the material. Each grain is a distinct crystal defined by a highly ordered atomic lattice (Image 7.5). The boundaries where different grains meet, called **dislocations**, are much less ordered. Within the crystal, the atomic bonds that hold the lattice together are loose, and allow flexibility; so the atoms can be pulled apart and they will spring back together. On the other hand, at the dislocations, the atomic bonds are often complex and entangled, defining a rigid structure that can't be pulled apart easily. These and other imperfections in structure also define the weak points in the material. So, the pattern, size, and geometry of the grains and dislocations define a material's basic properties. Generally, a larger grain means greater malleability; and a metal with small grains and many entwined dislocations will generally mean increased hardness.

This basic principle has allowed for great flexibility in the production of automotive steel. For example, because crystals grow simultaneously from molten metal as it cools, allowing the metal to cool slowly provides more time for larger crystals to form. The resulting large grain pattern with minimal dislocations produces a more malleable, tougher metal. Not hard, but tough, able to absorb energy and deform without fracturing. On the other hand, quickly quenching a hot metal will produce smaller grains and multiple dislocations, for a harder metal that deforms very little under stress. Work hardening can have the same effect. As a metal is mechanically deformed, grains are broken apart, defining a complex and entangled web of dislocations that produce a firmer metal. If you've ever fiddled with

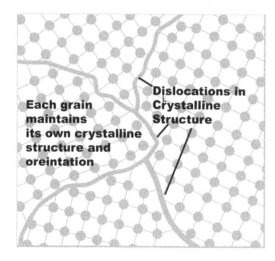

IMAGE 7.5
Metal grain structure.

[6] J. Bowyer, S. Bratkovich, K. Fernholz, M. Frank, H. Groot, J. Howe and E. Pepke, Understanding steel recovery and recycling rates and limitations to recycling. Dovetail Partners, March 23, 2015.

a paperclip until it breaks, you've experienced this effect. It becomes ever harder to bend, right up until it snaps.

Combining differing metals to define an **alloy** can have a similar outcome. Adding one or another additional metal to steel can make different sorts of crystalline structures more likely to form and so change the nature of the material. By varying manufacturing methods and alloying, steel's strength can vary more than any other common building material. These variations allow carmakers to define what they need for each application. The art and science of putting all this together has developed over the years, but very recent innovations have dramatically added to the strength and versatility of steel.

In fact, it's not an exaggeration to say that ingenuity in steel manufacturing has revolutionized the typical BiW. Improvements in production processes now allow for steel to be manufactured to much higher standards. For example, advances in control processes and improved material cleanliness, called **inclusion control,** help remove impurities from molten steel; and **vacuum degasing** uses low pressure to allow contaminating nitrogen and hydrogen to escape from molten metal. This allows contemporary steel to be about 20 times more pure than steel of past generations, making today's metal a stronger and more reliable material. But the modern transformation of steel goes well beyond this.

The transformation really began with so-called first-generation advanced steels that emerged around the 1980s, providing new alloys with greatly improved strength. This shift was heralded in the 1990s, when a series of studies by the American Iron and Steel Institute developed an **ultralight steel auto body** (ULSAB) that demonstrated the potential for a BiW that was a fifth lighter than conventional body with greater strength and improved crashworthiness. The new alloys and production methods that followed provided new steel products with improved performance profiles. High-strength steels (HSS) that had been developed in the 1980s became increasingly used in the automotive industry by the 1990s. For example, **high-strength, low-alloy** (HSLA) steel improved upon conventional steel to produce materials with fine grain microstructures. Adding very, very small amounts, like a few thousandths of a percent, of alloying elements such as vanadium and titanium defined impurities that displaced crystalline patterns and cause lattice strain within the metal's structure. By resisting the movement of dislocations, the material's yield strength was significantly increased (from about 200 to 300–550 MPa).[7]

Metallurgical innovations that began to more precisely structure metals at the molecular level combined with new manufacturing processes to define remarkable new possibilities. For example, **martensitic steel** uses extremely rapid cooling to prevent carbon atoms from diffusing and results in the formation of a carbon-saturated, highly strained tetragonal crystal structure with a very large number of dislocations. The result is an extremely strong material, defined as an **ultra-high-strength steel**, with a tensile strength of 900–1,700 MPa, and very little elongation before breaking.[8] Similarly, **dual-phase steel** combines exact alloying with precision manufacturing to define dual grain morphologies within one material. The steel combined a general matrix of softer crystalline structures defined by an atom at each corner of a cube and one atom in the center of the cube, called a **body-centered** structure or **ferritic** steel, with much harder **martensite** structures as a secondary phase. Balancing the combination by controlled cooling, and strain hardening, enables precisely defined formability and strength characteristics. Alternatively, the

---

[7] P.K. Mallick, Advanced materials for automotive applications: an overview. In J. Rowe (ed) *Advanced Materials in Automotive Engineering.* Woodhead Publishing, Cambridge, 2012, 5–27.

[8] P.K. Mallick, Advanced materials for automotive applications: an overview. In J. Rowe (ed) *Advanced Materials in Automotive Engineering.* Woodhead Publishing, Cambridge, 2012, 5–27.

development of **transformation-induced plasticity** (TRIP) steel offers a defined micro-structure that can be thought of as a sort of molecular strain hardening that produces a strong metal while maintaining ductility. A close cousin of TRIP, **complex phase** (CP) steel exemplifies the use of precise and fine micro-structuring of the grain to produce an advanced material. CP steels include small amounts of martensite and pearlite in a dislocation-dense steel matrix with an extremely high grain refinement defined by micro-alloying of key elements such as titanium or niobium.

These so-called **advanced high-strength steels** (AHSS) have offered impressive moves forward in automotive safety. AHSS has allowed much greater crush resistance, for example, defining far improved materials for the passenger safety cage or front-end structures. And they can enable lightweighting. While the weight per volume of these new steels is not greatly changed, their added strength means a downsizing of thickness is possible while still providing improved strength. Typically, AHSS allows the weight of any given member to be reduced by 10%–20% or more. As a result, the proportion of mild steel used in automobiles has dropped notably since the early 2000s in favor of AHSS alternatives.

In fact, the automobile chassis has become a precise assembly of differing steel products that allow for specifically engineered strength, ductility, and weight where and as needed (Image 7.6). So, designing a safety cage and the surrounding chassis is no longer just about building a rolling enclosure out of carbon steel. The steel used to maintain a critical survival space and prevent intrusion in a crash will generally be a product that offers very high strength and relatively little elongation under stress. This might include B pillars, for example, to protect against side impact. On the other hand, key roof features and A pillars may require more of a balance of strength and formability, providing energy absorption as well as durability in a steel that is less strong and is capable of greater elongation.

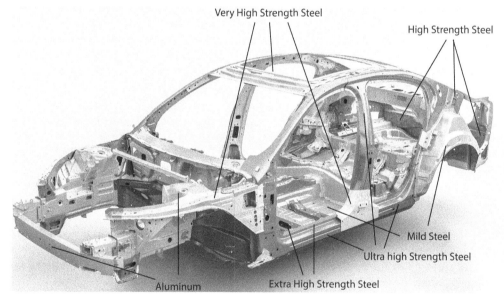

IMAGE 7.6
Steel safety cage.

Various types of steel are used to define the safety cage, including what are commonly called very-, extra-, and ultra-, high-strength steel.

Image: Volvo

The major body stampings used to define the shape of the car, on the other hand, might prioritize formability and offer less strength.

So, HSLA steel might be used to reinforce a door structure for side impact and moderate energy absorption. Or dual-phase steel might be used since it can be tuned to different yield strengths and ductility, making it useful for crumple zones, beams and cross-members, as well as fasteners, doors, and other body components. Extremely strong and hard martensitic steel might be used for door beams or roof sections to provide intrusion protection or roll-over protection when energy absorption is not viable. With high strength and ductility, TRIP steel may be used for frame rails, roof rails, seat frames, B pillars, and other components.

One of the challenges that remains true even with these advances is the trade-off between strength and ductility. The problem is certain automotive applications could benefit from a material that is both strong and flexible, able to provide a sturdy structure but with a capacity for energy absorption. This is where the second generation of advanced steels comes in. As you might guess, the key to the development of this steel is in the molecular structure of the material. The starting point is an atomic organization defined by a series of cubes with five atoms on each face, one on each corner and one in the middle. Think of dice with fives on all sides. This defines what is called an **austenite**, or **face-centered cube** (FCC), and presents the most tightly packed structure theoretically possible (Image 7.7). This is the typical structure of some metals, notably aluminum, but conventionally this sort of structure was only possible at very high temperatures in steel. However, the addition of about 20% manganese in TWIP makes this possible at room temperature. The resulting material is very strong; but the real difference comes when this structure is deformed under stress. As it's pulled apart, the crystalline structure is distorted along slight shifts in the alignment of the stacked crystalline planes, called **stacking faults**. This defines crystalline mirrored planes on either side of the deformation called **twins**. Under stress, the crystal lattice at these twin intersections is changed from a cubic form to a martensite; that's right, the same sort of structure as in very strong martensitic steel. As a result, the continued formation of these fine twin boundaries creates a strain hardening that makes the material very, very tough to break. Here's the punch line: if you try to pull TWIP steel apart, it's going to be hard and get harder; and it might elongate to twice its length before it actually snaps.

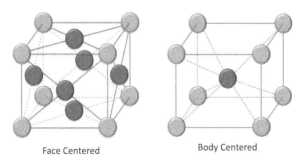

Face Centered   Body Centered

IMAGE 7.7
Body- and face-centered structures.

A metal's general characteristics are partially defined by its crystalline form. A face-centered cubic structure (FCC) has tightly packed atoms and includes metals such as aluminum and gold. A body-centered cubic structure (BCC) is less densely packed and includes conventional iron and chromium. However, iron can take on either form depending on its temperature (called allotropy); and steel alloys can be manufactured to have either form and so define variations in strength and malleability.

The multiple possible application of this strong but malleable metal in automotive design might be obvious to you. For example, this offers an exceptional option for side impact protection. It can provide a strong structure that is still capable of deforming and absorbing the energy of an impact, but growing ever harder as it does, so providing improved protection for the occupants.

The challenge with TWIP steel, and other so-called second-generation steels, is two-fold. First, cost. TWIP steel is very expensive to manufacture, and simply unaffordable for high-volume production cars. Second, these steels, as impressive as they are, are difficult to work with and hard to join together. The effort to design newer variations that can provide the strength and ductility of second-generation steels but at a reduced cost that can allow use across the automotive industry is underway, and defining a third generation of advanced steel (Image 7.8).

The development of third-generation steel builds on the research of first- and second-generation AHSS. The idea is to enable a mixture of martensitic and austenite structures to produce precise microstructural designs with targeted strength and ductility while preserving affordability. One interesting technique is a thermomechanical treatment called **quench and partitioning.** The process begins with a precise quenching of very hot steel to a critical and exact temperature to produce a refinement of grain structure, with a targeted balance of strong martensite and formable austenite crystalline structures. This is something like a blacksmith tempering steel when making a sword to get the toughness of some austenite and the hardness of martensite as well. The quenching is followed by a slight reheating and cooling to cause a partitioning of the carbon atoms from the martensite to the austenite in order to stabilize the structure at room temperature. Ever more complex combinations of processes and treatments are being explored including variations in tempering, rolling, bake hardening, and stamping to develop a targeted arrangement of martensites and austenites and yield a material that has superior strength, toughness, and formability. In essence, the aim is to separate the qualities of strength and ductility by independently shaping the micromechanical characteristics responsible for each of them.

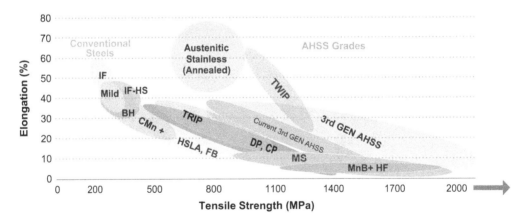

IMAGE 7.8
Advanced steel strength and ductility.

A wide array of advanced steels offer combinations of strength and ductility that were unimagined just a generation ago, and offer a wide range of performance and safety possibilities to automakers.

Image: WorldAutoSteel

Perhaps one of the most interesting characteristics of new steel innovations is the capacity to define ultra-fine grain structuring and reinforcement at the nanometer scale, causing variants of third-generation materials to be named **nanostructured steels**. With grains in the 10–100 nm range, they are defined by assemblies of crystals which can at times have only a few dozen atoms each, orders of magnitude smaller than conventional steel. With careful control of crystal development and grain growth, defects including the grain boundaries that degrade conventional steel's strength can be controlled or excluded. The fine grain structure also impeded the motion of dislocations, an effect that defines grain boundary strengthening (Image 7.9). The smaller the grain, the stronger and tougher the material becomes (called the Hall–Petch relationship in material science). In this case, the results are metals with extraordinary properties, well outside the performance boundaries of even the most impressive existing AHSS. Future third-generation AHSS can offer nearly twice the strength of many second-generation steels and about eight times the strength of mild steel.[9]

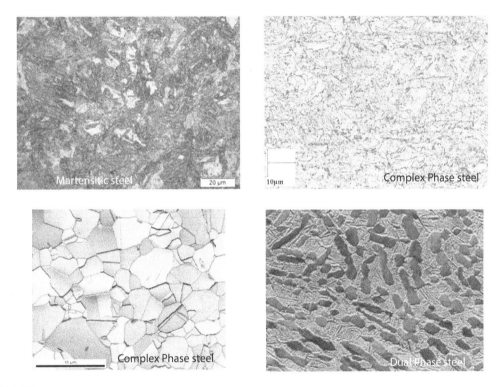

IMAGE 7.9
Steel under the microscope.

These photomicrographs of differing steels clearly reflect important variations in their crystalline structure. The Martensitic steel (upper left) reveals highly strained and dense small crystals. The high strength of Complex Phase steel (upper right) is reflected in its extreme grain refinement. The greater ductility of Twinning-Induced Plasticity (TWIP) steel (lower left) is reflected in the larger crystal structures. And the combination of a ferrite matrix and hard martensitic islands of dual-phase steel (lower right) can be plainly seen.

Image: WorldAutoSteel

---

[9] D. Chatterjee, Behind the development of Advanced High Strength Steel (AHSS) including stainless steel for automotive and: An overview. *Materials Science and Metallurgy Engineering* 4(1), 2017, 1–15; and Advanced High-Strength Steels Application Guidelines Version 6.0. World Auto Steel, April 4 (1), 2017.

However, innovations in the application of steel aren't limited to new steel chemistries; they are also defined by new applications and uses of existing steels in creative ways. A great example of the new manufacturing possibilities available through advanced methods is laminated steel, often referred to by the brand name **Quiet Steel**. A viscoelastic (viscous and elastic) polymer core is bonded between two thin sheets of steel to produce a sandwich. The result is a sheet material with great bending strength for its weight. But it's great advantage is its capacity as an NVH dampener and its ability to be tuned to attenuate specific frequency ranges. Because of this, conventional dampening and sound absorbing materials that are usually fitted around the car can be reduced or left out all together; this in turn can allow considerable weight savings.

Perhaps one of the greatest advantages of steel is that it's been the dominant material for quite a while. Engineers know how to use it. Plants are tooled for its use. Workers are trained in its fabrication. Indeed, a driving force for the development of third-generation steels is to create a material with improved characteristics that can be produced in existing facilities. It is often said that one of the great advantages of steel is its ease of manufacturing; notably, it can be formed and welded easily. While there is truth to this, it might be more correct to say its great advantage is in the *familiarity* of manufacturing. Designers know and understand steel, and can design for crashworthiness with known and existing manufacturing methods and facilities. For another material to replace it, there needs to be very good reason. And, as we'll see, for many applications, those reasons seem to have arrived.

Alternative Metals

Overall, the age of advanced steel products hasn't really been represented by a drop in vehicle weight. Instead, with increasing demands for more features and capacity adding weight to the vehicle, the use of AHSS has enabled curb weights to stay about the same rather than increase (Image 7.10). It is clear that advanced steel offers the most impressive weight savings for each dollar spent.[10] However, there is a clear limit to its promise. The typical 10%–20% decrease in component weight savings enabled by steel innovations may not be enough to continue to allow manufacturers to meet weight targets. This is particularly true as weight becomes more critical with rising fuel economy expectations and the demand for increased range from electric vehicles. But a more dramatic weight drop may only be available with a turn to alternative metals, polymers, and composites.

Until relatively recently, the use of aluminum in automobiles was pretty much limited to engine castings and wheels. However, aluminum has enjoyed a notable rise in automotive componentry in the past decade.[11] The great advantage is its specific weight, about 60% lighter than steel by volume. However, an aluminum component needs to be about 50% thicker to provide the same bending stiffness. So, replacing a steel body panel with an aluminum panel can typically save about 50% in weight.[12] Like any other material substitution, a switch from steel to aluminum also normally requires significant redesign.

[10] *Automotive Metal Components for Car Bodies and Chassis: Global Market Study*. Roland Berger, Munich, Germany, 2017.

[11] Aluminum content in North American light vehicles 2016 to 2028: Summary report. Ducker Worldwide, July, 2017.

[12] P.K. Mallick, Advanced materials for automotive applications: an overview. In J. Rowe (ed) *Advanced Materials in Automotive Engineering*. Woodhead Publishing, Cambridge, 2012, 5–27.

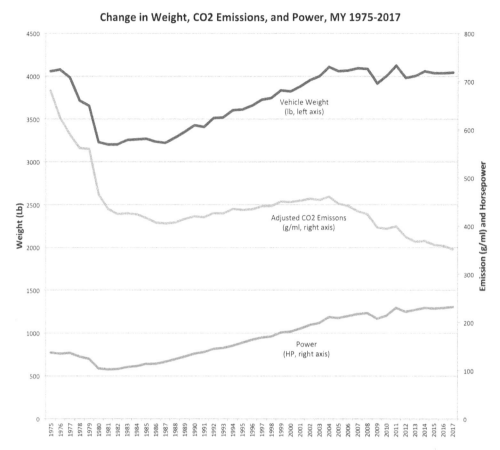

IMAGE 7.10

Change in weight, CO_2 emissions, and power, MY 1975–2017.

Although CO_2 emission have declined significantly while power has increased, continued improvement will require a continued reduction in vehicle curb weight, which has remained relatively steady for more than a decade. Continued moves toward multimaterial design is seen as necessary to achieve further weight reduction.

Data: US Environmental Protection Agency

Among other concerns, aluminum cannot maintain the same concentrated loads as steel, so a redesign may be needed to ensure loads are more evenly distributed across the component.

Like steel, variations in alloying and treatment can produce variations in mechanical properties, allowing varieties of aluminum to be used across the automobile. So-called 5000-series aluminum is alloyed with magnesium to produce a material that combines strength and formability. However, this aluminum does not respond to heat treatment; so the strength of this particular alloy is limited. Adding silicon will define an AlMgSi alloy. The resulting 1%–2% or so magnesium silicide that's produced will make heat treatment possible. So, this 6000-series alloy provides a stronger material. And its improved surface appearance and strength make this alloy preferable for body panels. The effect can be significant. For example, Jaguar's use of 6000-series alloy allowed side body panels to be reduced from 1.5 to 1.1 mm, with no decrease in strength and a 27% weight reduction.[13]

---

[13] I. Adcock, "Jaguar's Lightweight Challenger". *Automotive Design* September, 2014.

If even greater strength is needed, a heat treatable alloy of AlZnMg produces a very high strength 7000-series material that defines the strongest aluminum alloy in production. Suitable for bumpers, intrusion beams, and other ultra high-strength applications, this alloy can be two to three times stronger than conventional aluminum.[14]

Perhaps the most compelling example for the use of aluminum is the European Union-funded Super-Light Car (SLC) project. Coordinated by Volkswagen but including 38 automotive industry players, this project examined the materials, fabrication and joining technologies and costs for a potential ultralight BiW that utilized only technologies and materials suitable for high-volume production. Using a VW Golf chassis, the result produced a 34% weight reduction with no compromise to performance or safety. As you might expect, the BiW was just over half aluminum. A bit over a third was steel, with the rest largely magnesium and plastic.[15] It's not so much that aluminum is revolutionary; in fact, the opposite may be its greatest advantage. Aluminum is familiar, with a well-developed history of production and alloying possibilities, allowing a feasible, affordable, yet substantial drop in weight.

Interest in aluminum has been growing among carmakers for a while. The Acura NSX offered an early and inspiring example, with the first aluminum semi-monocoque body in 1990. The current model uses an advanced space frame design that combines aluminum and some steel coupled to a polymer and carbon fiber body (Image 7.11). Cast aluminum nodes provide strong joining points for the extruded aluminum frame members and serve as rigid mounting points for the front and rear suspension and powertrain. Similarly, Audi's ASF was developed in the 1990s and offers a key example for carmakers, with a frame that the carmaker claims is 40% lighter and 40% stiffer than traditional steel.[16]

IMAGE 7.11
Honda's aluminum innovation.

Honda defined the first aluminum semi-monocoque body with the Acura NSX. A range of aluminum manufacturing techniques are used to produce the highly innovative NSX body, defined in general by advanced aluminum castings used to join aluminum extrusions and stampings. In key locations where conventional casting methods would be too brittle to provide needed energy absorption characteristics, a method of aluminum sand casting pioneered by Honda was used. Ablation casting uses high-pressure jets to quickly wash the sand off the casting while rapidly cooling the material, producing a ridging component that preserves the frame's energy-absorbing capacity.

---

[14] T. Summe, Insider's Look: 7000-series aluminum alloy innovation. Novelis Aluminum Blog. Available at http://novelis.com/7000-series-aluminum-alloy-innovation/6/26/18

[15] J. Hirsch, Aluminum in innovative light-weight car design. *Materials Transactions* 52 (5), 2011, 818–824.

[16] Audi technology Portal. Available at www.audi-technology-portal.de/en/body/aluminium-bodies/

The list of carmakers turning to aluminum is growing. Such as the Mercedes SL Class adoption of an aluminum space frame in 2012. Or the Chevy Corvette's move to an aluminum unibody in 2006, with 3 mm thick panels to ensure needed rigidity. BMW 5- and 6-series cars use an aluminum firewall and forward subframe rails, riveted and glued to the steel unibody. Called GRAV for *gewichtsreduzierter Aluminium-Vorderwagen*, or lightweight aluminum front-end, the entire structure weighs about 100 pounds. By drawing weight off the front end, the result is not only a lighter car, but also significantly enhanced handling. However, the true leader in the incorporation of aluminum across varied platforms may be Jaguar Land Rover. Using unibody construction, the carmaker has relied heavily on aluminum in vehicles across their fleet, from the Range Rover Sport, to the F-Type to the XJ luxury sedan. As a result, the carmaker claims it has reaped weigh savings, fuel efficiency, improved crash safety and improved vehicle dynamics across its fleet.[17]

The move toward aluminum has been somewhat less compelling and notably different among US manufacturers. In fact, rather than a high-end performance sedan, the Ford F150 pickup may be the poster child for the increased use of aluminum by American carmakers. In 2015, Ford's hugely popular truck was reintroduced with major aluminum body components. This, along with other lightweighting efforts, resulted in a 700 lb diet for the F150, with a notable fuel economy boost. Though, in a market where perceptions of toughness are vital, the switch has not been without controversy; and Ford has relied heavily on the dubious moniker 'military grade aluminum' to preserve the truck's manly bonafides. Still, we may be likely to see the use of aluminum grow in the American SUV and truck market where major weight reduction is needed to achieve efficiency goals (Image 7.12).[18]

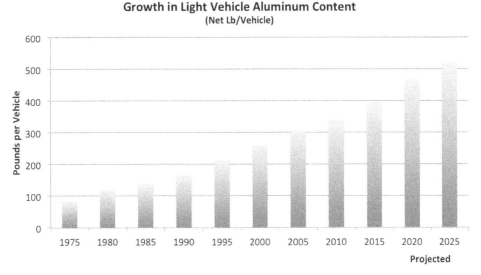

Growth in Light Vehicle Aluminum Content
(Net Lb/Vehicle)

IMAGE 7.12
Growth in light vehicle aluminum content.

The use of aluminum in automobiles is projected to continue to increase as carmakers seek cost-effective light weighting options.

Data: Ducker Worldwide

[17] *Lighter, Stronger, Cleaner*. Available at https://media.jaguarlandrover.com/2016/lighter-stronger-cleaner
[18] *Nicolai Müller and Tim Ellringmann, It's time to reassess materials for automotive lightweighting. Automotive World* Special report: Vehicle lightweighting. Penarth, UK, 2016. *Automotive World*, 2016.

Of course aluminum is not appropriate for all components, so its use necessitates a **multimaterial design** approach. For example, the Jaguar XE uses steel closing panels for the doors and trunk, the B posts are composed of aluminum, steel and structural foam to define an impact-absorbing sandwich, and high-strength steel underpins the rear cabin.[19] The trick isn't really in the fabrication of these differing components, it's in joining them. The melting point of aluminum is significantly different from steel, making conventional welding difficult. Audi relies on complex aluminum die-castings at high-stress joints. More generally as the variety of metallic and non-metallic material use increases, alternative joining methods, self-piercing rivets, high-strength adhesives and specialized fittings have been developed. While this can add complexity to the manufacturing process, it enables extrusions to be thinner since there is no need for thickness at the ends to enable welding. And, the increased use of mechanical joining techniques simplifies the incorporation of polymers and composites, further facilitating material variation.

More generally, it's often noted that aluminum presents some production challenges in relation to steel. Again, it's not particularly harder to work with than steel, but the requirements are significantly different. So, the up-front cost of the required changes in production processes and methods presents a significant barrier to a switch. It's not only less weldable but also has different requirements and capacity for bending and stamping. So, the transition can be tricky, even for apparently straightforward components and processing. However, the ease of extrusion of aluminum compared to steel can actually make certain parts much easier to manufacture. While a steel component may require multiple steps in the manufacturing process, stampings, welding, and so on. An aluminum extrusion may fit the same purpose with one manufacturing step.

Production challenges aside, we're likely to see more use of aluminum in the future. Its growing use by carmakers is decreasing cost and increasing manufacturing expertise. And continued advances in aluminum alloys are making aluminum stronger and more easily used in manufacturing. The result is a rising adoption rate for a wider variety of components. Increasingly, a turn toward aluminum for lightweighting is making more sense for mass-market models, not just high-tech space frames and luxury sedans.

We may well see a similar but more modest pattern with another lightweight metal, magnesium. One of the great advantages of magnesium is very low density, about two-thirds of aluminum, with a higher strength to weight ratio. However, the modulus of elasticity is much lower than either steel or aluminum, making it more difficult to ensure needed stiffness. And, of course, you can't manufacture magnesium as you would steel or even aluminum. For example, it can't be easily stamped. But it can be cast in very thin sections. And high-pressure casting can be used to produce complex parts that can at times actually simplify componentry. This makes it a viable option for instrument panels and other interior structural components for example. Alloyed with manganese and aluminum, it can provide strong parts that provide good crash resistance. In fact, magnesium overall has several manufacturing advantages over aluminum, including easier and faster casting and higher machinability.

Given these advantages, you might expect magnesium to surpass aluminum; but magnesium also presents some serious problems. First, it is expensive. And its production is extremely energy and carbon intensive, making it a hard material to justify in environmental terms, an increasingly vital concern in the industry and to humanity. And its production is largely concentrated in one country, China, raising supply concerns. On top of all this, a perennial challenge is corrosion. Indeed, the metal's propensity for rapid corrosion is the

---

[19] I. Adcock, "Jaguar's Lightweigth Challenger". *Automotive Design* September, 2014.

TABLE 7.1

Table Comparing Metal Materials

Material	Relative Weight (Compared to Steel)	Cost of Weight Reduction	Advantage	Disadvantage
Traditional Steel	100%		An available and known material	High weight
Hot Formed Steel	75%–85%	Low	High strength and attractive cost for light weighting	Initial investment and limited weight savings
Aluminum	50%–60%	Medium	Good formability and lower weight	Needed changes in production methods and joining
Magnesium	45%–55%	High	Very light	Threat of corrosion
Composites	Up to 25%	Very High	Excellent performance to weight	Expensive and not recyclable

Source: Roland Berger Global Market Study.

central reason the 'mag wheels' that were so poplar a few generations ago, are now hardly ever seen. Magnesium is highly reactive, and will form a redox couple with steel and other metals, with the steel acting as cathode to the magnesium's anodic role. The result is the magnesium is very quickly corroded unless very particular joining techniques are used. Because the effect is less pronounced with aluminum, aluminum hardware can be used to fasten magnesium parts. This helps, but the challenge is still there, and could generally constrain the metal's use to limited bolt-on components (Table 7.1).

The ideal solution for many of these challenges would be titanium. It has a very low density, high strength-to-density ratio and excellent corrosion resistance, making it an idea metal for automotive application. So, why don't we just make cars out of titanium? Well, it's very, very expensive. While aluminum sells for just under 2 dollars a pound, and carbon steel at about 50 cents a pounds, titanium will cost you about 25 dollars for each precious pound.[20] So, as a major component, titanium won't be competing with steel or aluminum anytime soon; but, for specific components where strength, weight or corrosion resistance really matters, titanium is a viable choice. For example, with a very low modulus of elasticity and excellent durability, titanium can be a good choice for spring coils. The deflection of a titanium spring is less than it would be if it were made of steel. This allows fewer coils to be used, which means less weight in a critical component and an increased natural frequency for improved road holding.[21]

As we consider innovations in metals, we might recall the metal matrix composites discussed in Chapter 1. The idea is to reinforce a light material by adding very hard ceramic particles. A variety of options—silicon carbine, aluminum oxide, ceramic particles or short fibers—can be added to aluminum, titanium or magnesium to produce a material with toughness, durability and wear resistance. And the material maintains low density and a high strength-to-weight ratio. While these materials are seeing the most interest in drivetrain components, there are some possible promising applications in other areas.

---

[20] Metal Miner. Available at agmetalminer.com/metal-prices/

[21] P.K. Mallick, Advanced materials for automotive applications: an overview. In J. Rowe (ed) *Advanced Materials in Automotive Engineering*. Woodhead Publishing, Cambridge, 2012, 5–27.

An interesting example is the development of **syntactic foam**. Hollow ceramic microspheres are incorporated into a metal matrix to produce a foam-like composite with about half the density of the metal used for the matrix and a very high capacity to absorb impact energy. This metallic foam can be used to reinforce frame sections and set the energy absorption capacity of crumple zones.[22] It can also be used as a very lightweight core material in a sheet sandwich structure.

Along a similar line, but with far greater micro precision, **nanocomposites** contain much smaller reinforcing particles, such as carbon nanotubes or nanoclay. These can be placed in a variety of matrix materials, including plastics, ceramics and metals. Because the surface area to size ratio for these small particles is very high, they are more solidly fused with the matrix; and a relatively small amount of nanoparticles can have a large impact on the material's properties. Nanotubes in particular seem to offer real potential for increased strength and decreased thermal expansion. Research is ongoing, and these technologies are not yet fully deployed in the industry; but the promise of highly tuned nanomaterials for specialized applications is growing.

Manufacturing Metal

Just as important as the composition of new materials are new fabrication techniques and possibilities. Remarkable new production methods now offer a more uniform material, with precise sheet thicknesses, better surface qualities, and cleaner and stronger joints. For example, **laser beam welding** (LBW) provides a highly focused and deep joining of metals. The resulting precision makes more complex joints possible and enables overall lower heat application, which means greater manufacturing flexibility, lighter components (since they do not need to be designed to manage the heat stress of welding without distortion) and stronger bonds.

Advancements such as precision laser welding have made **tailored blanks** possible. Rather than the simple stamping process of the past that require the thickness and composition of a part to be uniform and so defined by the greatest needed thickness and strength at any particular spot, tailored blanks combine different grades and thicknesses of material into one stamped part. Parent materials are laser welded together to provide thickness where needed and less material where it's not needed, combined into a single stamped part. Similarly, variable rolling can be used to vary sheet thickness, patches can be added, and localized coatings and heat treatment applied, to define a more optimized product with defined variations in thickness, material and coatings.[23] This can allow more exact and varied strength, impact resistance, stiffness, along with less waste, a lower overall parts count, and improved crashworthiness.[24]

Similarly, the use of **hydroforming** instead of stamping has allowed for more complex shapes that are stronger and lighter than conventionally stamped parts. The basic principle is to use a one-sided mold, and apply a high-pressure fluid to the other side of the material

[22] A. Macke, Metal matrix composites offer the automotive industry an opportunity to reduce vehicle weight, improve performance. *Advanced Materials & Processes* 170 (3), 2012, 19–23.

[23] M. Merklein, M. Johannes, M. Lechner and A. Kuppert, A review on tailored blanks—Production, applications and evaluation. *Journal of Materials Processing Technology* 214(2), 2014, 151–164.

[24] G. Sun, J. Tian, T. Liu, X. Yan and X. Huang, Crashworthiness optimization of automotive parts with tailor rolled blank. *Engineering Structures* 169 (August), 2018, 201–215.

to conform the metal to the die shape. This process can be used to shape cylindrical parts from extruded tubes as well, by applying pressure to the inside of the tube and expanding it to a surrounding die. The material cost can increase, but by avoiding the dual stamping and welding that used to be needed for such cylindrical parts, overall manufacturing costs can be reduced and the resulting part is stronger and lighter than previously possible. Moreover, once again, since traditional forming led to thinner edges and ends and so required thicker than needed overall size to allow ends strong enough and thick enough to enable joining and welding, hydroforming can enable the use of minimum thickness, allowing overall lighter components.

Perhaps the new process that has most significantly transformed the automotive industry is **hot stamping**. Once a marginal technique for specialized components, carmakers have moved toward the broad use of hot stamping in high-volume production, making this probably the fastest growing form of tailored products. The basic idea is to heat a blank of boron steel to a temperature where its fixed ferrite crystalline structure changes to a softer austentite structure. The material is then stamped and very quickly quenched to form a hard martensitic structure that ensures high strength. Once again, because of the resulting high strength, components can be made thinner and lighter; and parts can be made with more complex shapes without worrying about springback. And the added strength can reduce the need for some of the welding, stiffeners and reinforcement, simplifying component design. Volvo's X90 is about 40% hot stamped, providing a strong and lighter safety cage. But Volvo's not alone, Mazda, Honda, Volkswagen and others are quickly adopting this technique in their BiW production (Image 7.13).[25] In fact, hot stamping is likely to quickly become the standard for structural body components.[26]

IMAGE 7.13
Hot stamped outer door ring.

Steelmaker ArcelorMittal has developed an inner and outer hot stamped door system for the Acura RDX. The outer ring provides improved safety while enabling a panoramic roof opening. Material thickness varies from 1.2 to 1.6mm as needed to ensure rigidity and strength.

---

[25] *Automotive Metal Components for Car Bodies and Chassis: Global market Study.* Roland Berger, Munich, Germany, 2017.
[26] Automotive Metal Components for Car Bodies, 2017.

Plastics

Plastics have a long history in automobiles. But until recently their use was limited to a few accessories and cabin embellishments. It is only lately that plastics have been used for structural components. In fact, the use of plastics has grown dramatically in vehicle manufacturing and is likely to continue for two reasons: versatility and cost. Plastics are a remarkably versatile group of materials. Think of the plastics you encounter in everyday life, from water bottles to furniture to elastics; they can be designed for an incredibly wide variety of applications and characteristics. They can be made chemically inert, ductile, durable, and strong, and are highly moldable. And they are notably inexpensive.

Plastics are made of repeating units of carbon compounds called **monomers**, and are a part of a large group of compounds with repeating monomer units called **polymers**. Other polymers include starches, which are repeating units of sugar, proteins, which are repeating units of amino acids, or even wood, which is made of cellulose, repeating units of the monomer glucose. The long chains of carbon compounds that make up plastic polymers can be put together in a wide assortment of configurations, thus explaining the great variety of properties that can be produced. They can be branched or linear, they can be interconnected or individuated, and the monomers themselves can be made of a wide variety of chemical compositions, like ethylene, propylene, styrene, and vinyl chloride. All this allows for the production of a seemingly infinite range of characteristics.

You can imagine these polymer chains as a cluster of boiled spaghetti at the microscopic level. When you try to pull or bend a polymer, it's like pulling these spaghetti strands past one another. If the chains are long and entangled the resulting material is more likely to be firm, imagine stirring a pot of three-foot long entwined spaghetti strands. If the chains are loose, or if we add a lubricant that lets the strands slip past each other more easily (a plasticizer), the material can be soft and pliable. Imagine one-inch-long strands in olive oil. If we physically lock the strands together wherever they cross over each other, called crosslinking, they form an interconnected three-dimensional structure and define a material that can be very firm. The pattern of the strands can assume crystalline or amorphous characteristics, increasing the possible material variations.

These physical characteristics fall into two major categories of plastics: thermoset and thermoplastics. Thermoplastics tend to be the materials you are most likely to think of as consumer plastics. They are generally pliable. If you heat them, they get softer as the strands have an easier time sliding past each other. As they cool, resistance to this sliding increases so they get firmer, and they're likely to return to their original shape. They're easy to mold and extrude, and fairly easy to recycle since the process of heating, softening, and reshaping can be repeated more than once. Thermosets, on the other hand, are defined by the strong crosslinking of the strands with powerful chemical bonds, sometimes called **vulcanization**, through a hardening process or **curing**. The resulting interconnected network of strands is fixed, and will not soften with heating. Think of an epoxy or a typical fiberglass component. Fixed, rigid, and unchanged by temperature, they maintain their strength under changing conditions; but they can lack durability and can be brittle compared to tougher thermoplastics. And recycling thermosets is challenging at best.

One of the most common polymers being used in automotive applications is the thermoplastic **polypropylene** (PP). It offers low density, good mechanical properties, low cost, and good impact strength. The plastic bumpers, body trim, washer fluid tanks, carpets and many other parts of the car are likely to be PP. **Polyurethane** may be a somewhat distant second, as it's the material principally used in foam for seats, insulation, and cushioning.

While one of the less common consumer plastics, polyvinylchloride (PVC) is also a frequent material used for automobiles due to its lightweight, high strength and toughness, though its environmental effects are problematic at best.

So-called **engineered thermoplastics** define high-performance polymers with chemical compositions that are more precisely designed to produce targeted mechanical properties and heat and chemical resistance, making them useful in automotive manufacturing. Acrylonitrile Butadiene Styrene, or ABS, polyacetals, polycarbonates, nylon, and others are defining new options in specialty plastics with designed durability, strength, and chemical stability for automotive applications. They can be used for precise gears, cams, or bushings, exterior trim, bumpers, headlight lenses, display screens, handles and much more.

The challenge with polymers is their lack of strength. While the tensile strength of polymers used in automotive construction range around 20–50 MPa, and can approach 200 MPa for nylons, this compares very unfavorably to steel's tensile strength typically in the range of 400–1,500 MPa, or even aluminum's tensile performance of up to 500 MPa or so.[27] So, while replacing metal parts with plastic can reduce weight, it can also negatively impact crash worthiness, as plastics can break not only more easily but in a brittle way, not absorbing the energy of a more ductile metal. So, how do we take advantage of the manufacturing flexibility and lightweight of polymers while achieving the strength of materials with characteristics closer to metals? We combine the two to define **fiber-reinforced plastics** (FRP) composite materials. Basically, we encase strong fibers in a polymer resin that forms the body of the material, or the **matrix**. We might also add fillers or chemical modifiers to tune the properties of the material.

Combining high strength, high Young's modulus materials such as glass or carbon with a polymer to form the matrix, allows some of the best of both (Image 7.14). You may be

IMAGE 7.14
BMW's composite life module.

The passenger compartment, or 'life module', of the BMW i3 is assembled of resin transfer molded carbon fiber composite components. The superior rigidity means the car does not require a B pillar. With the rear doors swinging rearward, this provides a large opening for easy entry and loading.

Image: BMW

---

[27] M. Ashby, *Materials and the Environment*, 2nd Edition. Elsevier, Oxford, 2014.

thinking that glass is a strange material to turn to as a strengthening component. After all, it's pretty easy to shatter glass. True. But instead of shattering it, try pulling glass apart. You'll find it's pretty hard. In fact, the tensile strength of glass FRP can exceed 1,000 MPa.[28] So, using these fibers to reinforce a material can provide the toughness of plastic with the tensile strength of glass. The great advantage of these composite materials is their potential for reduced weight. They can be a third lighter than aluminum and, of course, twice that advantage over steel.[29]

The basic idea of fiber composites is not new. In fact, fiberglass has been around for quite a while. Most notably, the 1953 Corvette was produced with an all-fiberglass body. Today, glass fiber composites are not uncommon in interior applications, and are increasingly finding their way to body panels and even semi-structural members. The most common form of FRP in automobiles, probably representing over 90% of uses in the industry, is generally short fibers of a few millimeters to perhaps 25 mm, randomly oriented and encased in a thermoset resin matrix.[30]

These FRPs need to be designed and evaluated carefully, as the energy absorption capacity varies by type of fiber, matrix material, geometry of the component, fiber architecture, and ratio of fiber to plastic matrix.[31] The strength to weight ratio is notably higher than most metal alternatives. And, with a tough polymer matrix, they offer good fatigue strength. However, composites require careful consideration and engineering to ensure crashworthiness. They tend to be more brittle than steel; so rather than an energy-absorbing ductile buckling, they still tend to crack and shatter, with the plies and fibers breaking from the matrix. Designed properly, the controlled brittle failure of FRP can offer viable energy absorption and general crashworthiness in a lighter and more rigid package (Image 7.15).

In fact, the use of FRP has a few advantages over steel beyond their light weight. The typical production configuration uses a large roll of sheet FRP, called **sheet molding compound** (SMC), also called **prepreg** as in pre-impregnated material, that can be compression molded into components on site. The simplicity of this process can offer lower tooling costs than those required for making a component of stamped steel. And, because FRP allows for a single component to potentially replace multiple metal parts, the parts count and associated cost are reduced. So, for example, a single SMC radiator support with two components might be able to replace 20 or more steel parts that make up the support assembly.[32]

However, this is nowhere near the full potential for FRP. Fibers can be used in a variety of lengths and orientations to produce targeted characteristics and precisely engineered components. Orienting all the fibers in one direction can produce a material that is very strong in that direction. Typically, the fibers are oriented in line with the major axis of stress, and layers can be added and crossed at 45° or 90° to tailor the mechanical properties to the application. Changing the length of the fibers matters as well. Fibers can be as short as 3 mm or as long as 75 mm or longer. So, unlike metals, orienting the fibers in a particular direction, or a designed set of overlapping directions, can define specific strength in the

[28] Y.J. You, K.T. Park, D.W. Seo and J.H. Hwang, Tensile strength of GFRP reinforcing bars with hollow section. *Advances in Materials Science and Engineering* 3 (October), 2015, 1–8.

[29] M. Pervaiz, S. Panthapulakkal, K.C. Birat, M. Sain and J. Tjong, Emerging trends in automotive lightweighting through novel composite materials. *Materials Sciences and Applications* 7(1), 2016, 26–38.

[30] Polymer composites for automotive sustainability. European Technology Platform for Sustainable Chemistry (SUSCHEM), n.d.

[31] G.C. Jacob, J.F. Fellers, S. Simunovic, J.M. Starbuck, Energy absorption in polymer composites for automotive crashworthiness. *Journal of Composite Materials* 36(7), 2002, 813–850.

[32] P.K. Mallick, Advanced materials for automotive applications: an overview. In J. Rowe (ed) *Advanced Materials in Automotive Engineering*. Woodhead Publishing, Cambridge, 2012, 5–27.

IMAGE 7.15
BASF plastic wheel.

The chemical company BASF has developed a polymer wheel for the Forvison lightweight concept vehicle it developed jointly with Daimler. Nearly the entire wheel is manufactured of plastic composites fortified with long glass fiber, allowing a weight reduction of over 6 pounds (or 3 kg) per wheel. Given the heavy loads and rotation wheels must endure, this is no easy feat.

Image: BASF

direction needed. This can be supplemented with seamless reinforcement for precisely designed performance. For example, a very thin polymer body panel might be augmented with aramid honeycomb or foam to increase thickness and rigidity at key locations.

This enables FRP to provide a design efficiency that sets it apart from metals. Conventional metals have identical properties in all directions. The tensile strength of a steel component, for example, is the same whichever way you pull it, called **isotropic**. However, FRP can be **anisotropic**, meaning their physical characteristics are directional. This allows for a more efficient structuring of the material's properties, using material and weight to produce strength only in the directions needed. The resulting capacity for efficient and precise engineering enables a high strength-to-weight ratio. Along with its inherent durability and corrosion resistance, this can make FRP very attractive.

For many high-end applications, carbon fibers are replacing glass fiber (Image 7.16). Carbon fibers are stronger and can allow lighter components. This allows for much greater weight reduction. For example, the BMW M6 roof panel made of carbon fiber is thicker than a comparable steel panel, but also more than 10 pounds lighter.[33] This is particularly welcome high on the vehicle, as it will drop the center of gravity.

The potential performance benefit is great, but the manufacturing process can be cumbersome, expensive, and fiddly. Typically, the process begins with a precisely woven SMC fabric that is pre-impregnated with a resin. Fabric sections are cut to shape and often hand placed with precise orientations in molds to be cured under heat and pressure. Individual parts can be defined with oriented carbon fiber and then brought together into a final assembly in a closed mold. Epoxy foam, and polymer or metal components

[33] P.K. Mallick, Overview. In P.K. Mallick (ed) *Materials, Design and Manufacturing for Lightweight Vehicles.* Woodhead Publishing, Cambridge, 2010, 21.

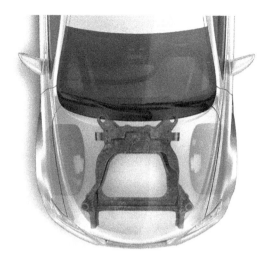

IMAGE 7.16
Carbon fiber can simplify design.

The Canadian automotive supplier Magna and Ford are testing this carbon fiber subframe for future use. The prototype replaces 45 steel parts with 2 molded parts, and weighs in at a third less than a stamped steel alternative.

Image: Magna

can be integrated at key locations for stiffness or attachment points. The whole process is called **resin transfer molding** (RTM). An impressive example of what is possible with RTM is offered by Lamborghini's Aventador composite monocoque. With a painstaking process, Lamborghini defines a precise body shell with tolerances of a tenth of a millimeter and a structure that demonstrates very high torsional rigidity and excellent road performance. But the most remarkable thing is that the whole monocoque structure weighs in at just about 325 pounds, or just under 150 kg.[34] Continuing the carbon fiber theme, the Aventador's seats and interior trim incorporate a patented woven carbon fiber fabric that is saturated with a special epoxy resin which strengthens and stabilizes the fabric while maintaining pliability, producing a soft and flexible seat and console cover that is half the weight of leather.

Lamborghini may be exceptional, but it's not alone. More than three decades ago Ferrari and Porsche incorporated carbon fiber panels on their racecars, as well as limited production cars to meet FIA requirements. Later McLaren and Ferrari developed carbon fiber monocoque cars. More recently the Ford GT supercar utilizes a carbon fiber monocoque. So does the Porsche Carrera GT, which weighs in at about 3,000 pounds, a remarkable achievement for a car with a 5.7-liter V10 powertrain.

Perhaps the most remarkable feature of FRP is their flexibility of application. They can be used as skins and combined with a foam or honeycomb core to produce a stiff but lightweight semi-structural component or a rigid energy- and noise-absorbing sheet. Long strand thermoplastics can offer structural use for cross-members, bumper beams and other applications. Components from cooling fans to wheels to valve covers to roof frames are being developed.

However, these composites also faced a couple of key obstacles. First, they present manufacturing complexities. They can be a challenge to join with metal structures as they

---

[34] A. Jacob, Built in Italy: The Lamborghini Avendator. *Reinforced Plastics Magazine* 57 (5), 2014, 29–31.

are vulnerable to concentrated forces and can't be welded; though advances in alternative joining are largely addressing this concern. A chief obstacle in the use of composites of the past was their more complex and time-consuming manufacturing requirements. Laying precisely oriented reinforcing fiber in a resin mold takes time and increases production costs. However, SMCs, RTM, improved injection molding, knit fiber fabrics, and the use of robots have begun to address some of these problems. So, although FRPs are still likely to remain more time-consuming and complex than metal manufacturing for a while, their increasing use not only in high-end performance cars but in high-volume production and especially electric vehicles is likely to continue to grow.

A key challenge to this growth is price. While the potential weight reduction can mean improved range or reduced battery capacity requirements for electric vehicles and so off-set some of the cost, the current price of FRP won't be fully offset by savings on a few battery cells. At present, the high energy required to produce the fiber is the principle cost determinant, and there are a number of research efforts being pursued to address this challenge. A noteworthy example comes from the US Department of Energy's Oak Ridge National Laboratory. ORNL has been working to develop a higher speed, lower cost process for carbon fiber production, and have recently made a new production method available that has high promise. Carbon fiber is typically produced through an energy-intensive and costly series of heating and stretching of a specifically produced and pre-cisely made polyacrylonitrile. The production of this raw material represents a significant potion of overall cost. However, the researchers at ORNL have shown they can make the carbon fiber using a simple acrylic fiber, the same stuff carpets and clothes are made from, reducing cost by nearly half and energy use by much more. If this or other similar efforts bear fruit, the use of FRP could become much more attractive very quickly.

One of the key remaining challenges of composite polymers using glass or carbon fibers, is their poor recyclability. Both glass and polymers are recyclable, of course; but with the dissimilar materials tightly bound, and separation therefore difficult, the potential for effi-cient recycling of the composite material is limited. This makes it difficult to meet the European Union's end-of-life-vehicle (ELV) requirement that a full 85% by weight of a vehicle must be recyclable. Efforts are being made to identify recycling possibilities for carbon and glass fiber composite. One of the more successful examples among carmakers is BMW, which has had some success in the reprocessing of carbon fiber in the making of the i3. The process is not perfect, and might be more correctly considered downcycling rather than recycling. But about 10% of the i3's fiber is made from waste.[35] And despite this challenge, the lifetime carbon footprint of the i3 is about half that of a similarly sized typi-cal car, making it one of the greenest EVs on the market.[36]

However, a few more environmentally friendly alternatives exist when the strength of carbon fiber is not strictly needed. The first is **self-reinforcing polymers** that use the same thermoplastic material for the fiber and the matrix. In addition to being easily recycled, because the fiber and the matrix are the same material, the bond between the two is strong, since they have the same chemical makeup, yielding a more durable material. While not as strong as glass fiber reinforced composites, self-reinforcing plastics do offer stronger materials than unreinforced polymers for a greater variety of applications in interior and exterior automotive use.

---

[35] J. Pellettieri, "The Case for Recycled Carbon Fibre in Auto Lightweighting," *Lightweighting World* October 4, 2016.

[36] Environmental Report BMW i3 BEV. Available at www.bmwgroup.com/content/dam/bmw-group-websites/ bmwgroup_com/responsibility/downloads/en/2016/Environmental-report_BMW-i3.pdf

Another path toward a more environmentally friendly composite material is the use of natural fibers, replacing glass with fibers extracted from the leaves, stems and fruits of various plants, such as hemp and jute. These biological fibers can also lighten the material, as the densities of most natural fibers are well below that of glass. However, the strength is not as high as glass, and some natural fiber composites can degrade with moisture and time. So, for the present, manufacturers are limiting the use of natural fiber reinforced polymers to components with no significant structural load, principally interior trim. When Mercedes replaced conventional FRP panels inside the E-Class with flax/sisal fiber reinforced panels, weight was reduced by a fifth and passenger protection actually improved.[37]

Suspension

So far, we've examined the capacity to produce a strong, rigid and precise architecture for a vehicle. But to make the chassis useful, it needs to connect to the road. The trouble is, the road is not as strong and rigid; it's often broken and pliable. As a result, the ride is not as smooth, controlled and consistent as we might like. If we simply bolted the chassis to a set of wheels, the potholes, bumps, crevices and occasional curbs the car encounters would make a ride unbearable at any speed above 15 mph. So, we connect the chassis to the road through a flexible, energy-absorbing mechanism, a **suspension system**, that can allow controlled semi-independent movement. Ideally, we'd like the wheels of the car to travel in close contact with the ground, adapting to every bump and divot to ensure positive road contact, or what's called **road holding**. At the same time, we want the chassis to do something very different. Whatever the road conditions, we'd like the chassis to ride smooth and even, maintaining a level posture as centrifugal forces push it to one side or the other. This is sometimes referred to as a **skyhook** analogy. Imagine the chassis suspended just above the ground by a hook, traveling smoothly and evenly across the terrain no matter what bumps may be beneath. That's our idealized car.

The problem is these two aims, a smooth ride and high friction for road holding, aren't always in full agreement. Optimal contact between the tires and the road generally requires a firm suspension that can effective work against any upward movement of the wheel and push the tire toward the pavement, ensuring handling and safety. As the road bumps the wheel upward, the suspension needs to push it back down, decisively and reliably. And we want the system to provide stability to support cornering and acceleration comfortably, without allowing the chassis to sway or tilt under centrifugal force. This similarly requires firmness. However, at the same time we want the suspension to provide a comfortable ride, or **road isolation**, soaking up shocks on the straightaways to give us that skyhook ride. This requires somewhat softer energy absorption. To complicate the matter, we want a car to maintain the same ride quality whether there's only a driver on board, or the thing is packed down like a clown car. And this ability to handle excess weight and maintain ride height requires very firm springs. So, once again, we face a trade-off, and the proper tuning of a car's suspension for any particular purpose requires achieving the right balance.

---

[37] M.W. Andure, S.C. Jirapure and L.P. Dhamande, Advance automobile material for light weight future: A review *International Conference on Benchmarks in Engineering Science and Technology ICBEST*, 2012.

The typical suspension uses springs to absorb road bumps and accommodate vehicle movement. As you might remember from your middle school science class, any mechanical spring exhibits a tendency toward harmonic oscillation. That is to say, if you compress or stretch an elastic body, it will move to restore it's original shape, but overshoot, and so be drawn back somewhat in the direction of the initial displacement, then rebound back, overshoot a bit less, and move back once again, and so on, defining a diminishing sinusoidal motion until it has used up its energy. If you like crazy carnival rides, this might be a fun way to ride in a car. But for the typical driver, bouncing down the road is not pleasant; and extreme **wheel bounce** tends to ruin handling and stability. So, we dampen the spring's oscillation with the poorly named **shock absorber.** These shock absorbers don't really absorb much shock, they dampen the springs' oscillation to prevent an otherwise wavy ride. Typically composed of a cylinder filled with oil or gas, and a sliding piston that works against the fluid viscosity, they provide proportional resistance to movement. The greater the speed of the movement between car and wheel, the stronger the resistive force. These two features, a spring and a dampener, define the dynamics of the connection between the **sprung mass** above (the body, chassis, and drivetrain) and the **unsprung mass** below (wheels, tires and brakes). A key concern is to keep the unsprung mass as small as possible, as the larger the mass, the greater the upward force imposed on the suspension with every bump, and the more challenging it is to absorb.

All of this also requires some form of mechanical articulation that can allow relative motion of the wheels and chassis while maintaining wheel orientation to preserve vehicle control, stability and tracking. Exactly how this is done varies significantly. Older systems included **dependent suspensions** that had a stiff axle housing connecting the two wheels and keeps them in a fixed relative orientation. With the exception of trucks, these systems have been largely replaced by independent suspension that allow each wheel assembly to move independently, and so better accommodate uneven road conditions and improve handling. Though, to accommodate larger cabin space or reduce cost, a semi-independent system can sometimes be used on the rear axle, with the two wheels impacting each other but allowed to move largely independently, such as a twist beam axle. Nevertheless, just about every contemporary car has independent front suspension and a growing number also have independent rear suspension.

Most contemporary vehicles use a variation on the MacPherson strut and **wishbone** or **multilink** suspensions to provide suspension that ensures that any given wheel's movement is unaffected by the movement of any of the three other wheels (Image 7.17). These struts combine a coil spring with a dampener, acting as a structural component, a wheel location device, and a shock absorbing and dampening mechanism. A double wishbone is defined by two semi-parallel wishbone-shaped control arms that allow articulation. The result does a fairly good job of keeping the tires perpendicular to the ground and generally offers good dynamics and load handling. **Multilink suspensions** use a more complex geometry defined by multiple control arms and linkages. This complexity offers greater latitude for variation and precise movement dynamics. While the cost and space requirements of these systems is greater than the simple axles of the past, the resulting improvement in handling, active safety, and ride quality is increasingly considered well worth it.

Typically, putting all this together requires careful design and certain trade-offs. Depending on the design priorities—that is to say the desired capacity for cornering, road holding, road isolation, or load control—differing geometries, springs and dampeners are selected. But the result is inevitably a compromise. Modest dampening and soft springs provide good road isolation, but terrible road holding, and firm springs and dampening offers good road holding, but a very stiff ride. We've seen this sort of thing so often in

IMAGE 7.17
Performance suspension.

The Volvo V60 utilizes a wishbone front and multilink rear suspension to provide excellent handling while maintaining ride quality for precisely engineered driving dynamics.

Image: Volvo

other systems that you might be able to predict the next point. What if, instead of settling on a mechanically fixed **passive suspension** system, we use digitally controlled actuators to define a dynamic system that can change with driver preference and road conditions, what if we used an intelligent and **active suspension**?

Chassis Control

The most modest form of an active suspension could be defined by controllable dampeners. A conventional shock absorber relies on multiple disks covering ports in a moving piston to define a dampening force. So, varying the disks and ports can define different sorts of dampeners, but typically once they're in place, they can't be changed. Replacing these disks with controllable valves can allow the port to be varied and the dampening effect to by adjustable, manually for some, or remotely in other cases. This sort of dynamic valve control can simply offer predefined settings for the driver, sport or cruise modes, or it can vary dampening based on driving conditions, with an ECU using inputs from steering angle sensor, speed sensor, yaw sensor and others to adjust accordingly. A system might provide stiff dampening in cornering, braking and acceleration to reduce roll and pitch; and on straightaways, the system can automatically switch to softer dampening. The result is a more even ride, less squat and dive with acceleration and deceleration, improved braking, and improved road holding and cornering.

However, mechanically changing the size of an orifice on a piston that is located in a cylinder filled with fluid is not easy. A more sophisticated solution comes by leaving the port unchanged and simply changing the viscosity of the fluid itself. This ingenious innovation is made possible with the use of a **magnetorheological** (MR) fluid. An MR fluid is defined by micrometer-scale metallic spheres suspended in a carrier oil. Applying a magnetic field causes the particles to align with the magnetic flux, and resist movement. So, varying the field strength varies the effective viscosity. By placing a variable magnetic field around the port in the shock absorber, an electronically variable dampening effect can be produced, defining an **electromagnetic dampener**.

It's worth noting that this MR effect has more than one application. Similar systems can be used to provide dynamic drivetrain mounts. First appearing in the Porsche 911, the system works like suspension. In normal driving, the mounts are forgiving, allowing the NVH of the engine to be absorbed for a smoother and quieter ride. But, in hard driving, the slight movements of these heavy components can negatively influence performance, so the system tightens up, and momentarily locks the engine and transmission in place when rpms rise. The basic idea of variable engine mounts is not entirely new, hydraulic systems have been around for a while and are used to help accommodate cylinder deactivation, but MR fluids make this a more elegant option.

Strictly speaking, the sorts of systems we've discussed so far can be called semi-active suspensions. They adapt to road conditions, but do not apply any dynamic force to the suspension mechanism to counteract these conditions. So, they can't really be called fully active, and they can't completely cancel roll and pitching movement. An active system, on the other hand, can do just that, provide force through an actuator to regulate road forces and provide a more precisely controlled separation of sprung and unsprung mass.

The simplest form of active suspension systems focus on controlling roll. In tight cornering, a vehicle will sway outward, lifting the inside wheels, compressing the outside suspension, reducing road grip and defining an uncomfortable sway. The higher the center of gravity, the greater the leverage of the list, making this an even bigger concern in SUVs with rollover potential. To address this in conventional suspension systems an anti-roll bar or sway bar is installed. This provides a rubber-mounted connection between right and left wheels. The basic idea is to increase roll stiffness by connecting the two sides, so a rise at one side induces a rising action at the other, counteracting the roll tendency in tight turns. The effect is limited and blunt. And it means that any bump that hits one tire in level driving effects both sides, increasing the impact's effect. But, recently, the conventional sway bar has been improved with the use of a digitally controlled actuator to counter body roll tendencies. With input from sensors like the steering angle sensor, yaw sensor, and acceleration sensor, the mechanism can either be placed at the center of a split anti-roll bar, or at the end links on either side of the bar, and can produce forces to counter the effect of body roll.

The details of these systems vary. BMW's Active Roll Control (ARC) uses a hydraulic rotary actuator at the center of the rear anti-roll bar. Toyota's active Power Stabilizer Suspension System utilizes an electric motor for motion control and can respond in milliseconds, improving on the generally slow response time of hydraulic control. Audi's electromechanical active roll-stabilization system used in the diesel SQ7 relies on a small electric motor to drive a three-stage planetary gearbox that can couple or uncouple the sway bars. On a rough road, the bars can be completely uncoupled to allow each wheel to adapt to irregular road conditions and enable a smoother ride. But, when turn-induced roll is a concern, the bars can be fully coupled. Porsche's Dynamic Chassis Control, on the other hand, uses hydraulic actuators on either side of the anti-roll bar, exerting counteracting torque on the bar to accommodate driving conditions. Variations on this theme are growing and include a variety of hydraulic, electromechanical, and pneumatic systems.

However, we can move beyond roll control to a fully active suspension that can exert precise control at all four wheels. A basic system places an actuator in series with a spring and dampener for what can be called low bandwidth control. Why bandwidth? The natural frequency of oscillation, or **resonant frequency**, of the unsprung mass and sprung mass is different. So, we have two frequency ranges to deal with: what we might call the body rattle frequency of 10–15 Hz and the much lower wheel hop frequency of about 1–4 Hz. Low bandwidth systems focus on accommodating body rattle, and rely on passive systems to continue to address the bounce of unsprung weight. Because these systems are limited to low bandwidth dynamics, they generally do not fully address vehicle oscillation, but they are significantly less expensive than full bandwidth control.[38]

Fully active, high bandwidth control generally requires the use of a highly controllable actuator in parallel, rather than series, with the spring. A rapid control force can be produced to address roll and pitch motions and provide chassis stability. Such systems can require a lot of energy and generally need a 48-volt system. However, the use of a permanent magnet electromechanical actuator could potentially allow energy to be absorbed by the actuator and stored, allowing this to be used for actuator motion later. This could significantly reduce the energy demand of the system.

Such an active suspension system can accommodate driver preferences more completely. For example, Volvo's active chassis allows a 'comfort' mode for high-speed stability, an 'eco' mode that lowers the vehicle for improved aerodynamics, a 'dynamic' mode that provides firmer suspension, and an 'off road' mode that increases ground clearance by an inch and a half or so (40 mm).[39] Drivers can even define their own individualized mode.

But these systems need not be simply reactive. Systems might identify road conditions by scanning the road ahead and provide preemptive accommodation for a superior response. An ingenious example is provided by the ClearMotion system initially developed by Bose. An ECU uses road sensor data to expand and retract electric actuators that accommodate changing road irregularities. The system can actually lift a wheel to skip over a pothole. The goal is simple, whatever road irregularities and bumps, whatever the driving conditions, tight turns or hard braking, the system adapts and the car stays level, a skyhook.

While this sort of full reactive control is not yet in production, proactive suspension tuning certainly is. Mercedes-Benz Magic Body Control is a notable example. An accelerometer is paired with a forward stereo camera at the windscreen to identify road conditions ahead and preemptively tune the suspension accordingly. Similarly, Audi uses a front mounted camera to detect bumps in the road ahead; and an electronic chassis control system adjust the suspension accordingly. An electric motor that can provide nearly half a ton of force at each wheel does the adjusting, and makes a 48-volt system necessary. These systems don't do much for road holding, but they can increase passenger comfort considerably.

In fact, passenger comfort is a major concern, and likely to get much more important as automated driving systems become more common. Systems such as these could become even more vital with the development of self-driving vehicles. As passengers read, sleep, or chat while riding, the propensity for motion sickness will increase. So, systems that can counter vehicle motion may become indispensable.

---

[38] X.D. Xue, K.W.E. Cheng, Z. Zhang, J.K. Lin, D.H. Wang, Y.J. Bao, M.K. Wong and N. Cheung, Study of Art of Automotive Active Suspensions 2011. *4th International Conference on Power Electronics Systems and Applications*, 2011.

[39] Y. Elattar, S. Metwalli and M. Rabie. PDF versus PID controller for active vehicle suspension. Conference: AMME-17, At Cairo, Egypt, 2016; and J. Ekchian et al., A high-bandwidth active suspension for motion sickness mitigation in autonomous vehicles. SAE Technical Paper 2016-01-1555, 2016.

Bringing It All Together

Increasingly, the addition of advanced **Vehicle Control Systems** (VCS) like active suspension, electronic stability control, antilock braking systems, as well as additional systems that we'll discuss in Chapter 9, require harmonization. The conventional model of having each system independently controlled by a distinct ECU increasingly does not make sense. An integrated system is needed to synthesize steering, suspension, throttle and braking functions and ensure that they behave in complementary ways.

The addition of active steering systems can make integration even more necessary. With a history dating back to the introduction of electronic power steering (EPS) in the 1980s, these systems have become increasingly sophisticated, intelligent, and common. Early on, EPS provided greater efficiency, improved steering feel and reduced vibration or kickback from the road. Four wheel steering (4WS) has been around for a while too. At low speeds, these systems can turn the rear wheels the opposite direction of the front and significantly improve maneuverability. At high speed, the rear wheels turn slightly in the same direction as the front and significantly improve stability. Snaking at low speed, crabbing at high speed. More recently, **active front steering** (AFS) adjusts the relationship between steering angle and steering inputs to accommodate driving conditions and speed. And now, the active four wheel steering available in a few performance cars produces stability and control at high speed and ease driving at low speed by intelligently modifying steering control at all four wheels. Nissan's 4-wheel active steer (4WAS) is probably the most notable example, with variable active front and rear wheel steering.

These systems can be integrated with active suspension, ESC and ABS, to produce what is variously called **Integrated Chassis Control** (ICC), Unified Chassis Control (UCC), or Vehicle Dynamics Management (VDM). The aim is improved synthesis of multiple control functions and sensor data to enhance stability, handling, passenger comfort, and NVH management. The result for the driver is not only improved safety and a more comfortable ride, but also the ability to customize a car's performance easily, as the basic parameters of the system can be effortlessly adapted to suit driver preferences.

An innovative example of VCS coordination is provided by Nissan's ingenious **intelligent ride control** system. Rather than rely solely on active suspension to address vehicle pitch, Nissan brings brake and throttle control into the game. You may have already noticed that depressing the accelerator can cause a nose up pitch called **squat**, and releasing the accelerator produces a nose down pitch called **dive**. Hitting the brakes can cause a similar dive. So, why not use these dynamics to accommodate road irregularities? Coordinated with active dampeners that moderate road motion, intelligent ride control can counteract the pitching caused by road bumps by varying engine torque or applying braking ever so slightly. So, as the Nissan Micra passes over a bump, for example, the wheel speed fluctuates slightly, vehicle motion and wheel speed sensors identify the shift, and engine torque is quickly and slightly modified to reduce the effect. If the pitching is more significant, slight braking is applied, not enough to slow the vehicle significantly, but just enough to help calm the pitch motion and maintain a more level ride.

In sum, the simple mechanical connections of the past are being replaced with intelligent, active actuators that don't just move mechanical components; they determine driver intent and modify system operations accordingly to produce the targeted motion. So, for example, steering is no longer about aiming the front wheels through a mechanical connection. Effectively, the steering wheel allows the driver to express an intention, and the system determines the mechanical actions that will best produce it. That may mean a change in steering angle; it may be the application of differential braking or torque, a throttle position change, selective roll dampening or suspension adjustments to achieve the designated maneuver as safely and comfortably as possible.

The overall aim is real-time monitoring and control of every dimension of the car's motion and orientation. The vehicle's six degrees of freedom—lateral, longitudinal and vertical motion as well as roll, yaw and pitch—need to be known and synthesized continuously to define the interoperation of vehicle controls. The goal is optimal control of the vehicle along every axis, beyond the trade-offs and toward a full synthesis of peak handling and optimum ride quality.

Modularity

In case it's not obvious by now, the fabrication, assembly and operation of the vehicle chassis is getting increasingly complex. Manufacturers are using polymers, reinforced composites, and multiple alloys of aluminum, steel, and even magnesium. As discussed, the process requires a variety of creative joining methods, innovative fasteners and adhesives. With each step in the diversification of materials, the level of difficulty and the complexity of the challenges increase. Add to this advanced active suspension technologies and sophisticated chassis control systems, and this starts to get very particular, tailored, and extremely expensive. As a result, carmakers are now looking for opportunities to simplify and standardize their manufacturing processes and chassis designs in order to reduce costs.

For an increasing number of carmakers over the past decade, the answer has been in modular design. This is not entirely a new idea, of course. The notion goes back to the body-on-frame system's ability to accommodate a variety of body types. And automotive manufacturers have long relied on standardized platforms for different models in the same class. Chrysler's K cars from the 1980s are a hallmark example, but far from alone. The Volkswagen Beetle and Karmann Ghia shared the same platform; and today the Audi TT and Volkswagen Golf use largely the same framing. The idea is to maximize the number of common parts, increase economies of scale, and simplify supply lines and manufacturing flexibility.

However, this isn't just about a common undercarriage anymore; the modular concept is now being expanded to define assemblies and componentry that are scalable, with variable structural dimensions (Image 7.18). So, a common architecture can define vehicles in multiple segments with differing wheelbases and track widths.[40] This is not a fixed platform, but rather a set of assemblies and practices that allow for universal assembly methods and componentry. Volkswagen's Modularer Querbaukasten (MQB), or modular transverse matrix, provided one of the first major examples, defining a full modular design for a common front-engine, front wheel drive chassis. The system permits variation in nearly every longitudinal dimension, allowing it to underpin everything from a subcompact model to an SUV. To differing degrees, nearly every major manufacturer has moved toward some version of a modular assembly system.

---

[40] J.F. Lampón, P. Cabanelas and V. Frigant, The new automobile modular platforms: From the product architecture to the manufacturing network approach. Munich Personal RePEc Archive Paper No. 79160, 2017. Available at https://mpra.ub.uni-muenchen.de/79160/

IMAGE 7.18
Toyota new global architecture.

Toyota is one of many carmakers who have moved toward modular design and shared architecture. Toyota New Global Architecture (TNGA) was adopted by the carmaker in 2015 in an effort to more fully integrate the development of major chassis and powertrain components and facilitate the strategic sharing of parts and components. The aim is to simplify production and the development of new models, and especially to reduce costs.

Image: Toyota Motor Corporation

There are two key reasons this is appealing to carmakers: cost savings and manufacturing flexibility. The most obvious cost advantage comes from the economies of scope and scale in manufacturing. However, for the major carmakers, an additional advantage is the capacity to outsource entire modules rather than just basic componentry. This pushes the cost and risk of product development, engineering, and design to suppliers, significantly reducing the exposure and investment of the carmaker. With the increasing sophistication, complexity and cost of suspension, steering, and chassis modules, carmakers have good reason to want this shift. In addition, a modular system offers them operational flexibility. Production can be quickly shifted from one plant to another rapidly and globally. And in general, vehicles can be assembled more quickly, relying on just-in-time delivery of major modular assemblies, reducing idle inventory and allowing faster response to market demand. Conceptually, vehicles could actually be made to order and delivered to a dealer within a week.

This modular idea may take on an entirely new scope with the expansion of electric vehicles. With drive-by-wire systems quickly becoming more universal, the links between the cabin, body, drivetrain and chassis are electrical, not mechanical. For electric vehicles in particular, this allows the production of a common rolling platform, a fully functional powered chassis that only need be connected to the passenger cabin and body through a wire harness. This universal rolling platform could then accommodate multiple body types as a bolt-on structure. The poster child for this is Tesla's Model S rolling chassis, or skateboard, that could once be seen at any Tesla showroom (Image 7.19). Composed of a basic frame, suspension, wheels and of course an electric motor, the striking characteristic is the absolute simplicity of the thing. Just bolt on a body, a couple of seats and some batteries and you're ready to go. BMW's i3 provides a similar example, built in two modules:

IMAGE 7.19
Tesla skateboard.

a rolling chassis dubbed the 'drive module' and the so-called 'life module' defining the cabin and body.

Variations on this theme—a basic electric platform that can accommodate multiple body types—are being explored by several carmakers. The idea of an essential platform, optimized for crashworthiness, handling, and performance, and tunable with simple software modifications, that can serve as a universal base for multiple models and vehicle types is more than a little appealing. In addition, with such a system the possibilities for control open up dramatically. Unlike a conventional ICE system and drivetrain, it's possible that each wheel could be controlled independently. And, with a wheel hub system, steering and torque at each wheel might also be selectively controlled. The increased unsprung weight might be adequately managed by an advanced active suspension system. This opens up the possibility for what can be called eight variable control.[41] Forget basic four-wheel steering and traction control, the force, position and direction defined by each wheel could be selectively determined. The range of motion possible could define new capacities for handling, maneuvering, and even parking. This may mean the already-remarkable shift to electric drive will be that much more revolutionary and extend well beyond the drivetrain.

---

[41] M. Abe, Trends of intelligent vehicle dynamics controls and their future. *NTN Technical Review* 81, 2013.

8

The Power of Shape

Aerodynamics has long been the redheaded stepchild of automotive design. True, cars have been visually streamlined for quite some time. The fins of the '59 Eldorado or the art deco flow of a '30s Delahaye are a couple of favorites, and they all reflect an appreciation for flow and movement. But these lines were about style, not function. In fact, often these sorts of streamline features get in the way of reducing drag and noise. Through the 1980s, with the exception of racecars, aerodynamic design largely remained an afterthought, to be added to a defined body near the end of the design process when convenient. No surprise, the typical sedans were boxy, with ill-defined aerodynamics, and sometimes moved through the air like a brick through molasses.

More recently things have changed. True aerodynamics, not just streamlining for stylistic charm, has taken a more prominent role in car design. There are three reasons for this: first is the importance of fuel economy. Faced with fuel economy regulation in Europe and the US, automakers are keen to squeeze out every last drop of efficiency; and they have noticed that small changes in shape can produce significant changes in highway economy for minor changes in production cost. Second, the need to reduce cabin noise and flow, a prerequisite for the ride quality expected by today's consumers. Third, while this factor is decreasing a bit, the top speed of a performance car is still a salient selling point, particularly in the European market. And aerodynamics has a big role in defining a car's top speed. Top speed can serve as a metric of the vehicle's overall capacities for performance buyers, whether they will ever reach those speeds or not, much like American buyers tend to fixate on 0–60 times.

The results are not negligible. Very often, a 10% decrease in drag can result in a much higher reduction in fuel demand than a similar reduction in vehicle weight; and modest changes in body shape can potentially reduce drag with almost no increase in cost. So, we can expect this transformation to continue, redefining the shape of our daily drivers to enhance efficiency, performance, and ride quality.

As a result, we have seen increasingly sophisticated aerodynamic features in high-end production cars; and we've seen these features trickle down through the fleet. In fact, a concern and awareness for aerodynamics has literally reshaped the modern car. Curved rooflines are now ubiquitous. Equally common are the sharp rear corners that ensure an aerodynamically sound clean break for airflow. Slots, breathers, and louvers that manage airflow around wheel wells are so common and desirable that many times they are added simply as an esthetic feature.

As you might expect by now, the cutting edge of innovation goes well beyond these common examples. Like so many fields we have examined, automotive aerodynamics has been defined by trade-offs. Most notably, racecars and high-end performance cars had to negotiate two priorities: the reduction of lift, allowing the car improved road holding and cornering, and the reduction of drag, enabling higher top speeds and improved high-end acceleration. And, very often, reducing one meant increasing the other; so compromises were unavoidable. However, once again as you might expect by now, the use of digital control has increasingly allowed us to have our cake and eat it too. By quite literally changing the shape of the vehicle as it travels, we are able to continuously adjust a vehicle's aerodynamics to manage drag and lift, optimizing each as driving conditions demand. But before we get to that, we need to understand some of the basics.

The Nature of Drag

The defining element of automotive aerodynamics is **drag**, defined formally as the resistive force to the flow of an object through a fluid. In this case, the fluid is air and the object is a car. So, drag is the force resulting from airflow that resists the forward movement of the car. At highway speeds, up to 75%–80% of a vehicle's resistance to motion can be defined by aerodynamic drag.[1] This makes drag a focal point of aerodynamic design in production cars, and a key to improved highway fuel economy.

Sometimes when folks imagine drag they tend to think about the roughness of the surface of the car. The thought is that defining a smoother surface will allow improved slipperiness through the air and so lower drag. However, in reality, this sort of drag, called **surface drag**, is a relatively minor component of overall automotive drag. So, that bottle of wax that promises to improve your fuel economy isn't *entirely* wrong: over the course of a few hundred thousand miles, it should save you a gallon or so of gas. The most significant form of drag is defined by the shape of the car, not surface friction. Though, this does not mean that we can ignore skin friction; as we'll see, it interacts with the vehicle's other characteristics to define overall aerodynamic performance.

The more substantial component of drag is the result of the shape or form of the car, so we call it **form drag** (or sometimes pressure drag). As a vehicle travels through air, the air needs to accommodate the car's movement, to move around it. This results in differences in airspeed over the surface of the car, which in turn defines different zones of varying pressures. Higher pressures in the front of the car have a component that pushes backwards on the car; and lower pressures at the rear of the car similarly has a component that *pulls* it backward; both therefore resist the car's movement. These pressures vary with the shape of the car. So, as we'll see shortly, cars with identical cross-sectional areas can exhibit significantly different drag.

The energy required to displace that air is calculated similarly to other forms of mass in motion. Here's the basic equation for the energy of a mass in motion, called kinetic energy:

$$KE = \frac{1}{2}mv^2$$

In this case, instead of a solid mass, we have a displacement of air; so, we can use air density and the car's cross-sectional area to define the size of that displacement, substituting those two variables for the mass. If you think about it, it makes sense that the amount of air that needs to move will be defined by the cross-sectional area of the vehicle. A boxy van, with a large front cross section, will need to displace more air than a sedan, even if they are the same length and weight. And it makes sense that air density is a key factor; the heavier the air, the more energy required to move it. It's worth noting quickly, that since air density changes with humidity and temperature, the actual drag on a car will change a bit with weather conditions, improving with warm dry weather when the air is thin.

[1] W.H. Hucho, Introduction to automobile aerodynamics. In W.H. Hucho (ed) *Aerodynamics of Road Vehicles: From Fluid Mechanics to Vehicle Engineering*, 4th edition. SAE International, Warrendale, PA, 1998, 1–46.

What's really worth noting here is that drag is a function of the velocity *squared*. This is important. It means aerodynamic drag grows exponentially, and so gets increasingly important as speed increases. You might remember the graph in Chapter 3. A small increase in speed while modesty driving around town might only increase a vehicles' drag by a tiny bit; but that very same incremental increase at highway speeds would result in a much greater increase in absolute drag force. Similarly, improved aerodynamics of a car might have a major impact at high speeds, but no discernible effect at all when cruising around town. This is why aerodynamics has been so important to racecars, but historically discounted in many production cars.

The implications for efficiency and general performance should be clear. While aerodynamics may not matter much in defining city mileage, it can have a much more significant impact on highway fuel economy. And, while the shape of a car may not impact its 0–60 time, it can have a major effect on that car's top speed. In fact, conceptually, aerodynamics can delineate the top speed. As we saw in Chapter 3, the speed at which the drivetrain's capacity to produce acceleration is met by the increasing aerodynamic drag will demarcate a maximum theoretical speed. Though, in reality, it's more likely that software will limit the car's top speed based on regulatory or mechanical limits and safety.

To really understand how this all comes together, we need to look at the shape of the car. More precisely, how efficiently an object moving through air is able to split the air in front of it, redirect the air around it, and allow that air to reconnect behind it, and so minimizing the air's resistance to its movement.

The Power of Shape

Here's an interesting fact that early car designers often missed: Cars are not airplanes. You've probably already noticed this, and maybe even cursed this fact the last time you were stuck in a traffic jam. What this means in terms of aerodynamics is that the air that travels across an airplane's wing in flight stays relatively ordered, smooth, and parallel to the plane's movement, in aerodynamics we call this sort of shape a **streamlined body**. However, the air that travels across a car's surface is generally much more unsettled, with the streamline flow breaking from the surface and tumbling into disorganized, erratic air movement. The smooth, ordered air, called **laminar** flow turns to disordered eddies of **turbulent** air. We call shapes like these **bluff bodies,** or sometimes blunt bodies. So, while early car designers often relied on cues from aircraft to shape their designs, we now understand that shaping the aerodynamics of buff bodies like cars is fundamentally different from the aerodynamics of streamline bodies like airplanes.

Similarly, unlike the smooth flow over an aircraft wing that can be usefully thought of as two dimensional, flowing linearly over the plane's wing and parallel to its axis, the flow over a car is fully three dimensional, with turbulent surface flow, irregular eddies, and complex interactions. Air moves longitudinally down the car's length, laterally or sideways across the car's surface, vertically, and in multiple combinations and variations of the three. In fact, this is a major reason why form drag dominates automotive aerodynamics while friction drag is a much bigger concern in aircraft design. Because the airflow is so much more disturbed by the movement of a car, the effect of surface friction ends up being a minor player.

To be able to understand how efficiently a car's shape is able to displace air, and so compare and assess the drag characteristics of different shapes and features on a car, we give shapes a numerical score, called a **drag coefficient** (C_D). The less form drag a shape generates as it moves through the air, the lower the C_D. So, this number is a key way to understand the aerodynamic performance of a car. Though it is not the only metric; as you'll recall, the frontal area of the car matters just as much. These two observations allow us to complete the transformation of our formula for kinetic energy into a formula for form drag by incorporating a measure of the resistance of the car, or any other shape, as it moves through air (or any other fluid):

$$\text{Drag Force} = \frac{1}{2}\rho V^2 A C_D$$

where
  ρ, air density
  V, velocity
  A, cross-sectional area
  C_D, drag coefficient.

If we want to reduce the drag on a car, we don't have a lot of options. We could lower the velocity, but that sort of defeats the whole point. There are limits to how much you can reduce the frontal cross-sectional area of an automobile and still maintain interior space, room for the engine, and other design priorities. These limits are pushed pretty hard in a Formula One car, but in a sedan, there are major constraints on the shrinking of the frontal area. The surrounding air density isn't really our call. So, it's clear that the key variable in the aerodynamic design of a production automobile is reduction in the coefficient of drag.

The basic nature of drag coefficient is reflected in the shapes below (Image 8.1). All of the shapes have the same cross-sectional area, though their C_D and drag change significantly.

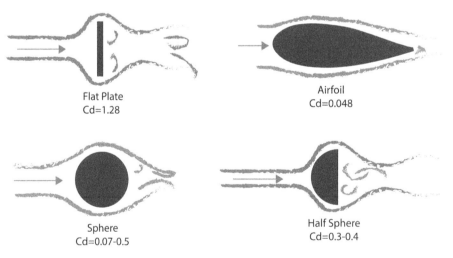

IMAGE 8.1
Coefficient of drag.

With all the objects having the same cross-sectional area, the drag produced still varies as a function of the object's shape. The more blunt the object's front, the greater the air pressure ahead of the object resisting movement. And the more the object is tapered in the rear, the less negative pressure pulling back on the shape.

Data: NASA and Hoerner, Fluid Dynamic Drag

Reducing the frontal positive pressure alone will reduce the C_D. That's why the half-sphere is an improvement on the flat plate. Changing the front and rear of our shape offers continued improvement. So, when we go from the half-sphere to a full sphere, we get a notable potential reduction in drag. This tells us that the back of the shape, where the air reconnects, can be just as important as the front. In fact, elongation that fills that trailing area of turbulence and allows the air to reconnect more smoothly, as a teardrop or airfoil shape, shows a major reduction in C_D. The key here is the distinction between the parallel, well-ordered lines of **laminar** flow and the swirling, irregular lines of turbulent flow. Ideally, we want to maintain laminar as much as we can. So, when we elongate the shape even more, stretching it into an airfoil, we get a continuing reduction in C_D by largely avoiding the turbulent wake.

Boundary Layer

A key to understanding aerodynamics is knowing when airflow will erupt into turbulent flow as it breaks away from the object's surface. When air travels across a surface with little turbulence, we call that **attached flow**. In this condition, as the airflow gets closer to the surface, it slows. This is because as we near the skin the air is increasingly affected by surface friction. We can imagine that right at the surface, at an impossibly small distance from the car's skin, the air is pulled along with the car itself, and there is zero relative movement. As we move away from the surface, the speed increases until we get to the maximum speed of the air, or rather the speed of the traveling car. We call this thin layer of flow at the surface the **boundary layer** (Image 8.2). Within this zone of slowed air the internal friction, or viscosity, of the air and surface friction of the moving car define a sheer stress that draws out an air velocity gradient, like the image below. As we move downstream,

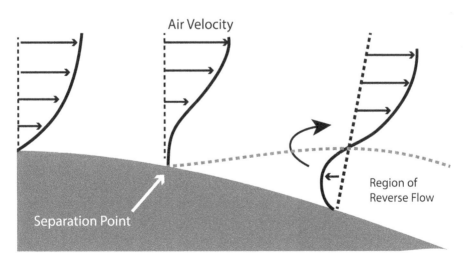

IMAGE 8.2
Airflow separation.

The boundary layer is defined by the slowing of the air near the surface due to surface friction and the viscosity of the air. As the surface drops away from the airflow, the air detaches from the surface, defining a zone of turbulent reversed airflow and increased pressure drag.

the thickness of this layer tends to increase. So, while surface friction is not itself a major component of overall automotive drag, it shapes this thin layer which strongly shapes the entirety of the flow around the car and the determination of drag.[2]

Typically, on the backside of a car this gets tricky. As the back surface slopes down, it pulls away from those molecules of air that were being pulled along the surface. So, as the rear window, for example, drops away, it defines an area with decreasing air pressure, a vacuum, which can pull some of the flow down and define a swirl of turbulent air. Where the airflow breaks away from the surface is the **separation point**. What follows separation is a disturbed flow defined by eddies of irregular movement that increase drag, noise and even surface soiling and water deposits in its wake. The sooner separation happens, the larger the area of low pressure and so the greater the drag. This is why our teardrop and airfoil shapes had a much lower C_D than the sphere; they delayed or avoided separation. So, a key to automotive aerodynamic design is to delay or avoid separation.

As mentioned, the creation of turbulence due to early separation can also increase aerodynamic sound and contribute to overall noise, vibration, and harshness (NVH). Because drag is closely associated with NVH, designers can sometimes be doubly concerned with separation, as it also increases cabin noise. However, aerodynamic drag and NVH are not always clearly associated, and the reduction in one does not always mean the reduction in the other. Noise reduction tends to be most closely related to managed airflow at the A pillar, for example, while drag reduction is much more dispersed across the vehicle surface. So, identifying priorities and assessing the value of various design options can be difficult.

The Shape of a Car

With all this in mind, we can start thinking about the particulars of how the air moves around a moving automobile, and what impact that has on the car's performance. Airflow over and around a car is determined by the car's shape, and in turn defines a topography of air speed changes across the vehicle's surface. Where the surface protrudes outward, the airspeed must increase to get around the obstruction (we'll see why in a bit). Where the surface retracts, airspeed slows. These variations in airspeed define variations in pressure. And, as we've seen, these changing pressure gradients tear the flow away from the surface, defining separation, but can also allow the flow to re-form as attached laminar flow. Mapping where and how this happens across the cars surface is the starting point of aerodynamic design.

Fundamental to this effort is an understanding of the relationship between air speed and pressure. In fluid dynamics, **Bernoulli's Principle** says that when a fluid increases in speed, it decreases in pressure. So, if air is forced through a narrow channel, called a venturi, it will speed up as the channel narrows. With a smaller cross-sectional area, molecules need to speed up to make it through the restriction; much like you might be pushed quickly through a narrow doorway as a slow-moving crowd exits a stadium. The upper surface of a car functions in a similar way. By forcing the air to travel the longer distance over the top of the car, the speed increases. And, according to this principle, as the speed increases the pressure decreases (Image 8.3). Since the bottom of the car is flat, this effect causes a net negative pressure on the top of the car resulting in a lifting force on the car.

---

[2] D. Hummel, Some fundamentals of fluid mechanics. In W.H. Hucho (ed) *Aerodynamics of Road Vehicles: From Fluid Mechanics to Vehicle Engineering*, 4th edition. SAE International, Warrendale, PA, 1998, 47–82.

However, a venturi is a pretty simple shape, and a car is not. So, the way the topography of pressure lays out on the surface of the car isn't quite as regular as it is through a venturi (Image 8.4). For example, the positive pressure on the front of the car defines a major source of drag. If you've ever put your hand out a car window, you've felt this drag. We might call

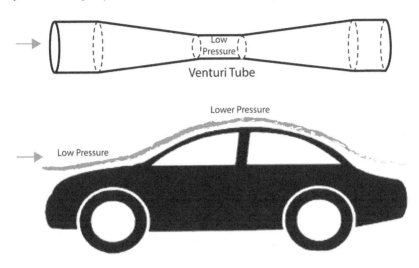

IMAGE 8.3
Airflow venturi.

Just as airflow passing through the restriction of a venturi must speed up and therefore lose pressure, the air traveling over a car must also speed up; and the reduced pressure defines lift on the car's upper surfaces.

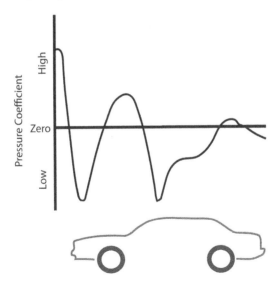

IMAGE 8.4
Pressure change over the automobile.

As air flows over the automobile, it defines zones of differing pressure. High pressure is formed at the front of the car. As the air travels over the hood, it gains speed and drops in pressure. A more modest high-pressure area is defined as the base of the windshield. The lowest pressure is typically found at the top of the car, where airspeed is highest. Pressure then rises as the air slows toward the back of the car, with a drop in pressure when the airflow begins to separate from the surface.

this Newtonian drag, as it's the simple result of Newton's third law of motion: for every action, there is an equal and opposite reaction. As the car hits the air, it pushes against it, and the air in turn pushes back.

As this high-pressure air spills onto the hood, it speeds up, defining a zone of lower pressure and associated lift as we discussed. Hitting the lower windscreen, another Newtonian high-pressure point is defined. And, once again, as the air speeds over the top of the car, it defines a major source of low-pressure lift. There may then be a small rise in pressure on the trunk deck as the air slows, and a final drop into turbulent flow off the back of the car. Understanding the innovations that allow us to shape and control these forces requires that we look at each of these pressure regions with a bit more detail.

The Front of the Car

The high pressure at the front of the car is the largest single component of form drag. Our ability to avoid this drag is limited since, as we've noted, the frontal cross section of the vehicle is defined by vehicle type, engine requirements, and other design requirements and cannot be easily modified for the sake of airflow. However, lowering the front hood and bumper to allow more air to easily move over the car can reduce the size of the high-pressure air pocket at the front. As a result, in general the nose of many cars has moved downward over the past few decades. However, global safety standards and the need for cooling airflow have limited this trend. In particular, European-defined safety standards have led to a rise in some hood designs to reduce likely pedestrian injury in a low-speed impact.

Tapering or rounding the front sides can allow the high-pressure, slow-moving air to flow around the front corners smoothly, reducing frontal pressure while also avoiding the increased turbulent flow along the side surfaces that might otherwise take place as this high-pressure air tumbles around a sharp corner. With smoother curves along the front of the hood and fenders, the air is able to reattach quickly. So, the resulting zones of separated turbulent flow, or **separation bubbles**, around the front of the car are minimized. Even a small radius curve can have a notable effect, and is reflected in the general pattern of front-end smoothing and rounding found on recent production cars.

The front of the car must also accommodate intake paths for cooling and induction. Once this was as simple as mounting a radiator in the path of oncoming air. However, now this has grown much more sophisticated, with defined pathways for engine cooling, brake cooling and potentially an oil cooler, turbocharger and intercooler, or in the case of EVs, battery, motor and power electronics cooling. So, the total cooling requirements on a modern vehicle have grown greatly, and require much more precise design. Balancing these cooling requirements with a need to adopt a smaller blunt cross section to address drag, while maintaining required space under the hood, can be difficult.

An innovative solution now used by multiple manufacturers is **active grill shutters** (Image 8.5). At low speed, when engine cooling is needed and aerodynamic drag is low, the shutters are open, allowing airflow over the radiator and other components. However, at high speeds these shutters close, sealing the front of the car and encouraging more outflow over the vehicle's top and sides, reducing frontal pressure. An ECU can control shutter position based on vehicle speed and coolant temperature, closing all or some of the shutters, and improving not only aerodynamics but also engine warm-up and aerodynamic noise.

IMAGE 8.5
Active shutters.

Active Grille Shutters (AGS) allow for a balancing of the aerodynamic benefit of rerouting air around the engine compartment and the cooling effect of airflow when needed.

Image: Röechling

IMAGE 8.6
Grill-less EV.

The reduced demand for cooling airflow allows EVs to avoid the aerodynamic disadvantage of a large open front grill.

This technology is particularly useful for blunt fronted vehicles with a normally unobstructed grill. So, while an SUV might see a 5% drop in C_D with shutters closed, a sedan might see half of that advantage or less.[3] So, for example, the active shutters on the Chevy Cruze Eco and the diesel RAM 1500, are likely to offer a bit more bang for the buck than a similar feature on a more streamlined sedan.

Of course, it would be even more effective to eliminate the grills all together, now possible in EVs. Though EVs may require airflow for battery and power electronics cooling, the reduced need for airflow allows for a reduction or elimination of the front grill (Image 8.6). This has helped Tesla, for example, design cars with notably low drag coefficients.

---

[3] G. Larose, L. Belluz, I. Whittal, M. Belzile, R. Klomp and A. Schmitt, Evaluation of the aerodynamics of drag reduction technologies for light-duty vehicles: A comprehensive wind tunnel study. *SAE International Journal of Passenger Cars—Mechanical Systems* 9(2), 2016, 772–784.

The Model S defines a C_D of 0.24, an impressive accomplishment for a four-door sedan, even though it's a bit higher than the 0.21 the company had initially targeted.[4]

Though less significant, high pressure also forms at the base of the windscreen. This is a bit more difficult to release, since it's not as easy to curve or taper the lower screen. An increased windscreen rake angle helps; but this can also impact visibility and headroom. And while increased inclination reduces drag, this is only true to a point, after which a continued increase has relatively little effect.[5] Similarly, the line at the upper edge of the windscreen, where is connects with the roof, will also be reduced to avoid separation and associated form drag as the high-pressure air transitions to the low-pressure top. The Tesla above offers a striking example with a large curved windshield that is smoothly integrated into the roofline. But this is only one of the more notable examples of an industry-wide trend toward smoother integration of the windshield with a curved roofline.

Another avenue for the reduction of front-end drag is a lower body-panel extension that can redirect flow from under the car. A front **air dam** can help place more high-pressure airflow over the top of the car, and by slowing the airflow under the body it can reduce drag caused by the undercarriage (Image 8.7). However, as we'll see, with more streamlined underbody panels being adopted, this advantage is diminished. Nevertheless, an air dam can reduce front-end lift, and if properly designed, can also increase cooling airflow by redirecting it appropriately, though the overall impact is not major.

Beyond some modest possibilities, there are not many options left for continued improvement of the front of the ICE automobile. So, the back of the automobile has taken on much greater relative importance.

IMAGE 8.7
Active air dam.

The active air dam produced by Magna and used in the 2019 Ram 1500 pickup truck deploys automatically to redirect air around the vehicle offering a 7% drag reduction and improved fuel economy. It is the first of its kind on a high-volume vehicle.

Image: Magna

[4] D. Sherman, Drag queens: Five slippery cars enter a wind tunnel; one slinks out a winner. *Car and Driver* June, 2014, 86–91.
[5] D. Hummel, Some fundamentals of fluid mechanics. In W.H. Hucho (ed) *Aerodynamics of Road Vehicles: From Fluid Mechanics to Vehicle Engineering*, 4th edition. SAE International, Warrendale, PA, 1998, 47–82.

Addressing the Rear Wake

The importance of the rear of the car to overall vehicle aerodynamics is often underappreciated by the typical gearhead. About a third of the total pressure drag on a car is defined at the rear.[6] And while the frontal pressure on a car tends to be predictable and constant, the pressure on the rear fluctuates erratically and can be influenced by body shape much more than the front, making it an important area to improve aerodynamic performance.[7]

As airflow drops off the blunted rear of a car it defines a low-pressure wake. The core of this wake is a **recirculation zone** defined by relatively slow swirling eddies that define a turbulent bubble, a gurgle of disturbed air, with relatively little passage of air through the zone. This is sometimes called **dead water** since this same sort of effect is defined in the water behind a moving ship. This low-pressure turbulent wake is a major drag source. All things being equal, the larger the cross section, the larger the effect. So, a large squared back, say from a boxy hatchback or van, defines a larger low-pressure wake (Image 8.8). A sedan, with a more modest blunt back, will generally define a smaller wake. Think of the car like a piston drawing through a cylinder of air. As the piston draws away, it creates a vacuum behind it; and that relative vacuum pulls in all directions, resisting the movement of the piston that created it.

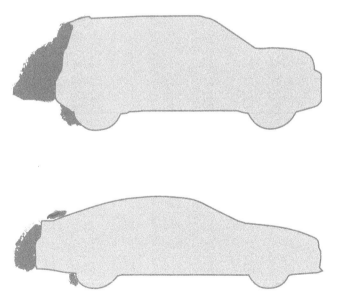

IMAGE 8.8
Dead water.

The blunt back of a boxy vehicle such as the Kia Soul demonstrates significant aerodynamic disadvantage when compared to significantly smaller dead water zone defined by a more fastback shape such as the Optima.

[6] G. Rossitto, C. Sicot, V. Ferrand, J. Borée and F. Harambat, Wake structures and drag of vehicles with rounded rear edges. *Proceedings of 50h 3AF International Conference on Applied Aerodynamics*, March 29, 2015—April 1, 2015, Toulouse, France.

[7] K.S. Song, S.O. Kang, S.O. Jun, H.I. Park, J.D. Kee, K.H. Kim and D.H. Lee, Aerodynamic design optimization of rear body shapes of a sedan for drag reduction. *International Journal of Automotive Technology* 13(6), 2012, 905–914.

Addressing drag at the rear can be more challenging than at the front of the car because it is complicated by the temperamental nature of separation. If the rear slopes more gradually, as a fastback sedan, the airflow is able to follow the body contour more easily while maintaining greater laminar flow. With a notchback, the separation may occur at the roof drop, so a smooth shape that allows the reattachment of the air and fills the vacuum could be key. In general, a drop-off angle, or slant angle, somewhere around 15° can allow for a maximum reduced drag coefficient while maintaining attachment or allowing reattachment at the rear trunk.[8] As the angle increases and defines a more severe slant angle, the drag generally gets larger, up to a critical angle of about 30° (Image 8.9). After this, a more severe angle drop from the roof to the rear window can trigger separation with no reattachment at the rear hood and the drag remains pretty much unchanged. Of course, the angle needs to accommodate other design priorities like rear visibility and trunk space. So, decreasing the angle drop-off from the roof to the rear window, perhaps accompanied with a raised trunk height to allow for the gradual rake, can reduce drag. However, if a gradual taper cannot be achieved within design parameters, the option may be a more severe drop that definitively separates flow and achieves a modest, though higher than minimum, drag.

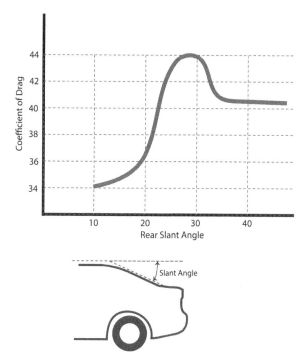

IMAGE 8.9
Rear slant angle and drag.

Generally, a small slant angle has the least impact on drag. As the severity of the drop-off increases toward 30°, drag increases significantly. Beyond that point, the drag decreases somewhat then remains relatively constant as the slant angle increases.

Source: Hucho, Aerodynamics of Road Vehicles

---

[8] H. Fukuda, K. Yanagimoto, H. China and K. Nakagawa, Improvement of vehicle aerodynamics by wake control. *JSAE Review* 16(2), 1995, 151–155.

The results of these dynamics on the general shape of production cars may be clear to you at this point. Trunk decks have gotten higher on cars (at least the ones that maintain a distinct trunk line). Fastback hatchbacks have become more common. And arched roofs are now nearly ubiquitous (Image 8.10). This arching can help reduce transition angles from the windcreeens to the roof. Too sharp a bow, of course, can have the oposite effect, triggering separation. But a suficintly gradual bowed roofline can reduce the presure peaks where the roof meets the glass in the front and help avoid early separation in the rear. This can have the advantage of also increasing head room and maintaining desierable body lines. Of course, if the bow increases the frontal cross section of the car too much, the advantage can be lost. A 5% decrease in C_D can't be paid for by a 10% increase in frontal cross-sectional area.

Of course, one way to reduce the dead water at the rear is to provide a more gradual rear slope by elongating the car. The longer the car, the more gradual the transition can be, and so the more laminar the airflow (Image 8.11). Think of the low drag coefficient of a teardrop shape with a long tappered back. However, there are pretty clear limits to this strategy. Nobody wants a 20 ft long car just to improve aerodynamics.

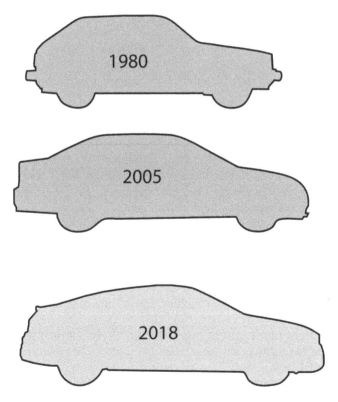

IMAGE 8.10
The evolution of shape.

The evolution of the Honda Civic's rear shape over the years reflects changing aerodynamic design. Maintaining a smooth and continuous roofline with an angle shift of less than 15° at the front and rear glass can be helped with a curved roofline. By starting the downslope before the window and raising the trunk deck the 2005 Civic pictured offered a notable improvement on the earlier Civic. By 2018, the roofline had gown more bowed with a fastback shape that reduced the blunted rear cross section and resulting wake.

IMAGE 8.11
Reducing rear separation.

The Mercedes-Benz 'Concept IAA' takes the reduction of rear wake a step further with the capacity to deploy a rear body extension and actively increase its length by over 15 inches (390 mm).

Image: Wikipedia

IMAGE 8.12
The teardrop.

The Toyota Prius, with an impressive C_D of 0.24 has clearly learned something from the teardrop.

With a rounded front, flat bottom and elongated tail, the aerodynamic shape we seem to be moving toward is something like a teardrop. This shouldn't be a surprise, after all the reason a waterdrop has the shape it does is because it is formed by the flow of the air around it, molding it with friction and shaping it into its lowest resistance form. In fact, the notion of a teardrop ideal has clearly influenced car design. With rounded fronts, curved bodies and tapered rears all clear trends in car design. In fact, if efficiency were our only concern, a teardrop shape would be perfect (Image 8.12).

If the teardrop shape is so slippery, you may ask, then why isn't this the shape of Formula 1 cars, and why don't the leading supercars adopt this shape? The answer requires that we put drag coefficient back in context. Remember, when we consider C_D, we're only looking at the characteristics of the overall vehicle shape; but there are a whole lot of other factors to consider. First, the racecar is also trying to produce downforce. And, second, the front cross-sectional area is just as important as the C_D. And decreasing and lowering the frontal area changes the shape options available. So, while the super sleek Koenigsegg Agera RSR maintains a low center of gravity and minimal frontal area, it only manages a C_D of 0.37. And a typical Formula One car may be orbiting a C_D of 1.0. The need for engine and brake cooling can also require a larger C_D in a performance car. In fact, the industry's lowest C_D's tend toward larger sedans. The Tesla Model 3 has a C_D of 0.24. The G30 BMW 5-series offers an impressive C_D of 0.22. Hyundai's Ioniq has a C_D of 0.24. And Mercedes CLA comes in at a remarkable 0.23, and the more environmentally friendly CLA 180 Blue Efficiency at a slightly lower 0.22.[9] None of these are supercars by any stretch. Remember, the C_D is a measure of the aerodynamic efficiency of the car's shape; it is a key factor in determining drag, but not a total measure of the car's drag (Image 8.13).

---

[9] Top 100 cars ranked by drag coefficient. Available at www.carfolio.com

IMAGE 8.13
The importance of the backend.

Volkswagen's experimental carbon fiber plug-in hybrid XL1 gets 230 mpg. Achieving the lowest drag coefficient in the industry at 0.20 is principally the result of modifications of the back, not the front, of the vehicle.

Three Dimensional Flow

So far, we've discussed the flow over the top of the vehicle as though it occurs along two dimensions, longitudinally following the length of the car with no sideways, or lateral, movement. But, as previously mentioned, the actual flow is more complicated. After all, cars are three-dimensional objects (the more astute among you may have noticed this already), A key element to this three-dimensional flow is the difference in pressure between the top and the sides of the car. Remember that due to Bernoulli's principle, the pressure over the top of the car is generally lower. As a result, the relative high pressure on the sides of the car tends to move toward the low pressure on the top. This movement of air from the sides to the top creates vortices off the end of the roof as the car moves, defining conical high-energy spirals trailing the car (Image 8.14).

IMAGE 8.14
Rear vortices.

The lower pressure above the car and relatively higher pressure on the sides of the car cause a lateral movement of air that, when drawn by the wake, define spiral vortices trailing the car.

The effect is that the resulting high downwash on the back slant of the car can actually decrease the separation of the upper boundary layer and results in decreased drag.[10] But this depends greatly on how you shape the back of the car. There is a limit to the advantage, of course. If the rear slope angle is increased, drag reduction drops and vortex drag increases, defining an overall undesirable effect.

The same sort of logic that reduces wake along the top contour of the car can be used to design the side geometry that shapes wake vorticies. The intensity and shape of the three-dimensional vorticies that trail the car are principally determined by rear end geometry. So, like the top, tappering the rearward sides of the car to reduce the rear cross section and manage the transition of side flow toward the rear center can be advantageous (Image 8.15). Called boat tailing, this narrowing of the rearward sides can reduce the volume of the rear wake with only a small change in vehicle styling. Even a modest side tappering can have a significant effect, shaping beneficial vorticies, increasing the rear pressure and decreasing the wake area. Because the incremental advantage of additonal tapering drops significanlty, a long extended boat tail offers diminishing returns. So, we do not need to extend the rear of the car dramatically to reduce rear wake drag.

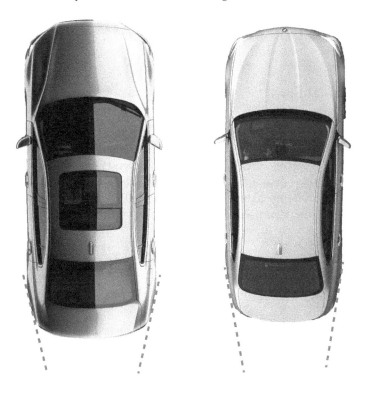

IMAGE 8.15
Boattailing.

This BMW 5 Series (left) and Volvo S60 (right) demonstrate the sort of boat tailing that is common in production cars.

Images: BMW & Volvo

---

[10] A. Thacker, S. Aubrun, A. Leroy and P. Devinant, Effects of suppressing the 3D separation on the rear slant on the flow structures around an Ahmed body. *Journal Wind Engineering and Industrial Aerodynamics* 107–108 (September), 2012, 237–243.

So, if we're not going to drive 20 ft long cars with extended pointy rears, at some point, the car needs to end and define a blunt rear face. But the best way to transition from gradual taper to a blunt end may not be what you expect. Recall that curving the corners at the front of the car allowed the high pressure to more easily move across the sides, reducing the size of the high-pressure bubble. But the flow geometry is entirely different at the rear, so we can't just apply the lessons of the front to the back. In fact, curving the rear shape can be problematic. A curved transition to a vertical rear panel can encourage an expansion of the low-pressure wake, effectively drawing the wake around the car's edges and increasing the resulting drag. Similarly, rounding the rear pillars or upright side edges can result in an overall increase in drag on the rear vertical surface.[11]

Alternatively, a sharp lip or edge on the rear trunk deck can force a clean break of the surface flow and weaken the downwash off the deck, reducing the negative pressure bubble on the rear surface. A small angular lift can tip airflow up at the end of the vehicle, and so help reduce the air being pulled downward into a turbulent, low-pressure wake trailing the vehicle. So, by defining a fixed separation point, the low-pressure bubble is contained by high-energy flow (Image 8.16). And by creating an air pocket above the rear trunk,

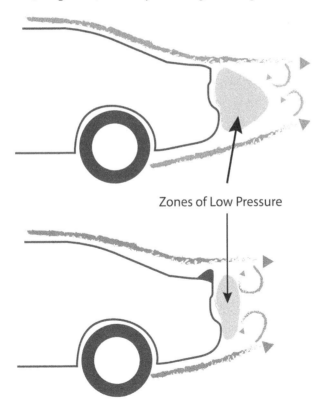

Zones of Low Pressure

IMAGE 8.16
Triggering separation.

A raised edge on the rear lip can allow a clean break of airflow that reduces the negative pressure on the rear surface.

[11] G. Rossitto, C. Sicot, V. Ferrand, J. Borée and F. Harambat, Wake structure and drag of vehicles with rounded rear edges. *Proceedings of 50h 3AF International Conference on Applied Aerodynamics*, March 29, 2015—April 1, 2015, Toulouse, France.

it can slow top flow and reduce lift while also reducing pressure drag.[12] A similar effect can be defined on vertical surfaces. For example, BMW defines a sharp-edged extension of the C pillar they call an 'air blade' to help with wake reduction. In fact, perhaps because the addition of a sharp lip or angular lift to trigger trailing-edge separation is a low-cost measure, it has become a common feature on nealry all production cars.

Vortex Generators

One of the more interesting ways to shape and trigger vorticies for reduced drag is lesser-used but worth examining as we think about the nature of airflow. The idea behind it is to intentionally trigger high-energy vorticies within the boundary layer as a way of avoiding premature separation. This may seem a bit counter-intuitive, triggering a form of turbulence to avoid separation into turbulence; but it works. We can do this with small blades on the surface, called **vortex generators**. This is not a new idea. Large aircraft have used vortex generators for decades to ensure airflow remain attached to the control surfaces of the wing.

Remember, the cause of airflow separation on a plane or a car is low energy air at the surface and the sheer force resulting from this friction. As the resulting pressure gradient grows more severe, the stability of the air under this sheer stress deteriorates, and at some point the airflow separates with the resulting eddies defining turbulent flow. Triggering stream-wise vorticies near the surface before separation can help energize the slow-moving air in the lower boundary layer by drawing in high-energy air from beyond the layer, relieving the sheer stress and delaying separation.

The same effect is used by golf balls. Yup, golf balls. Because they have no spin to stabilize them, we would expect golf balls to fly erratically, tossed one direction or another by the whims of fluctuating pressure over the surface. Knuckleballs in baseball exhibit this tendency. Traveling toward home plate with no spin they exhibit erratic and unpredictable motion as they're knocked around by the turbulent eddies they create. This is great if you're a pitcher, but not so good if you're a golfer. However, thanks to the dimples on the golf ball, they do not fly like knuckleballs. The dimples generate microvorticies of high-energy air along the surface, allowing attached flow across much of the back of the ball, and avoiding the erratic jumpiness of the knuckleball.

So, do we want a bunch of dimples all over our cars? It's been considered. But a more interesting option would be small blades placed just upstream of the separation point. While the ideal shape would be sharp with distinct edges, safety and esthetics dictate a compromise in production cars. So, the vortex generators that are used tend to be rounded.

The overall value of the vortex generators is probably modest, but not insignificant. Installed just upstream of the likely separation point they delay separation and generally increase the pressure on the rear surface. The result is a drop in both drag and lift.[13] As an added benefit, properly placed vortex generators can help maintain clean air for a rear wing (Image 8.17). In fact, while perhaps not distinctly defined as vortex generators,

---

[12] G. Dias, N.R. Tiwari, J.J. Varghese and G. Koyeerath, Aerodynamic analysis of a car for reducing drag force. *IOSR Journal of Mechanical and Civil Engineering* 13(3) Ver. I, 2016, 114–118.

[13] M. Jagadeesh Kumar, A. Dubey, S. Chheniya and A. Jadhav, Effect of vortex generators on aerodynamics of a car: CFD analysis. *International Journal of* Innovations *in Engineering and Technology* 2 (1), 2013, 137–144.

IMAGE 8.17
Vortex generators.

The vortex generator on a Honda Civic Type R can help delay separation and keep the rear airfoil in clean air.

features that trigger microvorticies can be usefully incorporated into overall airflow management to enhance the effectiveness of other features. For example, Ferrari claims that the vortex generators on the underside of the 488 helped increase downforce by 50%. And with very low additional cost, and increasing vehicle speeds, such features may make sense in a wider range of vehicles seeking high-speed performance.

Lift

As we've seen the airflow around a moving car is not the same on the top and the bottom. Air travels faster over the top of the car and, as we know, this means the entirety of the car acts like a wing and generates upward lift. This effect grows exponentially greater with speed. In fact, the formula for lift is essentially the formula for drag, except a coefficient of lift replaces the coefficient of drag. So, like drag, lift increases with the square of the vehicle's speed.

The tricky part is that the sorts of measures that can decrease lift often also increase drag. This is why racers often have to manage a careful balance of drag and lift reduction. You want low drag, of course, but you also need low lift to maintain road holding. So, aerodynamic modifications that may increase drag a bit may well be worthwhile if lift is reduced. Acceleration or top speed may drop a bit, but cornering speed would be increased. This explains the high drag coefficient of Formula One cars, for example. These calculations and trade-offs are no longer just about racing. While controlling wheel loads is vital for racing, it is also increasingly important for production cars. At highway speeds, the effect of lift can deteriorate performance and drivability. Front-end lift can impact handling; lift on the rear axle can inhibit stability.[14] But addressing this can be complicated. We want to reduce overall lift, of course. But we want to do it in a balanced way that addresses both front and rear axles. And ideally, we want to do this with the minimum impact on drag we can possibly manage. When the trade-off is unavoidable, the balance point needs to be carefully considered.

---

[14] W.H. Hucho, Introduction to automobile aerodynamics. In W.H. Hucho (ed) *Aerodynamics of Road Vehicles: From Fluid Mechanics to Vehicle Engineering*, 4th edition. SAE International, Warrendale, PA, 1998, 1–46.

The most notable mode of reducing lift is with an aerodynamic device, typically a flap, wedge or wing, at the rear deck often generically called a **spoiler**. More precisely, lift can be countered by modifying the top of the car in one of two ways: changing the flow over the top of the car to reduce lift production by use of a spoiler, or generating an opposing downward force by using an airfoil or wing. Unfortunately, each of these can have negative implications for drag. But, the trade-off is often worth it, and so we are now seeing more serious attempts to manage lift on production cars.

A spoiler can disturb the low-pressure flow over the top surface and so reduce lift. But, as mentioned, this comes with an increase in drag. The more common form of this on a production car is a raised lip at the rear trunk edge. You'll recall that a modest upturn of this rear lip can define a separation trigger, helping to define a clean brake of surface flow and reducing the rear low-pressure wake. This can also reduce lift a bit. As this angle is increased, it has an increasing effect on windward flow, and so greater lift reduction and an increasing impact on drag. A small amount of this reduced lift is the result of increased downforce from the Newtonian effect of deflecting airflow upward. The more significant impact is caused by the slowing of the air that occurs ahead of the spoiler. This slower, higher pressure air 'spoils' the lifting effect on the rear of the car. However, the drag created by this sort of spoiler can be very high.

Alternatively, a separate inverted airfoil or rear wing can produce downward force with less of a drag penalty (Image 8.18). Any airfoil produces lift perpendicular to its orientation. So, in the case of an aircraft wing, most of the wing's lift is drawing the plane upward. But a small component called **induced drag** is also pulling the plane backward, since the wing has a slight nose-up orientation, or angle of attack. In the case of an inverted wing on a car, most of the resulting force is directed downward. But a small horizontal component defines an induced drag penalty. As the nose-down orientation is increased, the generated downforce grows, but so does the induced drag.

The placement of the rear wing is important. For any airfoil to generate useful lift it must have access to non-turbulent clean air. So, if the airfoil is contained within the wake, it may have a spoiling effect on upward lift, but it will not be able to produce a downward force. A wing placed well into the clean air unaffected by the boundary layer is more likely to

IMAGE 8.18
Rear airfoil innovation.

The Koenigsegg One:1 rear airfoil utilizes an active rear airfoil on a cantilevered support to help keep the airfoil in laminar flow and minimize the effect of the support structure on the lower face.

produce a reliable downforce. The distance from the trunk deck need not be significant; somewhere just over a third of the distance to the roof line seems optimal.[15]

On the other hand, we may not always want the wing too far from the car. An inverted airfoil mounted closer to the body can promote reattachment of the flow over the body causing greater lift reduction.[16] More notably, a rear airfoil can help reduce the pressure at the back of the car, effectively drawing air under the vehicle and redirecting it, creating more downforce. This is due to something called the **Coanda Effect** that leads fluids to follow the contour of a surface. You may have noticed that water from a faucet can be redirected by a bowl, spoon, or your finger, as the water seems to be drawn around the surface of the object. In this case, the air is drawn around the lower body of the car and deflected upward, defining a stronger downward force. The key is to integrate the effect of the spoiler or wing with the overall airflow; and the result can be a total downforce that is stronger than what the wing itself could provide.

So far, we've focused on the back end, where lift control is more prominent. But, some cars drive with the front wheels, and all cars steer at the front. So, we need to address the front end as well. This can be delicate, as too much front-end downforce could also increase steering sensitivity. We've already seen how air dams can be used to reroute air away from the underside of the car, which, as discussed, can have a positive impact on both drag and lift. Similarly, front-end spoilers or splitters define a blade airfoil at the front of the car that can capture high pressure on its top surface and reduce underbody flow. Placement is important, as placing this airfoil too low could reduce airflow under the blade and lead to an upward force, and the general impact can increase drag too much.

So, once again we are faced with trade-offs as we attempt to manage lift. A more prominent rear wing or spoiler, for example, can increase downforce and road holding, but at the expense of increased drag, reduced rear visibility, and maybe an undesirable impact on styling. A more subtle spoiler may suit the style of the car, but leaves us with a lift problem. A prominent front spoiler may reduce lift, but at the expense of drag and road clearance.

A solution can be the use of an active spoiler that can deploy only when needed and change its orientation to suit driving conditions or driver preferences. These can be as simple as a lift-destroying device that deploys at a set speed. Or, increasingly, these can be more sophisticated systems with dedicated electronic control module that utilize speed, throttle position, braking, driver settings, and road conditions to determine when and how to deploy. We'll discuss active aerodynamics in greater detail later. First, we need to complete our assessment of the automotive surface with the last major zone of the car, the underside.

The Ground

You've probably already noticed that cars travel on the ground. This is another thing that makes cars fundamentally different from airplanes. Because one side of the car is very near the ground, the flow around a car has little in common with an airplane. In fact, the change in aerodynamic lift resulting from proximity to the ground has been well known

---

15 A. Buljac, I. Džijan, I. Korade, S. Krizmanić and H. Kozmar, Automobile aerodynamics influenced by airfoil-shaped rear wing. *International Journal of Automotive Technology* 17 (3), 2016, 377–385.
16 J. Katz, *Automotive Aerodynamics*. John Wiley & Sons, West Sussex, 2016.

by aircraft engineers for generations; and this 'ground effect' is accounted for and felt by pilots with every landing as it increases the lift of the aircraft just before touchdown.

Most notably for cars, the proximity of the ground has the effect of increasing drag. If the car were flying, the air might travel around the car with some degree of symmetry. However, because airflow around a car is constrained on one side by the earth, the top and bottom flow differ greatly, and is marked by particular dynamics: A higher pressure is defined at the front as air is constrained by the funneling effect of the narrow channel between the ground and the car. So, a larger proportion of frontal airflow travels over the car. And most notably the airflow under the car induces higher drag as it is constrained and buffeted while passing through a turbulent narrow channel with many obstructions and cavities. This increased drag caused by the interaction of streamlines from two objects, in this case the car and the ground, is called **interference drag**.

In this light, it is no surprise that reducing the airflow under the car will reduce drag significantly. This is generally done in two ways. The most direct is to simply lower the body. This has the additional advantage of lowering the center of gravity and resulting favorable impacts on handling. Of course ground clearance quickly becomes an issue. As we've discussed, a simpler option that can also be added to a lower chassis is an air dam in the front of the vehicle, redirecting the lower frontal flow around and over the car. The upshot is that the rough drag-inducing underbody is not exposed to airflow, and so drag is reduced dramatically.

This isn't the only advantage of diverting the frontal flow. Cooling can be improved by redirecting flow to needed cooling channels for the engine, brakes, or batteries. More notably, both dams and a lower body have the advantage of reducing front-end lift. This lift isn't principally the result of pressure differential but rather a sort of Newtonian force: as the high-pressure air strikes the undercarriage and is deflected, it imparts a reactive upward force on the car. The result can increase steering instability. Reducing the flow of high-pressure frontal air to the underside of the car can notably improve the situation.

Reducing the ride height can have a more significant impact on drag than might be expected. It effectively reduces not only C_D but also frontal area. Most simply, as the chassis is lowered, wheel exposure to oncoming air is reduced. As wheels are effectively blunt bodies with troublesome aerodynamics (we will discuss this in a bit), the effect is considerable. In addition, the reduction of underbody airflow reduces side outflow from the underbody and the resulting yaw airflow impact on wheels, which can be a major contributor to overall drag. Pitching the nose slightly downward can offer an additional advantage, since all the benefits of a lowered chassis are achieved and the slight underbody height increase toward the rear can provide a diffuser-like effect that eases overall underbody air flow and reduces undesirable cross-flow. Overall, reduction in ride height alone offers improvements generally around 5%–8%, and when combined with a front air dam can exceed a 10% reduction in drag.[17]

Yet again (as happens so often) we're faced with a trade-off. A lower chassis can provide significant advantages in aerodynamics and handling, but it also presents the challenge of low clearance. But (as also happens so often) a solution comes with advanced digital control. **Variable Ride-Height Suspension** (VRHS) can automatically adjust height based on driving conditions such as speed, terrain, or driver input. Typically, at lower speed the ride height will increase to allow greater ground clearance, and at highway speeds, an

---

[17] G. Larose, L. Belluz, I. Whittal, M. Belzile, R. Klomp and A. Schmitt, Evaluation of the aerodynamics of drag reduction technologies for light-duty vehicles: A comprehensive wind tunnel study. *SAE International Journal of Passenger Cars—Mechanical Systems* 9(2), 2016, 772–784.

automatic drop can help maintain an aerodynamic advantage and improved handling. The idea is not new. Famously, the Citroën 15CVH first utilized a hydro-pneumatic self-leveling suspension that allowed the driver to adjust the ride height back in 1954. More recently, adjustable suspension was initially adopted by manufacturers that hoped to offer off-road capacity in a vehicle with improved highway performance, such as Subaru and Range Rover. Now a common feature in many production vehicles with all-terrain aspirations, some variable suspension systems alter ride height several inches others less than an inch.

These systems are not limited to vehicles with off-road intentions. Mercedes-Benz Active Body Control, with no illusions of off-road exploration, has long utilized the system purely for the aerodynamic and handling advantages. Mercedes drops about half an inch (11 mm) between 37–99 mph (60–160 km/h). Audi's adaptive air suspension automatically drops the car nearly an inch (22 mm) at speeds above 75 mph (120 km/h). With a more acute interest in aerodynamic drag, and its impact on EV range, Tesla's model S offers their smart Active Air Suspension as an option. The system is linked to the GPS, so it will remember a driver-selected ride height at a given location, where a bump or rough terrain may be present, and automatically adopt that height on subsequent visits.

The effect on lift can be complicated. Potentially, reducing the ground clearance can slow the air under the car, increasing the pressure difference between the top and bottom of the car, and so increase the overall lift of the car. With increasing ground clearance, the speed under the car is closer to the speed over the car, reducing overall lift effect, though Newtonian lift on the front of the underbody can increase, and certainly overall handling is impacted by the raised center of gravity. A diffuser at the rear of the car can help manage and shape these dynamics and control lift. But before we get to that, we need to discuss a prerequisite for the underbody diffuser: cleaner underbody airflow.

Efforts to smooth the underbody of the car are relatively new. This may be largely because the process is tricky. Accounting for the suspension, drive shaft, exhaust plumbing, and any number of other possible protrusions and moving linkages makes a fixed cover on the underbody difficult. However, the benefits are clear. By maintaining less resistance to flow under the vehicle, and providing a more organized flow into the rear wake, an underbody shield has been found to reduce drag by nearly 10% in some studies, though their impacts are probably more modest in most cases.[18] However, even modest results can be valuable when combined with the possibilities for rear drag reduction made available with smoother underbody flow. And a drag reduction in the single digits is not a bad return on investment for the price of a few plastic panels.

With a smoother underbody, the air leaving the underside toward the rear of the car is aerodynamically cleaner; and this offers another opportunity for drag reduction. In general, the release of the underside airflow to the rear of the car can either increase rear drag or help address it. A diffuser that allows the air to define a more ordered release and entry into the rear wake can help make sure it's the latter. The effect is modest but not insignificant, with the likely total percentage of reduced drag in the high single digits.[19]

---

[18] J. Cho, T.K. Kim, K.H. Kim and K. Yee, Comparative investigation on the aerodynamic effects of combined use of underbody drag reduction devices applied to real sedan. *International Journal of Automotive Technology* 18(6), 2017, 959–971; and G. Larose, L. Belluz, I. Whittal, M. Belzile, R. Klomp and A. Schmitt, Evaluation of the aerodynamics of drag reduction technologies for light-duty vehicles: A comprehensive wind tunnel study. *SAE International Journal of Passenger Cars—Mechanical Systems* 9(2), 2016, 772–784.

[19] S.O. Kang, S.O. Jun, H.I. Park, K.S. Song, J.D. Kee, K.H. Kim and D.H. Lee, Actively translating a rear diffuser device for the aerodynamic drag reduction of a passenger car. *International Journal of Automotive Technology* 13(4), 2012, 583–592.

The same dynamics that shaped airflow on the top and sides of the car body, more or less persist on the underside. So, the angle and design of the diffuser is important. In general, something like a 5°–10° reentry is typical and generally desirable.[20] Longer diffusers are sometimes thought to be more effective, but the marginal benefits diminish with increasing length. In fact, a smaller diffuser with an increased angle can be just as effective as a long, gradual diffuser.[21] Of course, if the angle is too steep it essentially just defines a larger rear cross section; separation begins on the diffuser plane, and the drag reducing effect is lost.[22] At an extreme, a high-angle diffuser can generate confounding vortices that increase drag notably. The diffuser angle can also impact pitch, with a larger angle causing slower underbody air at the rear and so pitching down the nose.[23] Clearly, there are trade-offs. Potentially, an increased diffuser angle can increase underbody mass airflow with a resulting decrease in underbody pressure and lift, just as an increased ground clearance would but without the disadvantages of a raised chassis. A properly defined diffuser can help manage flow around and across the wheels, reducing wheel drag. This can also feed the wake region and reduce the base area of the rear pressure wake, reducing drag (Image 8.19).

IMAGE 8.19
Rear diffuser.

A rear diffuser allows underbody airflow to enter the rear wake more cleanly, reducing the low pressure zone.

[20] C. Lai, Y. Kohama, S. Obayashi and S. Jeong, Experimental and numerical investigations on the influence of vehicle rear diffuser angle on aerodynamic drag and wake structure. *International Journal of Automotive Engineering* 2(2), 2011, 47–53; and J.P. Howell, The influence of a vehicle underbody on aerodynamics of a simple car shapes with an underfloor diffuser. Vehicle Aerodynamics, *R.Ae.S. Conference*, Loughborough, UK, 1994, 36.1–36.11.

[21] D. Hummel, Some fundamentals of fluid mechanics. In W.H. Hucho (ed) *Aerodynamics of Road Vehicles: From Fluid Mechanics to Vehicle Engineering*, 4th edition. SAE International, Warrendale, PA, 1998, 47–82.

[22] C. Lai, Y. Kohama, S. Obayashi and S. Jeong, Experimental and numerical investigations on the influence of vehicle rear diffuser angle on aerodynamic drag and wake structure. *International Journal of Automotive Engineering* 2(2), 2011, 47–53; J.P. Howell, The influence of a vehicle underbody on aerodynamics of a simple car shapes with an underfloor diffuser. Vehicle Aerodynamics, *R.Ae.S. Conference*, Loughborough, UK, 1994, 36.1–36.11; and W.H. Hucho, Introduction to automobile aerodynamics. In W.H. Hucho (ed) *Aerodynamics of Road Vehicles: From Fluid Mechanics to Vehicle Engineering*, 4th edition. SAE International, Warrendale, PA, 1998, 1–46.

[23] T.C. Schuetz, *Aerodynamics of Road Vehicles*, 5th edition. SAE International, Warrendale, PA, 2015.

Wheels

Controlling underbody flow can make reducing drag around wheels easier; and this is no small matter. Wheels and wheel wells frequently represent a third of a car's total drag, and in a streamlined car, the wheels can account for half the total drag.[24] But reducing wheel drag is challenging, since airflow dynamics around the wheels tend to be the most complex of anywhere on the car. With relatively blunt frontal cross sections wheels define prominent bluff bodies, of course. As the wheel plows through the air, vortices roll off the top and the bottom and tumble outward. The effect can disturb flow along the side of the vehicle and can affect the car's trailing vortices and impact the size of the wake. But wheels are much more complex and problematic than a typical bluff shape. They are rotating at high speed, complicating flow and intensifying eddies. They are confined in a wheel well, making air movement much more complex. But the most notable characteristic is that the flow under the car is not one-dimensional, it spreads outward toward the sides of the car. This outward lateral flow means the wheels are not experiencing a symmetrical headwind or longitudinal flow; they are under an oblique cross-flow emanating from the underbody. That is to say, they are in yaw. This further complicates an already complex three-dimensional flow around the wheels and defines stronger drag. One effect of this is a separation on the outside of the wheels that draws out turbulent airflow across the outer surface and increases overall drag forces. In fact, even a small shift in yaw angle can induce a multi-fold jump in drag.[25]

There are a few ways to deal with this. Lower side panels can reduce underbody outward flow and so the yaw angle of the wheel. However, this is limited by the need for road clearance. Wheel air dams can be effective. These small strips of flexible material on either side of the underbody can redirect airflow around the wheels, though the impact is generally limited to 2%–3% reduction in drag.[26] Nevertheless, since the cost is minimal, the technique is appealing. Capping the wheel well with a fender skirt can help; but this runs against some stylistic priorities, and is problematic at the front axle due to steering.

Of course, wheel design can matter, and more aerodynamic wheels can have a notable effect. Tesla offers a notable example with the turbine blade wheel style of the 'aero caps' on the Model 3 (Image 8.20). These caps are designed to redirect air under the carriage, countering cross-flow tendencies and providing smoother flow along the sides of the vehicle and to the back wake. But, wheels are a prominent design feature. And aero caps aren't for everyone.

An effective approach that doesn't require compromised esthetics is to change the flow through and around the wheel wells with **wheel well venting** or **air curtains** (Image 8.21). As an advantage, the esthetic of the needed slots and louvers can enhance the high-tech look and aggressive styling of a car. In fact, manufacturers across the production spectrum have adopted these as stylistic adornments, sometimes with little aerodynamic effect.

[24] D. Hummel, Some fundamentals of fluid mechanics. In W.H. Hucho (ed) *Aerodynamics of Road Vehicles: From Fluid Mechanics to Vehicle Engineering*, 4th edition. SAE International, Warrendale, PA, 1998, 47–82.

[25] D. Hummel, Some fundamentals of fluid mechanics. In W.H. Hucho (ed) *Aerodynamics of Road Vehicles: From Fluid Mechanics to Vehicle Engineering*, 4th edition. SAE International, Warrendale, PA, 1998, 47–82.

[26] G. Larose, L. Belluz, I. Whittal, M. Belzile, R. Klomp and A. Schmitt, Evaluation of the aerodynamics of drag reduction technologies for light-duty vehicles: A comprehensive wind tunnel study. *SAE International Journal of Passenger Cars—Mechanical Systems* 9(2), 2016, 772–784.

IMAGE 8.20
Aero wheels.

Tesla claims their aero wheel caps can result in up to a 10% improvement in efficiency. A bold claim, though given the significant contribution to drag caused by wheels, it is likely that these innovative covers result in a significant contribution to drag reduction.

IMAGE 8.21
Air curtain.

High-pressure frontal air is redirected across the wheel's outer face to limit yaw airflow and wheel-well outflow.

The basic idea of an air curtain is implied in the name: a plane of high-speed flow that can redirect air movement across the wheel's outer face, limiting yaw and wheel-well outflow. Typically, intake slots at the front of the car allow incoming ram air to be redirected, possibly accelerated through a narrow channel, and released vertically across the outer surface of the wheel. The result encourages lateral flow on either side, reduced yaw from the underbody, less turbulence in the well, and a cleaner side airstream. The effect is sort of like fender skirts, or spats, on all four wheels but with better style options and wheel access.

IMAGE 8.22
Audi aerodynamics.

This Audi TT clubsport concept car offers a clear example of the use of breather slots and channels to help maintain clean airflow on the sides and back of the vehicle.

Image: Audi

Specifics of each application vary (Image 8.22). Breather slots on the backside of the wheelwell might be added to ease airflow egress and reduce back-well turbulence. Nissan adopts this prominently with the GT-R to achieve a remarkable C_D of 0.26 and a top speed of just under 200 mph. Channeling before the rear wheel-well can help direct and organize airflow before wheel well induction, as Honda has done on its Clarity plug-in hybrid. The Lexus LC500, on the other hand, adopts prominent venting on front and rear wheel wells as a definitive stylistic feature with aerodynamic benefits and improved brake cooling.

But clever aerodynamics isn't just for sports cars and fuel sippers. Larger vehicles, with larger exposed wheels, can tap these features to squeeze a few more miles out of each gallon. Inspired by the ongoing battle for the mantle of most fuel-efficient full-size truck, Ford's F150 uses horizontal slats under the headlight to channel frontal high-pressure air through ducting to the outer-front wheel well and across the outer wheel, reducing wheel wake significantly.

Bringing the Body Together

None of the dynamics that we've discussed so far exist in isolation. They are complexly interactive, and so the mechanisms to control or shape one of them have inevitable effects on the others. Front spoilers impact the rear wake. Boat tailing affects diffuser dynamics. Diffusers influence the effect of front-end spoilers, and so on. This is why the aerodynamic design of modern vehicles is an increasingly complex and precise undertaking, with thousands of hours of computer simulation and wind tunnel testing to evaluate even the most modest changes.

All these dynamics are brought together over the body contour. The shape and lines of the outer skin define the aerodynamic landscape on which patterns of flow, pressure, and speed interact. So, as we'd expect, innovations in basic body contouring are defining a new normal. While body lines, ridges, and edges were once almost purely about style, now few lines or reliefs on a modern vehicle are defined without a careful aerodynamic analysis. More varied and flexible manufacturing possibilities, more precise analysis, and an expectation of improved efficiency and speed, mean body contour details have become increasingly important. Such intelligent aerodynamic design can address the drag due to surface disturbances such as mirrors and door handles, called **excrescence drag**, improving not only basic aerodynamic performance but also NVH as well as surface soiling. Subtle contours on a hood, for example, can redirect airflow, reducing wind noise or front-end drag. Precise changes to a mirror stem can reduce side panel separation or decrease noise. Rear corners are precisely angled to define a subtle but definitive separation point so air breaks off cleanly reducing the rear wake and soiling. A precise crease on the front bumper can generate a front wheel air curtain effect.

Of course, sometimes the effects of body contours are not so subtle. Performance cars have used advanced modeling to define innovative new body options. The BMW i8, for example, integrates aggressive airflow channels into the C pillars to feed high-speed air into the rear wake and reduce drag. Of course, no car provides a more audacious example of aerodynamics body contours than the Ford GT (Image 8.23). Airflow passes through

IMAGE 8.23
Air channeling.

The air channels on the BMW i8 help ensure clean airflow around the car and into the rear wake. The Ford GT offers a more dramatic example.

Upper Image: BMW

the front breather, over the sleek cabin and through deep rearward channels that mini-mize airstream disturbance and feed high-energy air to the wake. The Ferrari 488 uses similar contouring to enhance the performance of the rear spoiler. A narrowed body chan-nel directs and accelerates airflow over a rear airfoil. Ferrari claims its so-called 'blown spoiler' is able to increase downforce by 50% over its predecessor.

Similarly, channeled airways can shape flow not only over the body, but also through the body. Frontal ram air can be taken in for cooling and redirected for aerodynamic effect. The Chevy Camaro, for example, takes in front end air at the grill, uses it for cooling, and expels it at the hood extractor. The increased upper flow decreases lift by increasing upper pressure, and by preventing the air from exiting under the car it lowers underbody pres-sure. Austin Martin offers a subtler example. In order to preserve the sleek lines of the DB11, the carmaker designed channels from the C pillar through the body and out the rear deck lid. The air exits in a stream along the rear edge, improving downforce. At higher speeds, a small spoiler is deployed to address downwash. Increasingly, modern vehicles aren't just accommodating airflow for improved performance; through precise shapes and contours, they are actively redefining flow patterns for precise airflow management.

Active Aerodynamics

As we have seen repeatedly, the options made possible with advanced digital control have redefined the automotive industry. Countless examples in engines, transmissions, electric drive, and other areas have demonstrated how compromises and trade-offs that were once unavoidable are being defied by computer control. Characteristics that were once thought to be invariable, fixed in the car's design, from compression ratio to suspen-sion, are now altered on the fly to accommodate driving conditions. These trends are true in aerodynamics as well. In the past, the aerodynamic game was defined by trade-offs. Reducing ride height could affect drivability. Improved rear shaping might impact vis-ibility or headroom. Redirecting frontal flow can impact cooling. And, of course, most notably, addressing lift often meant increasing drag, and vice versa. But, advanced active intelligent control defining a new normal.

The general solution is deceptively simple: change the shape of the car to suit driving conditions. When handling is a priority, increase downforce. When speed is the goal, prioritize drag reduction. We've seen versions of this already. Ride height variation for example. Or deploying an air dam or spoiler at a defined speed. However, newer systems have grown more sophisticated, taking in a growing number of variables and providing a broader array of variations in vehicle aerodynamics. **Active Aerodynamics** automatically shifts aerodynamic features for optimal performance based on speed, braking, throttle position and driver input, and are increasingly moving toward the inclusion of road condi-tions, weather, or learned driver behavior. Active control surfaces can provide reduced or increased drag, as well as lateral shifts in drag or lift to enhance targeted performance. So, we can often have our aerodynamic cake and eat it too.

The McLaren P1, a car that seems intent on redefining the epitome of slippery, offers an impressive example. In 'race' mode, a rear wing takes an aggressive angle of attack, providing maximum downforce. The increased drag can be more than addressed with the P1's 903 horsepower from a 3.8-liter V8 coupled with a 176 HP electric machine. Efficiency is not a priority in this mode. The carmaker claims the front and rear wings are capable of 1,323 lb (600 kg) of downforce in race mode; but when the DRS (drag reduction system) button is depressed on the steering wheel, the airfoil's angle of attack is reduced to zero and overall drag is dropped by nearly a quarter. When the brakes are aggressively applied, the wing folds forward to create maximum drag for air braking.

Some carmakers have focused on precision control of body lift rather than managing a rear airfoil. Small body-panel flaps placed on upper surfaces can be used to shift lift and drag asymmetrically. The small plate surfaces on the upper body tilt up, forcing dramatic airflow separation, destroying lift, and increasing drag. This can be used to assist not only braking; but because they can be asymmetrically deployed, high-speed handling can be enhanced. For example, the Pagani Huayra utilizes four hinged carbon fiber panels, two at the front and two at the rear, as spoiler flaps. As they open, they force airflow separation, destroying lift and generating some Newtonian downforce as well as considerable drag. An ECU tracks airspeed, yaw, steering angle, throttle position, acceleration, and other variables to continuously adjust downforce and drag for optimum performance. So, for example, deploying the inside flaps in a turn increases downforce on the inside tires and helps address body role. In hard braking, the rear flaps are deployed to counter vehicle dive and maintain stability.

Active aerodynamics can redefine the underside of the car as well. Recent work by Porsche has outlined the design of a simple active air diffuser that uses a hinged lower rear panel to actively adjust diffuser dynamics to suit driving conditions. Ferrari uses a similar, though more complicated, system to provide synchronized control of the lower diffuser angle and upper spoiler on the LaFerrari.

Perhaps the most advanced example of linking active upper and lower, front and rear aerodynamic features is the Lamborghini Active Aerodynamics (or *Aerodynamica Lamborghini Attiva*, ALA) system in the Performante (Image 8.24). An active front spoiler can function as a normal splitter, dividing front airflow either through the engine bay or under the car. This enhances cooling and can produce useful front-end downforce, as discussed, at the cost of increased front-end drag. However, a flap valve in this splitter can

IMAGE 8.24
Active aerodynamics.

The Lamborghini Huracan Performante exemplifies innovation in aerodynamic design. Active flaps can reroute air under the car; similar flaps can redirect airflow to the underside of the rear airfoil, all in an effort to produce the optimal combination of drag and downforce to suit driving conditions.

define an alternative route for airflow, opening a channel that sends front flow under the car. The escape of front-end pressure reduces drag but at the expense of downforce. The active rear wing is enhanced with a similar mechanism. Under normal conditions, clean air travels over the rear airfoil producing downforce and some associated drag. However, active shutter valves can direct ram airflow onto the rear surface of the wing, destroying the downforce and reducing drag. Varying a rear wing's angle of attack is, of course, another way of achieving control; but the ALA option has the advantage of simplicity and light weight, avoiding the heavy hydraulics used in most active rear spoilers.

The ALA's effect is remarkable. In normal driving mode and sports mode, all channels are open at modest speeds to reduce drag and closed at preset speeds to improve downforce and stability. With both valves open, frontal pressure is released under the car and the rear wing is essentially deactivated, and drag is minimized. However, in 'corsa' or race mode, the channels are continuously adjusted to optimize handling and traction. All channels are open when driving straightaways; in a corner, downforce is increased on the inside wheels, allowing improved road contact for better inside braking and so improved traction control. When braking, both valves are closed, this not only maximizes aerodynamic drag, it maximizes downforce on both axles for improved traction. Lamborghini claims this increases downforce more than seven fold, improving braking potential tremendously.[27]

We are unlikely to see a system like this applied to a high volume production car anytime soon. Lamborghini is special. But, as we've discussed, the general aerodynamic trends in automotive design are nonetheless remarkable. Modestly priced models that were once content with a shape roughly equivalent to a shoebox on wheels now boast carefully sculpted contours. Flow-though channels, diffusers, air blades, airfoils, and air curtains are defining engineering works of aerodynamic art. Every crease, corner, and contour carefully considered for its impact. They may not have the fins of an Eldorado or the flowing fenders of a Delahaye, but contemporary cars are remarkable nonetheless, beautiful in form and with precisely defined function.

---

[27] Focused on Performance: The New Lamborghini Huracán Performante, March 2017, media.lamborghini.com.

9

Smarter Cars

By now it should be clear that cars have gotten a lot smarter. We have seen examples of ECMs and complex algorithms used to manage everything from ignition advance to rear spoilers. Repeatedly, advanced digital controls have made all sorts of things feasible that were once only vaguely possible. As a result, today's car is more responsive, more efficient, more comfortable, and safer than ever before; and the possibilities for future improvements are only growing. As mentioned before, it's no exaggeration to say that the entire operation of the modern automobile has been redefined by digital control.

But the fundamental purpose of a car is not just driving, it's transportation; it is to get you from here to there. And it is in that undertaking of getting from point A to B that the changes made possible by electronic control are most visible and salient to the driver. Called Advanced Driver Assistance Systems (ADAS), integrated digital features that take on tasks normally required of the driver have increased dramatically over the past decade. Parking, lane changing, blind-spot monitoring (BSM), speed control, all those irksome duties you had to learn to earn a driver's license can now be automated, handled by a cluster of sensors and artificial intelligence (AI). Some find this cool, other find it creepy; but you've got to admit, the tech is amazing.

Whatever your personal feelings about your car helping you drive, there are multiple reasons driver assistance is desirable. The most pressing is probably safety. Every year about 37,000 people are killed in motor vehicle accidents in the US, and about one and a quarter million people are killed each year globally.[1] It's likely that well over 90% of these fatalities are caused in part by human error.[2] As distracted driving increases, the challenge of addressing these numbers grows more dire. Imagine a transportation system where driver error and inattention is not an issue. Distracted driving, human error, road rage, and road fatigue could all be things of the past.

However, enhanced safety is not the only thing motivating ADAS development. The current systems could lead to self-driving vehicles, also called autonomous vehicles. And a vehicle that does not require the constant attention of a trained driver could have many advantages. It could provide transportation for people with disabilities or the elderly. Such a system could reduce traffic delays since the faster reaction time of digital control would allow vehicles to travel at higher speeds much closer together. And an enhanced communication network would allow vehicles to adjust their routes to better avoid congestion. This is no small matter. A recent study found that traffic congestion in the US alone is responsible for wasting 3 billion gallons of fuel annually, cost travelers about 7 billion hours of their time and cost the nation about $160 billion, or $960 per commuter.[3]

Done properly, the deployment of self-driving cars could reduce emissions through efficiency and ease pressure on infrastructure. Automated driving facilitates ride-sharing

---

[1] 2017 NHTSA Fatal Crash Data and World Health Organization *Global Status Report on Road Safety 2015*, Geneva, 2015.

[2] National Motor Vehicle Crash Causation Survey, Report to Congress, NHTSA DOT HS 811 059, 2008.

[3] D. Schrank, B. Eisele, T. Lomax and J. Bak, *2015 Urban Mobility Scorecard*. Published jointly by The Texas A&M Transportation Institute and INRIX, College Station, TX, 2015.

services, which means less parking is required in urban areas. Imagine the possibilities for enhanced walkability in our urban areas, if we could reduce parking demand and traffic by half or more.

The need for precision is immense and the stakes are high. As the popular quip has it, a highly reliable personal computer would be praised if it only crashed once a year, but this would be a terrible record for a self-driving car. And the option of a quick reboot while speeding down the highway is not acceptable. We need to get this right the first time. As a result, tens of billions of dollars have already been invested in automotive AI alone, and much more is to come. And don't expect too much too soon. Getting it right will take time. The learning curve is steep, but the stakes are too high to get it wrong even once.

Smarter Driving

The spectrum of driver assistance features spans from simple driver notifications to fully automated driving. To make sense of the multiple variations and options between these two extremes, the Society of Automotive Engineers (SAE) has defined an international standard for self-driving capacity that is largely equivalent to a system adopted by the National Highway Traffic Safety Administration (NHTSA) or the German Highway Safety Research Institute (BASt). The SAE scheme begins with Level 0, defined as no automation with the driver in full control of the vehicle at all times, and progresses to Level 5 defined by full automation of all driving functions at all times (Table 9.1).

Notice that Level 0 does not necessarily mean no ADAS functions are available; it may mean that the ADAS technology is limited to informing the driver or facilitating driving operations without undertaking direct control of any driving functions. BSM, also called blind-spot warning or detection systems, offers a good example. BSM can use different sensors to determine when an object, like a car or pedestrian, enters the vehicle's blind spot. An alert is issued, often in the form of a warning on the corner of the side-view mirror, or perhaps a momentary seat vibration. Advanced systems can distinguish between a pedestrian and a vehicle, or offer a warning when backing out of a parking spot and a moving hazard appears from either side. But the driver remains in full control at all times.

In fact, efforts to provide the driver with more information or reduce driver distraction are probably the most common current form of ADAS. Heads up displays (HUDs) offer a good example. These systems utilize a high-resolution projector to display driving and navigation information on the windshield or a retractable screen. By placing the information in the driver's line of sight, HUDs ideally allow the driver's attention to remain focused on the road. Originally developed for military aviation, and dating back more than 30 years in automotive applications, this technology is now commonplace in new vehicles, and a regular aftermarket add-on. Such systems can be integrated with enhanced voice recognition or gesture recognition, allowing the driver to use a pinching or swiping motion like a smart phone to control infotainment features.

Alternatively, an ADAS feature might enhance general drivability or safety, such as adaptive lighting. With the goal of improved visibility and reduced glare, adaptive systems swivel headlights toward the roadway ahead, enabling improved night visibility on curved roadways. The idea is not new, famously the 1948 Tucker utilized a third central headlight, called the 'Cyclops Eye', which rotated with steering angle. However, newer

TABLE 9.1

Driving Modes

SAE Level	NTHSA Level	SAE Name	Automation Condition	Steering/ Throttle/ Braking	Monitoring of Driving Conditions	Fallback for Driving Tasks	Driving Conditions Capacity
0	0	No automation	Human driver controls all functions	Human driver	Human driver	Human driver	N/A
1	1	Driver assistance	Either steering or acceleration/ deceleration automated	Human driver with system	Human driver	Human driver	Some
2	2	Partial automation	Conditional part-time control of both steering and acceleration/ deceleration	System	Human driver	Human driver	Some
3	3	Conditional automation	Conditional automation of all driving functions with human monitoring	System	System	Human driver	Some
4	3/4	High automation	Conditional automation of all driving functions with no human monitoring	System	System	System	Some
5	4	Full automation	Full-time automation of all driving functions under all conditions	System	System	System	All

systems do not wait for a change in the steering angle; they sense the road ahead and turn to improve nighttime visibility and range before the driver turns. Some are able to adjust lighting for weather conditions. Others can adjust brightness for road conditions, for example, automatically lowering high beams when an oncoming vehicle is identified. An increasing number of cars provide an adaptive high beam that uses automated partial shielding of the beam to avoid glare on the oncoming driver without totally reducing the high-beam illumination.

Level 1 systems move beyond enhanced drivability and define a condition of partial automation, with the direct automation of a specific driving function. Typically this means either control of steering (lateral control) or speed (longitudinal control). The human driver would maintain control of the dimension or function not under automatic control and can typically take over full control seamlessly. So, a very basic foray into Level 1 would include electronic stability control, or a parking assist program that provides steering but requires the driver to control the throttle and brakes.

Level 2 allows temporary automation of both speed and steering, or longitudinal and lateral control, but only under specific conditions, usually consistent highway driving. The driver is required to monitor the vehicle at all times and be prepared to take over control if conditions move out of the system's capacity, due to maybe leaving the highway, faded lane markings, or inclement weather.

Because modern vehicles do not rely on simple mechanical connections for control functions, but have incorporated brake-by-wire and steer-by-wire technology, adding automated driving functions to basic ADAS systems is a natural step and neither difficult nor unduly expensive. Moreover, the use of a standard **Controller Area Network**, or CAN bus to connect microcontrollers and ECMs throughout modern vehicles makes the integration of functions relatively simple. This connection between systems can facilitate innovations from a start–stop function to collision avoidance that could be more difficult and expensive if each system operated independently.

As a result, Level 1 and Level 2 driving is now commonplace. Many times these features build on preexisting basic ADAS functions. So, multiple carmakers link BSM with driver intervention, added resistance to the steering wheel if a driver misses the warning and attempts to change lanes with the blind-spot occupied. Lane-keeping systems (LKS) offer another good example. A basic system provides **lane departure warning** (LDW), which will issue an alert when the vehicle drifts from the center of the lane. **Lane-keeping assist** (LKA) takes an additional step and applies corrective steering if the driver fails to respond to the warning and the turn signal is not on. Typically, the driver can overcome the system by applying extra steering force; so, reestablishing human control remains quick and seamless.

Collision avoidance systems offer a similar sort of graduated application. If the system identifies imminent danger of collision, it can automatically take actions to help avoid impact and reduce the risk of injury. A basic Level 0 system may issue a warning to the driver. It may also pre-charge the brakes, allowing for a faster and stronger braking application when the driver depresses the pedal. It could tension the seatbelts or adjust adaptive headlights to help with collision avoidance. Taking it to the next level, in this case literally, a system may take direct control of a driving function, such as an automatic braking system. Others are capable of controlling steering to avoid a collision, particularly at higher speeds. These systems are so effective at lessening the impact of a collision that the European Union has mandated that all commercial vehicles be fitted with at least a basic form of crash avoidance.

Another familiar example is **adaptive cruise control** (ACC). Based on mature and reliable technology, the system is simply a smart cruise control, using sensors to track vehicles ahead and automatically adjust speed to match traffic. Some systems are only designed to operate at highway speeds; others are capable of bringing the vehicle to a complete stop. A variant of this technology is able to manage vehicle speed in heavy traffic. Traffic jam assist systems are available from Audi, Acura, Mercedes, BMW, and others, allowing partial or full control of the vehicle in slow speed traffic conditions, when driver fatigue and inattention can be high (Image 9.1).

While an ACC system can be applied with lane keeping to provide limited automation, these Level 2 systems require continued driver supervision, as curved roads or weather conditions can cause failure, and most ACC systems are only suited for specific road conditions. Typically, these systems incorporate a form of driver feedback that ensures continued driver engagement. For example, the system might sense the driver's hands on the steering wheel, and allow hands off driving only for a very short amount of time, deploying a warning that tells the driver to put hands back on the wheel after that time, and disabling the system if driver does not comply.

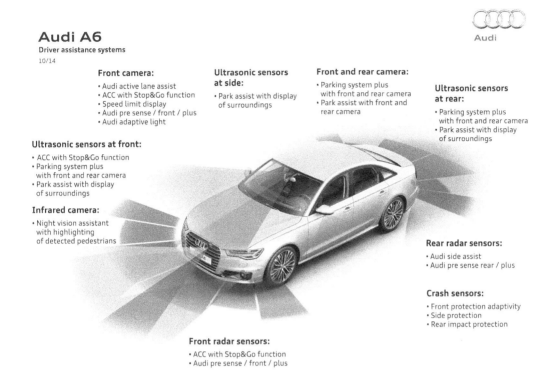

Audi A6
Driver assistance systems
10/14

Audi

Front camera:
• Audi active lane assist
• ACC with Stop&Go function
• Speed limit display
• Audi pre sense / front / plus
• Audi adaptive light

Ultrasonic sensors at side:
• Park assist with display of surroundings

Front and rear camera:
• Parking system plus with front and rear camera
• Park assist with front and rear camera

Ultrasonic sensors at rear:
• Parking system plus with front and rear camera
• Park assist with display of surroundings

Ultrasonic sensors at front:
• ACC with Stop&Go function
• Parking system plus with front and rear camera
• Park assist with display of surroundings

Infrared camera:
• Night vision assistant with highlighting of detected pedestrians

Rear radar sensors:
• Audi side assist
• Audi pre sense rear / plus

Crash sensors:
• Front protection adaptivity
• Side protection
• Rear impact protection

Front radar sensors:
• ACC with Stop&Go function
• Audi pre sense / front / plus

IMAGE 9.1
Audi ADAS.

Modern vehicles, like the Audi A6, incorporate a variety of sensors that enable multiple advanced driver assistance systems.

Image: Audi

Transitioning to Level 3 defines a system of extended self-driving under limited conditions. The system can maintain control of both steering and speed, as in Level 2, but for extended periods under defined traffic and weather environments. As in Level 2, the system can only function within specified conditions. But, if these conditions are maintained, the system can provide indefinite operation, allowing the driver's attention to go momentarily elsewhere. The key here is that the system also can recognize the limits of its control capacity, identifying when operational parameters have changed, and notify the driver when human control is needed. This allows the driver to pay less attention to the car's continuous control; but still remain ready to take over when changed conditions exceed the automated system's ability. There are a few Level 3 systems in production. Notably, Tesla's 'autopilot' in the Model S and X defined the first Level 3 system in a production car. Cadillac's super cruise system connects lane keeping and ACC for a similar form of reliable hands-free highway driving. Other systems, such as Mercedes-Benz Drive Pilot or Volvo's Pilot Assist require more regular driver intervention, and might be more correctly called high-performing Level 2 systems. Though both systems are being upgraded, and likely to achieve clear Level 3 capacity soon (Image 9.2).

Many point out that there are good reasons to be concerned about the transition to Level 3, since it rests in an uneasy middle ground. To be clear, providing information or

IMAGE 9.2
Autonomous driving brain.

All autonomous or semi-autonomous vehicles rely on a central processing unit to aggregate and interpreted sensor data and manage autonomous functions. In Volvo's case, this 'brain' is located under the rear cargo area, and supports the Pilot Assist functions that combine adaptive cruise control and lane keeping at speeds up to 31 mph (50km/h).

Image: Volvo

modest assistance to a driver that is in full control of the vehicle is not problematic; and a fully automated system that is safe is not a problem either. The problem is the in-between. A system that expects the driver to constantly monitor the car's operation but does not usually need human intervention presents a tricky situation. People often trust technology more than they should. The multiple internet videos of people doing foolish things in a Tesla when autopilot is engaged offers clear evidence of this fact. As drivers get comfortable with a system that rarely requires their attention, they may not be ready to take control when needed. In short, they may mistake a semi-autonomous car with an autonomous car. Indeed, Google decided to bypass this level for precisely this reason. And Ford and other manufacturers have decided to skip Level 3 automation in their self-driving development. Even if the driver understands the limits of the system, expecting a driver who has been reading emails and not paying attention to the road to suddenly take control in a critical situation is unlikely to provide a good back-up plan. This is sometimes called the **handoff problem**. Cadillac attempts to address this by monitoring the driver's head position when in hands-free driving, and issuing a prompt when the driver is not paying attention (more on driver monitoring later). Tesla simply requires the driver to keep a hand lightly on the steering wheel.

Levels 4 and 5 define fully automated driving functions. Self-driving may be a better term than autonomous for these vehicles, since every vehicle on the road already operates autonomously. Indeed the very term used to describe modern passenger vehicles implied this—*auto*mobiles. In Level 4, the vehicle manages all aspects of driving when conditions allow. So, for example, in severe weather or on an unimproved road the system may require human intervention. But, when under automated control, the driver does not need to pay attention, can focus fully on other tasks, and is not expected to provide back-up control. The steering wheel could even be designed to retract in normal driving, only extending for use the limited times it's needed. Level 5 has no such limit. The vehicle would operate independently under all road and weather conditions with no need for driver input. The occupant simply identified the destination and has a seat, no steering wheel needed.

The research and testing activity related to Level 4 driving seems to be expanding dramatically with every passing day. Level 4 cars are on the road in California, Pennsylvania, Texas, Washington, Arizona, and Michigan. Now a decade old, Google's self-driving car project, newly renamed Waymo, operates a fleet of driverless Pacifica minivans in Arizona *without* a safety driver. Apple operates a fleet of 30 self-driving Lexus SUVs in California. Lyft is operating self-driving BMWs in Las Vegas. Uber is field testing self-driving Volvos in Pittsburg. GM's self-driving subsidiary Cruise Automation is testing self-driving Bolts on the streets of San Francisco. And the list goes on. Ford, General Motors, Nissan, Tesla, and Mercedes are pouring billions into self-driving technology. And just about every major carmaker has promised a self-driving vehicle before 2021. Twenty-nine states have enacted legislation to accommodate or facilitate self-driving vehicle operation.[4] And federal regulators are looking at ways to adapt safety standards that were designed with human drivers in mind to an automated world.

Self-driving requires five basic capacities: perception, localization, navigation, decision-making, and vehicle control.[5] These are also the basic functions any driver undertakes anytime they get behind the wheel. And the various ADAS systems require varying degrees of these capacities depending on their function. Perception requires the use of sensors to scan and identify the surrounding environment. For a human driver, this is principally done with eyesight, and augmented with hearing. For a self-driving system, radar, lidar, optical cameras, and other systems are used. Localization simply means knowing where you are. This can be done through a combination of dead reckoning, satellite navigation, and sensor data. Navigation, also called path planning, means knowing how you are going to get to where you are going. Decision-making is pretty self-explanatory, the capacity to make choices based on route information, vehicle performance, the environment, possible hazards, and multiple other variables. Lastly, vehicle control is the process that connects this all with the car's operation, defining a real-time recursive process that responds to current road conditions and obstacles and gets you where you want to go safely.

Every driver does all of this all the time. The process can be tedious and exhausting, and this leads to mistakes, and that can lead to injuries. The basic idea is to take some, or all, of the load off the human driver. Designed correctly, driver assistance systems can consistently perform these tasks as well as the best possible human driver.

---

[4] Autonomous vehicles state bill tracking database, National Conference of State Legislatures. Available at www.ncsl.org/research/transportation/autonomous-vehicles-legislative-database.aspx

[5] Adapted from Cheng, H., *Autonomous Intelligent Vehicles: Theory, Algorithms, and Implementation*. Springer Science & Business Media, London, 2011.

Perception

There are a lot of ways human drivers can take in information from the surrounding environment, and there are countless variables and modes of information a driver processes constantly and unconsciously. Think of a typical moment in any usual drive. You need to be aware of where the curbs are, where the lanes are and what lane you are in, how the road turns ahead, traffic signals, road signs, road conditions, hazards, potholes, wet roads, traffic, and let's not forget the most important and unpredictable variables: other human beings in cars and on foot. Now stir in a multitude of other factors, a deer on the side of the road that looks like he just may jump out in front of you, the ball that just rolled across the street and the child that may follow it, and the list goes on. Compiling all of this requires that you know the size, movement, shape, and distance of all the objects around you, rain or shine, day or night. And, of course, it requires that you make sense of that information. We'll get to the making sense part later. For now, let's first focus on the sensors that can take in the raw information.

The most basic of automotive sensors may be **sonar**, or Sound Navigation and Ranging. The basic idea is pretty simple: send a sound pulse out in a defined direction and see if it bounces off an object and returns. If you know how fast sound travels and the amount of time it took for the pulse to get back to you, you can calculate the distance to the object. This sort of **echolocation** has been used by ships for a long time to identify everything from submarines to fish. The sort of sonar we're interested in is called active sonar, since passive sonar is just listening without sending out a pulse. Passive listening may work when tracking an enemy submarine in stealth mode, but it won't work as well for us.

Clearly, sending out sound waves might get pretty annoying, to say the least. You've probably seen old war movies with the submarine crew subjected to the 'pinging' of the enemy sonar as they try to escape. That sort of constant pinging from your car could get old fast. Particularly since the amount of sound intensity we need to produce a useful return would require a very loud ping. So, the acoustic wave that is emitted is in the ultrasonic range, above 20 kHz and beyond human hearing. In fact the sort of sonar used in cars is more specifically called an **ultrasonic sensor**. These high-frequency waves can be focused to produce a highly directed pulse, so we can send the signal in a fairly precise direction and know the location of the object more accurately.

This focused and intense wave travels like a pressure ripple through the air at the speed of sound, or just under 770 mph (or about 340 m/s). This may seem fast, but we'll soon see that it's pretty slow compared to other sensors. And, because the signal is an actual physical wave of pressure, it tends to lose energy quickly and dissipate, so it doesn't travel very far and doesn't offer great accuracy. If an object is closer to the transmitter, the returning echo will be stronger, of course. By setting a piezoelectric receiver to ignore all signals that are below the noise value of a response beyond a defined range, you can avoid false positives and provide a more definitive identification of a near object.

So, why use it if it's so limiting? The system costs are low and the signal processing is relatively simple. This makes it a useful sensor for close range detection when high-speed and precise resolution is not required. Other sensors, like cameras, are not great at close-in detection. So, when we don't really need detail, but just need to know whether an object is near the car or not, an ultrasonic sensor makes financial and technical sense.

So, when is it useful to be able to detect a general object at a range of 2–8 ft (1–2 m) without high processing speed? The most obvious answer is parking. A couple of sensors can be mounted at the rear, for example, and provide a warning when an obstruction is detected. The use of more sensors can enable a **parking assist** function. Properly focused, two to four sensors can identify a curb and provide data for automated parking. Different sensors can use different signal shapes to selectively identify and distinguish between a curb and a wall or a parked car, for example. First offered by Toyota in 2003, several manufacturers now offer some version of parking assistance. Systems like Mercedes's Parktronic use sonar to size up a parking situation and provide the driver with guidance, for example. Others, like Ford's Active Park Assist system, take over steering while parking, leaving brake control to the driver and defining a Level 1 system. Chrysler's Park Assist in its Pacifica, on the other hand, includes automated braking.

Ultrasonic sensors can be useful for other functions too. Activators for a smart trunk opener or other features that require proximity sensing can also use ultrasonic sensing, as can BSM or even gesture recognition. Still, while ultrasonic sensors are useful, they are very limited; we'll need to identify more precise and longer ranged sensors for more advanced functions such as ACC or ultimately self-driving.

Radar

Radar, or **RA**dio **D**etection **A**nd **R**anging, offers much improved range, speed and reliability over ultrasonic systems. Radar uses electromagnetic waves much as sonar uses sound waves to produce echolocation. The waves are in the radio portion of the electromagnetic spectrum and travel at the speed of light, about a million times faster than the speed of sound, allowing for much faster detection. Radio waves have much greater capacity for travel through the air and are far less effected by weather and indifferent to daylight. This makes radar the go-to sensor for inclement weather.

You might expect an automotive radar system to simply emit a narrow beam wave and measure the wave's travel to define an object's range. And, in fact, this is how radar units have functioned since World War II. However, this sort of approach is cumbersome, would be vulnerable to clutter and leaves something to be desired in accuracy. So, new automotive systems utilize continuous wave radar. Instead of measuring timing, the definitive measurement is frequency change. An uninterrupted wave is emitted with a frequency that constantly changes in some variation of a saw-tooth pattern, defining a **frequency-modulated continuous wave** (FMCW) radar. The frequency shift between the sending and received signal can be easily measured to determine distance; the greater the shift, the longer the distance.

This process is a bit trickier than it might appear because the distance between the transmitter and target may not be static. That is to say, the car and the thing it is sensing may be moving at different speeds. If they were moving at the same speed, the signal would return at precisely the same frequency it was sent. But there is a **Doppler effect** on the radar signal when there is a difference in speeds. If distance between the emitter and target is closing, the signal gets compressed by the relative speed, and the frequency is increased. If the target is moving away from the radar, the return signal is elongated, with a longer wavelength and decreased frequency. This effect must be accounted for, which adds a slight complication to the processing algorithm. But this frequency shift also allows us to gauge not only distance but also relative speed. This is useful in ACC and it is also useful in deciphering **clutter**. Roads are filled with competing forms and fixtures: signs, pedestrians, hydrants, lights, buildings, trees, and a whole lot of other objects, some moving and some stationary, and most will reflect a radar signal. So, allowing a clearer distinction between a stationary road sign and moving traffic, for example, can be challenging. Doppler can help.

A radar emitter sweeps in a circle, continuously emitting a changing frequency and receiving a reflected signal back. Not all materials reflect equally, and some will allow radar to pass right through. But those that reflect a strong signal, like metal, can be identified by orientation and range. The further the item gets from the radar, the weaker the orientation accuracy since the signal width expands as it travels.

The frequency and wavelength of radar is related to its range. Initial systems were used for relative short-range features, such as blind-spot detection or collision avoidance. Such so-called **short-range radar** (SRR) systems can detect objects to about 100 ft, or 30 m, and are used in obstacle detection, pre-crash sensing, and even parking assistance. **Long-range radar** (LRR) is useful for ACC or speed control in self-driving vehicles. With a detection range of 1,000 ft (300 m), these systems could identify road debris, the tail of a traffic jam, or other hazards in time to respond without emergency measures. The two are by no means mutually exclusive. In ACC systems, for example, typically two sensors are used, one to track the vehicles ahead and a short-range cut-in sensor to address cars that merge into the lane.

Automotive radar systems are undergoing great improvement in resolution and accuracy as a result of a regulated frequency increase from 24 GHz to above 79 GHz. As the frequency of radar goes higher, the ability to distinguish between objects, called **resolution**, and accuracy of range measurements improves proportionally. And, because when the frequency goes up the wavelength goes down, the smaller wavelength allows for a much smaller antenna. As we add an increasing number of sensors to the automobile, this is not an inconsequential concern. Lastly, the move reduces possible interference with satellite services and other transmitters, prompting European regulators to require a total phase out of the old systems by 2022.

The longer range and greater accuracy available with the shift will help enable radar use for ADAS functions. For example, cruise control systems can make good use of the next-generation 79 GHz systems. With a wider bandwidth of 77–81 GHz, next-generation 79 GHz systems will offer higher resolution, and an ability to distinguish small targets such as pedestrians or bicyclists at an extended range. Lower emission power requirements in this higher band will reduce mutual interference and decrease the size of the units, allowing multiple transmitters to be placed around the vehicle more easily. Typically, an array of four radar sensors, one on each side, can provide a 360° view of the environment without any significant changes to the vehicle's body design. Proper synchronization avoids interference between the sensors.

Even with these improvements, the great challenge of radar is resolution. To a radar unit, a tree can look very much like a pedestrian. Next-generation units may allow us to disaggregate multiple trees at a distance; but it may not allow us to easily determine which one is the pedestrian. One possible solution is the use of polarimetric radar that can emit pulses with defined horizontal or vertical orientations to provide much improved resolution.[6] Similarly, circular polarization with both left-hand and right-hand rotation capacity can provide differing bounce angles that help identify targets more precisely.[7] However, a

---

[6] S. Trummer, G.F. Hamberger, U. Siart and T.F. Eibert, A polarimetric 76–79 GHz radar-frontend for target classification in automotive use. *European Radar Conference (EuRAD)*, 2016 London, October 5–7, 2016.

[7] S. Trummer, G.F. Hamberger, U. Siart and T.F. Eibert, A polarimetric 76–79 GHz radar-frontend for target classification in automotive use. *European Radar Conference (EuRAD)*, 2016 London, October 5–7, 2016.

simpler solution that avoids complex radar polarimetry would be to combine radar signals with other sensors to gain greater resolution. The higher resolution available with lidar makes it a leading candidate.

Lidar

Operating much like radar, lidar sends focused beams of infrared light, lasers invisible to the eye rather than electromagnetic waves used by radar. The light can be more precisely focused, and reflects off any object that reflects light. So, all objects, wood, metal, stone, or plastic, return a signal. The emitter can send out millions of light pulses per second, as it revolves at several thousand rotations per minute. Measuring the return time and wavelength allows for a more detailed view of the world in real time and a disaggregation of clutter that surpasses radar. In fact clutter can make it difficult, if not impossible, for radar to distinguish between specific environmental features without help. Lidar has a much easier time with this.

The result of lidar scanning is a sort of topographical map of everything around the vehicle. Where an object has reflected the light, a **point cloud** is defined, providing a detailed contour of reflected light pulses that can clearly define the shape and movement of an object, regardless of material or color. Overlaying multiple point clouds, or **clustering**, provides a detailed shape of the object. With shape specifics, the object can be classified, enabling the distinguishing of pedestrians, buildings, vehicles, signs, etc. (Image 9.3). Predictive algorithms can then be used to define movement patterns and possibilities for each object based on its classification. For example, a pedestrian moves slowly, a car can move quickly, and a tree doesn't move at all. The end result is a dynamic 3D map of the vehicles surroundings in real time.

The increased resolution of lidar means that it can not only be used to identify the position of objects, but it can also be used to identify what they actually are, making out the details of a head or legs to distinguish, lets say, between a traffic barrier, a child, and a dog.

So, why isn't lidar fitted on every car? It's expensive, fragile, and has typically required a bulbous lump on the top of the car, none of which make it a good fit with production automobiles. However, lidar is thought by many to be indispensible to the development of self-driving vehicles. The preciseness of the data and the capacity to offer it in 3D makes lidar data relatively easy to decipher. And its capacity to identify a wide range of materials, from rock to plastic, make it a good fit with the needs of self-driving systems. In short, the precision and resolution lidar offers is the easiest route to defining a car that can identify and respond to its environment in real time.

However, it can't work alone. The most damming weakness of lidar is ironically that it depends on light. This means it does not penetrate things that block light very well, like snow or rain. And, while lidar can discern the outer shape of objects with great precision, it can't see around objects, and it can't identify non-shape details of object well. It may, for example, be able to identify a sign from an impressive distance, but it'll have trouble discerning what the sign says. So, lidar needs to be coupled with other sensors to refine the view of the surrounding environment.

So, lidar has its drawbacks. Among them is that even a modest lidar capacity could add 20,000 dollars to the cost of a vehicle. And the mechanical scanning lidar that is now in use can be fragile, perched precariously on top of the car, spinning at hundreds of rotations

IMAGE 9.3
Point cloud.

Each LiDAR laser pulse return provides the precise orientation and range of an object. As a collection, these defined points allow us to precisely map objects and surroundings in three dimensions. In fact, represented visually, the resulting point cloud offers a clearly recognizable image.

Image: Velodyne LiDAR

per minute, and requiring precise tuning and careful handling. This is a problem. To be mounted to the exterior of a working automobile, a system needs to be robust, able to handle the inevitable jostling of road driving, inclement weather, and temperature extremes. And, ideally, it would be discreet, not the large bubble that sits atop current autonomous vehicle prototypes. What's needed is a system that is compact, inexpensive and robust (Image 9.4).

So, what's the solution to lidar's fragile and costly nature? Well, it may be lidar. More particularly, a recently developed **micro-electro-mechanical system** (MEMS) may offer a lidar capacity that is much less expensive and far more robust than current systems. The basic innovation here is the capacity to make a lidar system much smaller and more compact, and the critical component that enables that is a micro-mirror a few millimeters across that oscillates at a very high frequency. Mounted on a chip, this tiny mirror is coupled with a laser diode that emits a collimated laser at the mirror as it oscillates a few thousand times a second or more. The laser is then redirected in an alternating arc

IMAGE 9.4
Solid-state LiDAR.

Magna and its partner, Innoviz Technologies, will be producing a compact lidar system for BMW's forthcoming autonomous vehicle production platforms. The unit offers a field of view of 73° horizontally and 20° vertically, a detection range of 150 m, and 20 frames per second in what they call a solid-state design.

Image: Magna

through a flat diffuser lens. The lens expands the beam vertically, creating a vertical slice of light that, like all lidar, is reflected off surrounding objects and returned back to the transmitting unit. The result is an ongoing arc scan by a slice of vertical light that then provides a full three-dimensional reflection of the environment. However, initial problems with vehicle motion upsetting the tinny mirror and scanning accuracy have identified needed improvement before this is ready for production vehicles.

Another option may be even more innovative. Quanergy has designed an alternative with no moving parts; this is a true solid-state lidar unit. The key is an optical phased array (OPA). The core component is a small plate with a matrix of phased controlled emitters. Each of these elements emits a controlled beam to form a precisely defined collective interference that defines and forms a resulting laser emission. By adjusting the relative phase of each emitter, the beam can be redirected. If all emitters are in phase, the beam goes straight ahead. But, for example, if an emitter shifts phase behind the others, the beam will turn toward that emitter. The result is a solid-state controllable lidar transmitter. With no moving parts, the system can refocus its scan in a microsecond, allowing the sensor to adapt quickly. A full 120° field of view is possible. This is not quite ready for production, but a very promising option for the possible near future.

Optical

A key sensor used in vehicle automation is not very different from the key sensor that has been used since the beginning of the automobile, vision. For many, optical sensors or cameras offer the best alternative for self-driving vehicles. Unlike the systems discussed so far, cameras are passive, receiving light from objects without emitting a signal. With rapidly improving resolution and decreasing cost, cameras are the go-to sensors for many ADAS functions. However, unlike the previously discussed sensors, they offer no range information, only a flat representation of brightness and color differences. The onus is then on the computational capacities of the system to make this data meaningful.

As the only sensor that can capture the color and texture of an object as well as its shape, cameras are indispensable for certain ADAS functions. For example, **traffic sign recognition**

(TSR) or traffic light detection can only be accomplished with a camera. These systems can identify certain road signs and warn drivers of hazards, notify drivers of exceeding the speed limit, or other services. Similarly, LKS, that require the identification of lane markings often with no major physical relief, could not be guided with radar, for example. And cameras can assist with functions that are principally accomplished with other sensors. So, while the all-weather capacity and range of radar may make it a good choice for ACC, a camera is better suited to distinguish small vehicles such as motorcycles, and so can complement the interpretation of a radar return. Similarly, cameras often work best in conjunction with other sensors and navigation system, as their capacity diminishes greatly with bad weather.

Relatively small and inexpensive, multiple cameras can be mounted on the car, each with a particular orientation, range and field of view. Often cameras are used in large numbers. Tesla is a key example here, as their autopilot feature relies on eight cameras that provide a full 360° of vision at over 800 ft (250 m) of range.[8] Tesla's intent is to continue to build on self-driving capacity to Level 4 and beyond relying principally on optical cameras (Image 9.5).

The basics of an optical sensor haven't change. Like all cameras for nearly two centuries, focused light comes through a lens to define a two-dimensional landscape of brightness and color on a flat receiver. In the past, that receiver was film, now it's a collection of tiny dots or pixels. These pixels operate like teeny solar cells that emit electrons when hit with photons of light energy. The incoming light hits the pixels with a certain color and brightness that is translated into an electric signal. These electron patterns can then be used to recreate that image on demand, defining your digital image.

However, the camera used in ADAS may not look much like the camera in your smartphone. A typical phone camera, for example, uses a **charge-coupled device** (CCD) as the sensor plate, allowing high resolution from a large number of pixels and moving the electron charge signal through the chip itself. However, automotive sensors are more likely to use **complementary metal-oxide semiconductor** (CMOS) sensors. The advantage is that the CMOS does not need to send the signal through the chip, making them easier and less expensive to manufacture. Instead, they utilize multiple minute transistors at each pixel to directly read and send light levels from each pixel. The space needed for the transistors means less pixels can fit on the surface, so resolution is lower. But the transmission of information from each pixel to the processor is simplified, reducing manufacturing cost and energy requirements. Unlike the camera in your cell phone, larger pixel size and lower resolution is fine for automotive sensor applications, we're not trying to get that perfect selfie for Instagram. What we do need is fast image acquisition and low light capacity. Larger pixel size helps with both.

The real technical challenge facing optical scanners is not the camera. Digital cameras are a mature technology, and resolution and cost continue to improve (Image 9.6). The real challenge is being able to take that image and extract data that an algorithm can utilize to actually identify routes and hazards and operate a vehicle. After all, what the computer sees is a flat plate of brightness and color variations. It's only your brain that turns that into a picture and renders it into a 3D understanding of the environment. Advanced detection, processing and algorithms are required to decipher these images and determine what they actually represent in the real world. This requires accomplishing two tasks: identifying *where* the object is, called **detection**, and determining *what* the object is, called **classification**. This is called **computer vision**, a subcategory of the broader field we've been exploring, **machine perception**.

---

[8] www.tesla.com/autopilot

	Number	Range (m)
Narrow Forward Camera	1	250
Main Forward Camera	1	150
Forward Radar	1	160
Wide Forward Camera	1	60
Forward-Looking Side Cameras	2	80
Reward-Looking Side Cameras	2	100
Rearview Camera	1	50
Ultrasonic Sensors	12	8

IMAGE 9.5
Tesla autopilot.

Tesla's self-driving features rely heavily, but not solely, on optical cameras, including eight cameras that offer a full 360° view, 12 ultrasonic sensors, and a forward radar for inclement weather. To enable the Autopilot feature, these sensors need to be able to identify and track the surrounding traffic and potential obstacles at significant ranges.

Images: Tesla

IMAGE 9.6
Enhanced optical sensing.

ZF and Mobileye have developed three-lens optical technology designed to support ADAS and automated driving. While a conventional lens can identify pedestrians, bicycles, and road hazards, a telephoto lens provides improved long-distance sensing and a fish-eye lens offers enhanced near sensing with a wide field of view.

Image: ZF Friedrichshafen AG

The process typically entails several undertakings. Multiple images from differing cameras may need to be integrated or stitched together. Filtering is applied to remove noise. Often the image is segmented to simplify the analysis of each portion. Within each segment, significant geometrically linear discontinuities in brightness are identified to detect edges, spaces, and corners, a process called **edge detection**. This provides the principle defining points in shape and **feature extraction**. And these features can then be compared to learned objects to provide object or image recognition. The algorithm can then identify objects and patterns to guide navigation. In the end, image processing doesn't really 'understand' pictures. It mathematically identifies geometric patterns and uses this to inform algorithms. With time, a system 'learns' data patterns and can more quickly and accurately recognize them. We'll explore this 'learning' process later.

The challenge with cameras is that they generally lack any distance information. When a human looks at a two-dimensional image, it's pretty easy to understand the three dimensions it represents. You brain will subconsciously sort things out. A building is generally much larger than a dog, so if a large dog is next to a small house in a 2D image, you'll automatically appreciate the three-dimensional implications, the dog is much closer to you than the house. And since you generally know how a dog and a house are shaped, you have a pretty solid three-dimensional intuitive understanding of the scene. This process is not so simple for a computer.

This can be improved by coupling two cameras to produce distance information. Much as the human brain typically utilizes **stereopsis** by using differences in information from each of two eyes to discern depth, comparing the relative images from two cameras placed side by side can enable range determination. As you look at an object with both eyes, the relative angles from each eye changes with distance. Your brain does the math, and gives you a sense of range. This same process can be done by computer if two camera angles are available of the same scene. The two cameras can define what's called a disparity map, identifying differences in object location and angle in each image, and calculate relative distances. Multiple manufacturers offer stereo cameras that can guide emergency braking, assist with ACC, or provide smarter BSM (Image 9.7).

There are a few challenges with relying on cameras. Even with stereoscopic capacity, the precision of distance measuring decreases notably with range. The complex algorithms

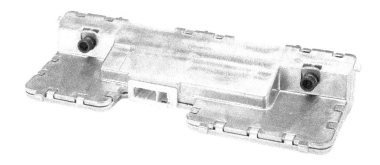

IMAGE 9.7
Stereo cameras.

By comparing the relative images from two perspectives, the range of an object can be calculated.

Image: Continental AG

that provide reliable distance data at 20 ft are much less precise when the distance is 300 ft and the resulting difference in the two camera images and angles is minimal. So, stereo cameras are very useful for blind-spot detection, but much less perfect for ACC. The challenge of nighttime driving can make this worse.

Some have pointed to infrared technology to solve the night issue. Similar to optical cameras, infrared cameras simply respond to a higher wavelength on the electromagnetic spectrum that is defined by heat, not light. Because it detects heat, it is not dependent on lighting and so could resolve the nighttime challenge of cameras. However, this idea is not without its flaws. Because it identifies heat differences, it would be difficult to see pedestrians if the ambient temperature were about the same temperature as a person, making the system less-than-perfect for, say, much of the summer in most of the country. Multiple systems were introduced by OEMs a decade ago, most focused on enhanced nighttime detection of animals and pedestrians most often by providing a warning at the instrument cluster. It's unlikely that these systems will be fully incorporated into a sensor suite for advanced ADAS.

Sensor Fusion

Providing useful localization and guidance requires a combination of different sensor inputs. Each sensor has strengths and weaknesses, so combining them produces complementary redundancy. Vulnerabilities to weather conditions by one sensor, for example, can be offset by other sensors that do not have this weakness. Or reliance on high-cost sensors can be reduced by providing supporting information from lower cost options. For example, an optical sensor could have great trouble identifying lane markings if the road is covered in even a light coat of snow. Radar, lidar or other sensors might be used to support or replace these sensor inputs in these conditions. Systems must also be adaptive. So, a lane-keeping function might be able to accept input from multiple cameras, at the front, side or rear of the vehicle, to accommodate conditions. A camera might allow a radar-guided ACC system to distinguish smaller targets such as motorcycles. Or yaw information can be used to adjust the radar sensor angle in a curve.

With robust redundancy, reliability can be further ensured. Data from one sensor can be used to verify the accuracy or calibration of another. Sensors can be used to verify each other's data, called **competitive fusion**. Or two different sensors can be used to derive information neither would be capable of providing alone, called **cooperative fusion**. In the end, multi-sensor fusion can offer high accuracy and reliability. Noise, uncertainty, and irregularities in sensor data can be resolved with advanced algorithms that can, for example, take real data over time and use it to identify a reliable pattern, correcting for noise and filling in missing data. This can be accomplished with **Kalman filters**, an algorithm that uses a series of data over time to estimate future patterns. So, for example, this might be used in lane tracking, when estimating future lane geometry based on past patterns can speed lane recognition and correct for momentary ambiguity. After all, this is how a human drives. No human driver is constantly searching for lane markings. You're in a lane, and use an unconscious filter of sorts to simply ensure the lane continues as it has. And if a lane marker is faded, you don't panic or turn suddenly, you know the pattern and you follow it. Similarly, **fuzzy logic** works through ambiguous data and uncertain information to define reliable and accurate overall patterns. Again, this is closer to the way the human driver's brain works. We take in information with various levels of uncertainty and aggregate the partial-truths to form greater certainty until we're confident with our understanding of the situation and can act. This makes precise control of a vehicle possible with imperfect sensor information, for both humans and machines.

Though differing projects have adopted differing sensor balance strategies, it is likely that a Level 4 and Level 5 future will require combined strategies and synergies, though the precise recipe is not yet clear. Mobileye, for example, has defined a 'sense and understand' approach to vehicle operation, relying heavily on on-board sensors. Google's Waymo has depended much more heavily on detailed maps streamed to the vehicle to orient sensor information, in what has been called a 'store and align' approach.[9] Both companies rely on lidar as a critical sensor. On the other hand, Tesla is moving toward a Level 4 capacity without lidar. While each initiative may end up with a different formula, it's unlikely that any single sensor technology or approach will resolve the whole puzzle.

Clearly, there is still needed improvement. Improving sensor performance in poor weather or low light is vital to the move to Levels 4 and 5. And improved sensor fusion to reduce uncertainty or over-reliance on a single sensor or technology is needed. Ensuring affordability of these systems is also a priority, as it would not do to have a system accessible only to the rich. In addition, as we plan for widespread adoption, passive sensors and low-energy detection capacity must be developed and improved to avoid signal density and interference of active sensors. Nevertheless, there is no question the capacity of the automobile to perceive and respond to its environment is here.

Driver Monitoring

In addition to the sensors that monitor what's happening outside the car, there is also a need to monitor what is inside the car. As long as there is some necessity for a human driver, there is an equal need to ensure an engaged and vigilant operator. As we've discussed, this

---

[9] M. Goncalves, I see. I think. I drive. (I learn): How Deep Learning is revolutionizing the way we interact with our cars. KPMG White Paper, 2016.

may be particularly true in Level 3 driving conditions. So, sensors that can monitor driver condition are certainly needed.

A simple system can use steering patterns to evaluate driver condition. Using input from the steering angle sensor and lane monitoring, the system can determine erratic or unresponsive behavior. Simple and inexpensive, this is now one of the most widely used driver monitoring systems. Typically, the frequency of micro-corrections is key to evaluating driver alertness. A more alert driver will provide consistent, small, and smooth corrections to address the effects of road conditions. A drowsy driver will provide larger and less frequent input. Assessing only the very small corrections allows the system to exclude the effects of necessary lane changes or curved roads. Tesla uses such a system with its autopilot feature, warning the driver if fatigue is sensed, and if repeated, eventually pulling the car to the side of the road and requiring a reset before autopilot can be used again. Mercedes Attention Assist adds patterns of acceleration and braking, and even time of day, to its steering monitoring to detect driver fatigue.

Alternatively, the driver can be monitored directly. A video system can monitor the driver's face, and use infrared eye reflection to measure the proportion of time the driver's eyes are closed, called percent eyelid closure or **PERCLOS**. Higher proportions of eye closure indicate drowsiness and can trigger a visual warning, seat or steering wheel vibration, sound, or a combination. GM's Super Cruise hands-free system uses a similar measure of eye position and head orientation. Looking away from the road or closed eyes will trigger a flashing disengagement warning meant to bring the driver's attention back to the road.

Localization

A certain amount of vehicle localization can be produced with sensors that measure the basic movement of the car using internal devices, sometimes called self-sensing or proprioceptive sensing in robotics. So, for example, using vehicle speed, acceleration, steering angle, and yaw can provide basic movement characteristics. An **inertia navigation system** (INS) using accelerometers and gyroscopes can enhance this sensing. This can allow **dead reckoning**, using direction and speed to estimate location. But the sort of inertial systems used in automobiles is not greatly accurate. Improved accuracy is possible, and used in military aircraft for example, but the cost is prohibitively high. However, even a basic INS or proprioceptive system can offer useful short-term navigation when outside sources are lost, say in an urban tunnel.

Moving toward a reliable and continuous localization capacity requires going beyond proprioception to external navigational inputs. In particular, **Global Navigation Satellite System** (GNSS) data can be integrated with INS information to produce precise and continuous location information. Once again, the idea is not new. For nearly a century, aircraft and ships have relied on coordinated networks of very low-frequency radio signals from terrestrial transmitters for long-range navigation. The basic idea was that if you know your precise distance from three known points, you could draw a circle on the Earth at that known distance around each point, and where the circles cross is your location. This is called **trilateration**, a form of triangulation that uses distances instead of angles. These systems worked well in the air or on the sea where interference was not a major problem. However, using a land-based system like this for a terrestrial vehicle in urban or mountainous terrain would be problematic. But this all changed as satellite technology became more commonplace. Placing the transmitters in space largely resolves the problem of reception and offers worldwide position trilateration. GNSS comprises multiple satellite networks, including Russia's Glonass, Europe's Galileo, and China's BeiDou navigation satellite systems, as well as the widely used US Global Positioning System (GPS). GPS was developed and maintained by the US military and is the most well developed and used of these systems.

The GPS satellite system relies on a network of 24 active orbiting satellites (and three spares). The satellites maintain extremely precise circular orbits at 11,000 miles above the Earth's surface, where they are largely unaffected by the Earth's atmosphere. The orbital constellation of satellites is defined by six planes, evenly spaced at 55° from each other, with four satellites in each plane. They circle the Earth twice each day with orbits that ensure that at any given time at least four satellites are in view at any location on the Earth, though typically more than six are visible at any time or place (Image 9.8).

Each GPS satellite sends a continuous signal that provides the precise time of transmission (TOT) and an identifying code. The receiver measures the time of arrival (TOA) of the signal and can therefore calculate its distance from the known location of the satellite. Any GPS receiver contains an almanac with the precise location of every satellite; and this almanac is updated by the signal from the satellite to ensure precision. The receiver finds the travel time of a signal by comparing what's called a pseudo-random code it generates with a code that should be precisely the same generated by the satellite. It moves its code back in time until the two are perfectly synched, and can then know the time of travel for

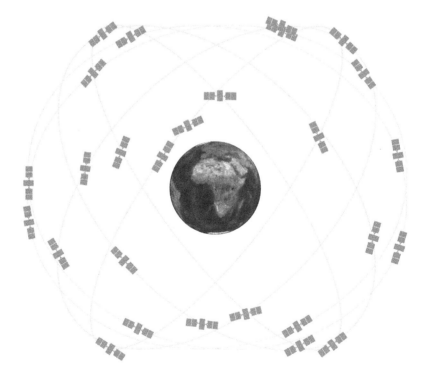

IMAGE 9.8
GPS constellation.

The US GPS system utilizes a total of 27 satellites, 24 active units, and 3 spares. They orbit in six planes, each spaced 55° apart, and offer complete coverage of the Earth.

the signal. The rest is simple trilateration. With signals from multiple satellites, a GPS unit can calculate what we can imagine as an arc of possible positions on the Earth's surface for each satellite position. So, with reception of signals from two satellites, two arcs could be defined, and possible locations on the Earth surface are the two spots where the two arcs cross. A third signal can define a third arc, and therefore only a single possible location.

The trouble with this idealized description of GPS is timing. Everything depends on an accurate measure of the time difference from the transmission to the reception of the signal. This is what allows us to define the distance measurements from each satellite. The satellite transmitters use multiple clocks that rely on the natural oscillation of a cesium atom as the timekeeper, called atomic clocks. These devices are extremely accurate, with an error of about a second every 100 million years. However, since a clock like this cost tens of thousands of dollars at a minimum, we can't put one in every car. This means the clock in a car's receiver is not nearly as accurate. Because the signal travels at the speed of light, or about 983 million ft/s, an error of one-thousandth of a second would place your location off by 983,000 ft or about 200 miles. That wouldn't be much help.

The solution comes, ironically, by relying indirectly on the clocks in the satellites. All the satellite clocks are perfectly synchronized. So, we need to determine only one number, the time offset of the receiver from universal time. If a receiver has access to three satellite signals, without knowing this offset, it can use trilateration to calculate a possible position but it won't be accurate. However, since the time offset from each satellite signal is the same, the position may be understood to be proportionally incorrect. The key is a fourth satellite. Information from that satellite would give use a fourth arc that will not coincide with the position defined by the previous three satellites. Since all the satellites' times are perfectly synchronized, we can now calculate the offset that would allow all of these arcs to cross at a single point. With four satellite signals, there is only one possibility. We can then apply that offset to correct our clock and track our location precisely. And the receiver can continuously evaluate and update its time as it continues to operate. Clever, right?

However, there is still another challenge. Since we know the time with great accuracy, we also need to know the speed of light with great accuracy, and this number is not constant. As the signal travels through the upper layer of the atmosphere, called the ionosphere, it can change speed and introduce error. Variations in moisture and pressure can also slightly vary the speed light travels as it nears the Earth. Computer modeling can help address these errors. A sophisticated receiver can use the relative speeds of different signals to correct for this error, called a **dual-frequency receiver**. GPS uses two frequencies one called L1 at 1575.42 MHz and a second called L5 at 1227.60 MHz. Comparing these two signals can allow direct measurement of the delay produced by the ionosphere for that satellite and enable a much more precise location. Initially very expensive, and only used when highly precise measurements were needed in industrial applications, dual-frequency receivers are now becoming much more available.

We're not done yet. Come on, you didn't think navigating with satellites in space was going to be easy, did you? There are other sources of error in the GPS system. Signals bouncing off buildings and other obstructions can introduce more error near the ground. And even with the precautions discussed, slight errors in the satellite positions can introduce further error. The solution comes from the use of **differential GPS** (DGPS). The basic idea is to add very precisely located stations on the ground. These stations receive signals from all visible satellites, and since the station is at an exact location, it can identify any error that may be present. It can then send information on the error and its rate of change for each satellite to receivers in the area to allow them to apply corrections for the satellites they are using. A similar system was developed by the US FAA to assist with aircraft navigation called **Wide Area Augmentation System** (WAAS); though the WAAS system transmits correction information via dedicated satellites rather than from ground-based stations, improving reception.

An alternative mode of enhancement for GPS utilizes not just the information contained in the satellite signal, but the actual phase of the electromagnetic wave that carries the pseudo-random code signal. Called **carrier-phase enhancement**, or CPGPS, the idea is pretty straightforward. First, remember that signals are carried within a carrier wave. That's sort of like the frequency of your favorite radio station. The frequency of the music changes, but the frequency of the station stays the same whatever the frequency of the signal it's transporting. So, the electromagnetic signal from the radio station or the satellite has a definite wavelength of a precisely known length that is used to deliver a signal. You might imagine this as a sort of repeating yardstick (or meter stick) extending from the satellite to the receiver with each yard precisely marked, but not numbered. So, if you could know your rough location, that is to say which yard you are on, you could refine that to a very precise location by measuring the remaining distance on the final yard. The basic idea of this sort of **kinematic positioning** is to multiply the carrier wavelength by the number of cycles and then add the phase difference remaining.

The result of all this is tremendous accuracy. Basic GPS can provide accuracy to within 16 ft (about 5 m) or so. This is fine for simple navigation, but you wouldn't want to maneuver a car down a city street with ±16 ft of precision. However, with a dual-frequency receiver or augmentation, that accuracy improves to less than an inch, or a few centimeters. And a CPGPS can bring that accuracy to less than a centimeter.

Mapping

So, GPS is clearly good. But any navigation system is only as good as the map. If you know your position with great accuracy, but don't know where the roads and buildings are, you're not really any better off. So, the move to self-driving cars could benefit greatly from precise mapping of roadway environments.

One way to ensure this is with a purpose-built high-definition map of everything, roads, buildings, posts, signals, signs, curbs, and all. This is a tall order and one reason why Google, Uber and others have driven millions of self-driving miles mapping the environment. Pre-driving specific roads allow a test vehicle to collect sensor data from optical, GPS, and lidar sensors, compare this to existing maps and data, and adjust maps accordingly. This process is tricky since temporary or minor discrepancies might be recorded as major obstacles. Imagine a few traffic cones being mapped as a fixed traffic barrier. So, multiple passes and continuous updating is needed. The system can face difficulty when distinct landmarks are not available, over a bridge, through a tunnel, perhaps driving through a desert or cornfield could present a challenge.

Nevertheless, the approach seen by many as the most viable is continuously updated high-definition mapping. A dynamic network of maps could be created and revised continuously through the sorts of communications networks we'll talk about next. Of course, the amount of data would be enormous. All roads would have to be mapped and continuously remapped, accounting for construction, temporary detours, moderate, and major variations in traffic patterns, and so on, endlessly. A cloud-based map could be updated continuously while accessed by drivers.

But this idea is not without challenges. The sort of detail required of maps to enable localization and planning is a tall order. And continuously identifying and updating changes quickly enough to enable ongoing adaptation is even harder. In general, this would be expensive and technically challenging; so some say it simply may make more sense to maintain basic maps of roadway information, and rely on sensors for the rest.[10]

Several carmakers see this as the most promising avenue toward Level 4 capacity. Mercedes, Audi, and BMW are developing a cloud-based high-definition mapping system that will enable their cars to see the world around them more accurately. Called HD Live Map, the system is updated in real time by cars in the network. The initial map is already in place, developed by the mapping company HERE, and using sensor data of cars already on the road. The highly detailed, multilayered map includes data on road conditions, traffic information, road hazards as well as highly precise roadway and infrastructure plotting. For now, the system only offers driver assistance, though Audi and BMW expect to have a Level 4 vehicle on the road and using this system by 2021. As previously mentioned, the fully self-driving Waymo car has adopted a similar approach, relying heavily on precise mapping (Image 9.9). In Japan, a consortium of carmakers led by Mitsubishi Electric have partnered to define the Dynamic Map Planning project in hopes of developing high-resolution 3D maps in time for a self-driving fleet during the 2020 Tokyo Summer Olympics.

While it may seem desirable to map the world in its finest detail to ease localization and planning on a predefined map that includes pixel-perfect annotations of lane markings and traffic signs, this comes with at least two significant disadvantages: First, it may not be possible to update maps to reflect changes in the environment quickly enough. And second, highly detailed maps are expensive to create, maintain, and transfer, since updates

IMAGE 9.9
High-definition mapping.

Image: Waymo, the subsidiary of Alphabet, Inc., that began as the Google self-driving car project, is one of several companies now undertaking high-definition mapping, though Google has been at it for more than a decade. This mapping supports the operation of their fully self-driving Chrysler Pacifica hybrid now being tested in multiple locations. Producing highly accurate maps of all roadways, and the continuous updating required, makes this a challenging way to support self-driving cars. However, several leading carmakers see cloud-based high-definition mapping as a key to self-driving capacity.

Image: Waymo

---

[10] W. Schwarting, J. Alonso-Mora and D. Rus, Planning and decision-making for autonomous vehicles. *The Annual Review of Control, Robotics, and Autonomous Systems* 1 (May), 2018, 187–210.

need to be constantly fed into and distributed by the system. Therefore, it seems advantageous to keep only a general map, and allow vehicle sensors to fill in the rest.

An approach which would rely much less heavily on preexisting maps is called **simultaneous localization and mapping (SLAM)**. SLAM allows the localization to happen on the road and with little a priori information. The vehicle would continuously observe the environment and adapt accordingly. Without relying greatly on established data and maps, the required computational capacity and sensor inputs are higher. A key, of course, is to differentiate objects, in particular static objects and moving objects, while correctly following road markings. The great advantage of this approach is that detailed prior knowledge of the road is not needed, making the vehicle more adaptable and dramatically easing the up-front investment and needed infrastructural support. Tesla's autopilot feature in the Model S relies on real-time SLAM processing and of course enables highway driving, with appropriate speed and lane changes. However, as the conditions become more complex, at an exit or intersection the human driver needs to take over.

Communication

Clearly, precision mapping will require connectivity. But, even if we rely on on-board sensors, communication is needed. The future of self-driving will not come from advanced sensors *or* communications; it will come from the convergence of the two. Even the very best sensors are inherently limited to what they can 'see'. With increasing sensor capacity, this is not severely limiting; but it leaves out all the obstacles, vehicles, or hazards that might be just beyond the line-of-sight or somewhere along the intended route. In a world of ongoing and extensive information sharing, this sort of limit isn't necessary. Instead, imagine the car as a node in a network of **connected vehicles**, sharing information with other vehicles and sensors on roadways, bridges, intersections and other infrastructure continuously. This could notably enhance and facilitate the sorts of functions we've been discussing. The ability to adjust vehicle speed to match traffic, or accommodate a sudden stop, could be simplified if cars shared their speeds and braking actions with each other. Or a car's ability to gauge road conditions can be helped by a sensor and transmitters built into the roadway. A complete system could include vehicle-to-vehicle communications (V2V), vehicle-to-infrastructure communications (V2I), and perhaps even vehicle-to-pedestrian connections (V2P), all generally referred to as vehicle-to-everything or **V2X connectivity**.

Using external sources to facilitate self-driving operation is not a new idea. In the 1950s, experiments in automated driving were enabled by a buried cable in the roadway. A truck with sensors mounted on the front and rear was able to follow the cable and receive signals transmitted through the embedded conductor. More recently, roadside permanent magnets have been examined as a guidance device. Detection would be easy, and unaffected by weather. Notably, the Swedish carmaker Volvo has experimented with magnets to guide self-driving cars in inclement weather and snow covered roads. As an added bonus, patterning the magnets' orientation, north up or north down, could allow them to provide a sort of binary signaling to deliver simple messages such as bike lanes, crosswalks, two-way traffic, or speed restrictions. Similarly, modified reflective lane marking tape has been proposed to enhance radar identification and roadway signaling. However, all of these would require extensive infrastructure modification at very high cost. An integrated communications network might be able to provide some of these advantages, with faster implementation and lower investment.

Imagine the possibilities. A child walks into a busy road and a sensor at the intersection issues a warning. This tells the oncoming car to apply brakes. A roadside sensor that sends road condition information to cars in the area helps determine the amount of braking needed. The braking car sends an instantaneous signal to the car behind it warning of the hazard and the vehicle's reduced speed. This warning signal cascades to the cars that follow, allowing them to adjust their speed slightly. The child clears the street. No one is hurt, the event is barely noticed by the cars' occupants.

Other proposed possibilities are even more dramatic. Sharing video imagery among cars on the roadway, for example, could allow each car to see what other cars are seeing. This could allow every car to know and see the position of other vehicles. A monitor on the dash could link to video from other cars on the road, allowing a driver to see the car hidden behind a truck or the pedestrian waiting to cross in front of a van stopped in the road. With shared video, you could see what's ahead in a traffic back-up. In essence, each car would be virtually translucent, allowing a driver to have an accurate and complete view of every car in the area and all relevant hazards.[11]

Cooperative road safety protocol might include differing notifications for the approach of emergency vehicles, warning of a slow vehicle, intersection collision warning and rerouting, perhaps even a warning to drivers of surrounding vehicles when a car makes a major driving error, turning the wrong way onto a highway, for example. If a car experiences slipping on icy conditions it can send a warning to approaching vehicles to adjust for adverse road conditions. An approaching car might reduce speed or even adjust all-wheel drive as a result. A message might warn of a fixed or moving hazard, such as a construction zone, oversized load, or recent accident. Messages might provide information on traffic and signal timing to improve flow. Even the detection of deer on the roadside by a vehicle sensor might trigger a warning to other vehicles to watch for deer. Or, in an area with a high deer presence, a ground unit might monitor the area and produce a caution signal when deer are near the road allowing vehicle systems to prioritize deer identification and avoidance.

Such a system would rely on mature technology, and could be implemented incrementally, a great advantage over other infrastructural guidance systems. While V2V technology requires similarly connected vehicles and won't reach its full potential until it is widely adopted, V2I connectivity with a focus on traffic management or safety can have a beneficial effect even if only used by a minority of the vehicles on the road. Indeed, the first steps are already in place. The Federal Communications Commission has defined a **dedicated short-range communications** (DSRC) system to serve as the basis of a two-way short- and medium-range vehicle communications capability. Seventy-five MHz of spectrum in the 5.9 GHz band is dedicated for use by what the USDOT calls **intelligent transportation systems** (ITS) and general V2X communications.[12] Similarly, ASTM International, SAE and the Institute of Electrical and Electronics Engineers (IEEE) have collectively defined standards for **wireless access in vehicular environments** (WAVE) adopted by the US Department of Transportation. These provide architecture, security standards, and management structure to enable high-speed networking between vehicles. Similarly, the EU has launched a 5G Action Plan to support transportation system connectivity and begun

---

[11] Y.F. Wang, Computer vision analysis for vehicular safety applications. *International Telemetering Conference Proceedings* 51, 2015, 1–31.
[12] www.its.dot.gov/

to develop models and build infrastructure.[13] Within a very short time, V2V signals might include blind-spot notifications, sudden braking warnings, collision warnings, approaching emergency vehicle notifications, and do not pass cautions. Likewise, V2I signals might offer warnings of traffic accidents, emergency vehicles, or more benign information on travel and traffic, parking, tolls, or weather.

These systems could build and enhance existing automated communications functions and enable the continued development of more sophisticated automatic collision notification (ACN) systems. Introduced two decades ago, systems such as OnStar ACN contact emergency response agencies in case of a collision. More recent systems rely on a variety of sensors to define an Advanced ACN (AACN) that automatically sends emergency information on the location of an accident, severity, type of impact, and even probability of injuries. The idea is to allow emergency assistance to be more quickly and efficiently deployed, saving lives.[14] The next step might allow the infrastructure to respond accordingly, re-routing traffic to facilitate emergency access, for example.

Decision-Making

Just about everything we've discussed so far requires some form of perception, decision-making based on that information, and adaptability to unforeseen conditions, and all this entails AI. This term is frequently surrounded my mystery and hype, and is often loosely used in popular culture to refer to rogue robots plotting to take over the world. So, it's worth setting the record straight at the onset: AI does not imply a form of consciousness. AI systems need not and cannot be aware of their existence, reimagine their purpose, or reflect on the nature of the human condition. Instead, AI is defined by algorithms that analyze data to identify patterns to information that can then be applied to specifically defined tasks. Their capacity to analyze very large amounts of data quickly allow them to be better than humans at some of these tasks; but it is a stretch to say that they are genuinely 'thinking', at least not anytime soon. More sophisticated systems are able to alter their processes based on past patterns in data in order to improve their capacity for future analysis. This capacity to improve performance based on experience is called **machine learning**; and this is a key to the development of self-driving vehicles.

There are many degrees and sorts of AI. In the vehicle, what is required for one sort of function, say managing vehicle speed and direction in traffic, may not be required for the operation of a traction control system or applying emergency braking. So, in current vehicles, and most certainly in future vehicles, a wide array of differing levels and types of AI will be implemented.

The typical popular view of AI programing is often something like a very long program that tells a computer what to do in any given circumstance. We can imagine this as a series of what are called 'if-then' statements. For example, 'IF the front sensor identifies an object in the road THEN apply braking'. This general approach can be called **rule-based** AI because every if-then statement defines a sort of rule. This type of system could work very well for a very basic ACC. It is fairly simple to define an algorithm that can maintain vehicle speed to ensure a set spacing with surrounding traffic. One can easily imagine a

---

[13] ec.europa.eu/digital-single-market/en/5g-europe-action-plan
[14] www.aacnems.com

series of rules that would enable this. Lane-keeping assistance could also be amenable to this sort of a priori programing.

However, as the operating conditions and requirements become more complex, and less predictable, a rule-based system quickly becomes unworkable. Even in the seemingly simple example above, identifying what it means to have an object in the road is very complicated when there is no human driver to decipher the nature of the object. If that object is a child, then the required response is very different than it would be for a cardboard box. But what visually defines a child? How is it different from a box, a fire hydrant, a dog, or even a cardboard cut-out of a child? What about location? What if the child is on the side of the road? What if the child is on the side of the road but a ball is rolling in the road? What if the child is wearing a large hat, or a mask, or riding a unicycle? The amount of coding needed to cover all possible iterations of a child near the road with a priori rules is mindboggling.

So instead, we rely on sophisticated machine learning, algorithms that can identify patterns and use them to integrate improved predictive capacities into future analysis. If, for example, the system is exposed to enough child-in-or-near-road scenarios, it can develop predictive patterns that will be able to project likely events and identify appropriate action faster and better than a human driver. The key to this is the process of data acquisition. To make this work, to define reliable patterns, an algorithm needs access to a very, very large number of scenarios. So, like a human driver, only so much can be gained by learning the basic rules, the real key to making a better driver is experience.

So, how do we give a machine experience? Machine learning can be supervised or unsupervised. In supervised learning the desired output, or target value, is known. So, for example, you could feed an algorithm hundreds of different images of a traffic signal, each with different angles, lighting, weather conditions, and so on. The algorithm identifies common patterns in the images and 'learns' to recognize a traffic signal. More sophisticated processes will require unsupervised learning, allowing the algorithm to refine its operation autonomously, without explicit programing. So, after an initial training, a self-driving system can use its experience on the road to refine its recognition capacity and responses. In fact, more generally in the AI world, functions that require changing solutions and adaptability to unforeseen conditions require a capacity for unsupervised learning; and adaptability is at the very core of driving. You might teach a system to brake when an object is ahead. That's pretty simple. But if you need the system to identify the object, its likely behavior, and appropriate response, you need adaptability; you need the system to learn on its own.

Because the dominant form of algorithms that can achieve this are modeled very loosely on the human brain, they are called **artificial neural networks** (ANNs). The rough idea of an ANN is to define a collection of nodes that very vaguely resemble the functions of neurons in a brain (Image 9.10). These work in a multilayered and complex network, with each node having an input and an output and a capacity for processing that takes place in so-called hidden layers. The communication between layers is basically a weighted value, used to shape the continued process of analysis in the next layer. Working together these layers can identify and quantify very complex patterns. And they maintain a capacity to adjust the hidden layers based on experience, called **backpropagation**, and so improve their operation. Of course, there are refinements and variation on this theme for particular purposes. Recurrent neural networks (RNNs) have the capacity to base decisions on previous learning and are useful in speech recognition, for example. Convolutional neural networks (CNNs) have restructured connections between nodes that helps avoid redundant analysis and so provide the ability to simultaneously recognize and detect objects in images, a useful characteristic for the high-speed recognition needed in automobiles.

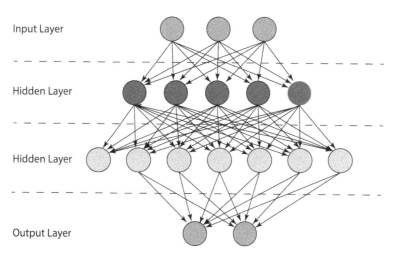

Input Layer

Hidden Layer

Hidden Layer

Output Layer

IMAGE 9.10
Artificial neural network.

This overall process is called **deep learning**. It's called deep because it requires a cascade of layered analysis. In image recognition, for example, imagine how our simplified model of ANN might identify a traffic feature through iterative steps. The first layer might identify a distinct object. The second layer may determine position; the object is in the upper-center of the roadway. The third layer then determines it is about $2\,ft^2$ in size. The next layer recognizes that it emits a red light. The algorithm then identifies it as a red traffic light, and applies the brakes. The real process is considerably more complex, as you can imagine; but the basics are the same. The initial layers determine basic characteristics, as the algorithm progresses, it moves through layers of analysis toward a more complex and precise analysis.

So, the creation of a safe Level 5 car will require the fusion of multiple sets and types of sensor data, a multilayered and complex processing capacity, and an ability to adapt the system with new information and accommodate unforeseen conditions, all very, very fast. The key characteristic is that learning will need to occur beyond initial programing or supervised learning. A system can be initially trained but will continue to refine its analytical capacity with increasing experience. With repeated exposure to new data, the system can correct for faulty predictions, improve its capacity to recognize patterns in the data, and essentially grow 'smarter'.

The challenge is developing the ability to organize and process information quickly enough to ensure driving safety. A massive amount of data from multiple sets of sensors needs to be continuously integrated and analyzed in real time. The identification of patterns, and a regression analysis to compare those to known patterns, requires a lot of computing capacity and time. A very simple AI self-driving system can require processing one gigabyte of data per second.[15]

Systems are improving. For example, the use of graphical processing units (GPU) now offers much faster image recognition capacity. Initially developed for videogames that demanded high-speed and complex imagery, GPUs can perform multiple concurrent

---

[15] A. Cornet, M. Kässer, T. Müller and A. Tschiesner, *The Road to Artificial Intelligence in Mobility—Smart Moves Required*. McKinsey Center for Future Mobility, New York, 2017.

calculations called parallel computing. A conventional computer is designed to complete distinct computations one at a time but quickly. So, an image is broken up into small bits, and feature extraction, detection and recognition take place sequentially. A GPU, on the other hand, can take in the whole image at once, or undertake a holistic analysis of a large data set, making it perfect for the sort of fast image recognition required for self-driving vehicles.[16]

However, the real capacity for deep learning may turn out to be more reliant on vehicle connectivity rather than isolated individual processors. Driving experience data could be stored collectively and streamed to vehicles in real time, allowing learned experience to become shared fleet experience and facilitating the maintenance of a much larger dataset. The system is likely to require constant upgrading and refinement, particularly early in its development. But a cloud-based system could enable not just intelligent cars, but an ITS.

The Road Ahead

Clearly, all the bugs haven't yet been worked out. But it is equally clear that change is happening fast. ADAS technology has fundamentally redefined the nature of driving. Self-driving cars are not a notion from science fiction. They are here, on the road, now. Level 3 driving is available in production cars. You can catch a self-driving ride share in cities across the country. Indeed, new systems are introduced as an option with little fanfare, as though a car that drives itself down the road is a normal, everyday thing; because it is.

Current research and development projects tend to emphasize differing technologies; but it is likely that continued progress in self-driving vehicles will rely on a combination of technologies and approaches. Merging rule-based learning with complex neural networks, deep learning and natural language processing to define a system that can manage the nearly unending possible iterations and variations in any given driving scenario. Radar, lidar, cameras, and other sensors will need to be seamlessly fused with enhanced GPS, mapping, and other technologies to define a capacity for a reliable, full-time 3D localization, and in the end will provide situational awareness and control that is far superior to anything a human driver could ever achieve. A self-driving vehicle must be capable of responding to situations it has never before seen and were not predicted. And it must do all this flawlessly. A one in a million error rate is absolutely not acceptable. We're not there yet.

A transition to self-driving will require some period of time when both human and machine drivers are on the road. This means the machines must be able to understand and predict human actions. Human drivers do this all the time. Someone is swerving in their lane as they fiddle with their phone, and you change lanes or speed to avoid them. A driver is coming up fast in your rearview mirror and you wait to change lanes until they pass. Or a driver clearly exhibits poor driving skills, and you give them a wide berth. Similarly, self-driving systems must be able to identify the driving patterns of other drivers and account for them in its decision making, accounting for human driver behavior and probable intentions even when they are neither rational nor legal.

The basics of this are relatively easy; but the devil is in the details. Giving a swerving driver a wide berth is fairly obvious. But driving is a social act. And the social cues and

---

[16] M. Goncalves, I see. I think. I drive. (I learn): How Deep Learning is revolutionizing the way we interact with our cars. KPMG White Paper, 2016.

norms that we take for granted while behind the wheel will not be apparent to a computer. Eye contact between drivers that indicates 'you can go ahead and I'll stay here' is not obvious to an algorithm. Often, a driving scenario that is unclear or ambiguous—navigating around a barrier, avoiding a dog in the road, managing a four way stop when all cars seemed to arrive at once—are resolved with a glance or a wave that signals intent. This may be a bit like a systemic projection of the handoff problem: with every car fully automated, these issues may not be a problem. It's the in-between stage that's tricky.

Similarly, critics point out that as a social activity, driving is also an ethical undertaking.[17] What do you do when there's a child in the road and avoiding a collision requires maneuvering that may damage your car? That's an easy one. But what about a puppy? A squirrel? The line is not the same for everyone. What about the often-cited ethical example of the trolley problem? Would you take one life to save five? Should a driver choose to hit a baby carriage in the road or swerve and hit a trolley filled with nuns? In the real world, these dilemmas are handled by improvisation or instinctual reaction that may later be subject to post hoc ethical reasoning. But, how do we manage this when the rules need to be deliberately defined ahead of time?

And the transition is unlikely to be without turbulence. Recent accidents of self-driving cars underscore this and have piqued public concern. The death of a pedestrian in Arizona by a self-driving Uber vehicle. The fatal crash of a Tesla on autopilot in California. The public is right to be concerned. Basic ethical principles, as well as the need to build public trust in self-driving vehicles, require that they work perfectly, all the time. Early failures could be disastrous for the slow upward ratcheting of public faith in self-driving technology. But it's also worth pointing out that the number of accidents per mile for self-driving cars nationally is well below the number we would expect if humans had been in control. Still, it is important that we not go too fast. The integration of self-driving technology needs to be deliberate and careful.

And there is much technical work to be done. Perception in poor weather needs to be improved. Sensor fusion and analysis needs enhancement, particularly in complex environments, and especially if a reliance on high-definition mapping is to be avoided. And, if not, a massive and detailed set of interconnected maps of all roadways and features is needed; and the infrastructure connectivity to continuously access, evaluate, update and enhance these maps will need to be developed. Alternatively an improved SLAM is needed that can safely and flawlessly accommodate all changing road conditions. A management and communication network that can facilitate cooperative traffic management and driving is required. The development of safety protocol and measures in case of sensor or system failure will be needed. System security will need to be addressed. And, absolutely not least of all these challenges, the policies and legal and ethical framework necessary to ensure the accountability, safety and security of vehicles and the system needs to be assembled.

Whatever happens, one thing is clear: the future of the automobile will be decidedly interesting.

---

[17] D. Bollier, Artificial intelligence comes of age: the promise and challenge of integrating ai into cars, healthcare and journalism. A Report on the Inaugural Aspen Institute Roundtable on Artificial Intelligence, Aspen Institute, Washington, DC, 2017.

Index

A

AACN, *see* Advanced ACN (AACN)
ABS, *see* Antilock braking (ABS)
Absorbed glass mat (AGM), 181
ACC, *see* Adaptive cruise control (ACC)
Acceleration, 70–72, 93, 95, 162, 251; *see also*
    Deceleration
ACN system, *see* Automatic collision
    notification (ACN) system
Acoustic tuning, in induction systems, 53
Active aerodynamics, 261–263
Active cooling, 112, 176
Active front steering (AFS), 229
Active grill shutter (AGS), 240, 241
Active inflation management, 98
Active roll control (ARC), 227
Active suspension, 226
Active torque vectoring, 94
Actuator(s)
    electromechanical, 77
    hydraulic, 77, 94
    mechanism, 75
Acura NSX, 212
Adaptive cruise control (ACC), 268, 269, 281
ADAS, *see* Advanced driver assistance system
    (ADAS)
Advanced ACN (AACN), 290
Advanced chain drive, 13
Advanced digital control technology, 39–40
Advanced driver assistance system (ADAS),
    265, 268, 278, 293
Advanced high-strength steel (AHSS), 206,
    209, 210
Advanced ignition system, 29
Advanced lead carbon (ALC) battery, 181
Advanced valve control, 50
AER, *see* All-electric range (AER)
Aerodynamica Lamborghini Attiva (ALA), 263
Aerodynamic resistance, 66
Aerodynamics, 233, 235–240
    active, 261–263
    Audi, 259
    body contour, 259–261
    drag, 234–235
    front sides, 240–242
    lift, 251–253
    rear wake, 243–247

    shapes, 235–237
    three dimensional flow, 247–250
    travel on ground, 253–256
    vortex generators, 250–251
    wheels, 257–259
Aero wheels, 257, 258
AFS, *see* Active front steering (AFS)
Aggressive shift logic (ASL), 87
AGM, *see* Absorbed glass mat (AGM)
AGS, *see* Active grill shutter (AGS)
AHSS, *see* Advanced high-strength steel (AHSS)
Air batteries, 193
Air channeling, 260
Air-cooling system, 177
Air curtains, 257, 258
Air dams, 242, 253
Airflow venturi, 238, 239, 261
Air-guided GDI system, 30
ALC battery, *see* Advanced lead carbon (ALC)
    battery
All-electric range (AER), 148
Alloy(s), 121, 205
    aluminum, 14, 211–212, 214
    lithium, 191
    magnesium, 15
    steel, 18
All-wheel drive, 95–96
Alnico magnets, 121, 122, 129
Alternating current (AC) generator, 104
Alternative metals, 210–216
Aluminum alloys, 14–15, 211–212, 214
Aluminum 5000-series, 211
Aluminum pistons, 18
AMC-SC1 (magnesium alloy), 15
AMT, *see* Automated manual transmission
    (AMT)
Angular momentum, 7, 8
Anisotropy, 121
ANNs, *see* Artificial neural networks (ANNs)
Antilock braking (ABS), 92
ARC, *see* Active roll control (ARC)
Armature(s), 102
    rotation, 125
    squirrel cage, 114, 115
    windings, 118, 130
Artificial neural networks (ANNs), 291, 292
ASF design, *see* Audi Space Frame (ASF) design